International Series in Operations Research & Management Science

Volume 324

The book series **International Series in Operations Research and Management Science** encompasses the various areas of operations research and management science. Both theoretical and applied books are included. It describes current advances anywhere in the world that are at the cutting edge of the field. The series is aimed especially at researchers, advanced graduate students, and sophisticated practitioners.

The series features three types of books:

- Advanced expository books that extend and unify our understanding of particular areas.
- Research monographs that make substantial contributions to knowledge.
- Handbooks that define the new state of the art in particular areas. Each handbook will be edited by a leading authority in the area who will organize a team of experts on various aspects of the topic to write individual chapters. A handbook may emphasize expository surveys or completely new advances (either research or applications) or a combination of both.

The series emphasizes the following four areas:

Mathematical Programming: Including linear programming, integer programming, nonlinear programming, interior point methods, game theory, network optimization models, combinatorics, equilibrium programming, complementarity theory, multiobjective optimization, dynamic programming, stochastic programming, complexity theory, etc.

Applied Probability: Including queuing theory, simulation, renewal theory, Brownian motion and diffusion processes, decision analysis, Markov decision processes, reliability theory, forecasting, other stochastic processes motivated by applications, etc.

Production and Operations Management: Including inventory theory, production scheduling, capacity planning, facility location, supply chain management, distribution systems, materials requirements planning, just-in-time systems, flexible manufacturing systems, design of production lines, logistical planning, strategic issues, etc.

Applications of Operations Research and Management Science: Including telecommunications, health care, capital budgeting and finance, economics, marketing, public policy, military operations research, humanitarian relief and disaster mitigation, service operations, transportation systems, etc.

This book series is indexed in Scopus.

More information about this series at https://link.springer.com/bookseries/6161

Raymond Bisdorff

Algorithmic Decision Making with Python Resources

From Multicriteria Performance Records to Decision Algorithms via Bipolar-Valued Outranking Digraphs

Springer

Raymond Bisdorff
Luxembourg City, Luxembourg

ISSN 0884-8289 ISSN 2214-7934 (electronic)
International Series in Operations Research & Management Science
ISBN 978-3-030-90927-7 ISBN 978-3-030-90928-4 (eBook)
https://doi.org/10.1007/978-3-030-90928-4

This Springer imprint is published by the registered company Springer Nature Switzerland AG
The registered company address is: Gewerbestrasse 11, 6330 Cham, Switzerland

This book is dedicated to my colleague and dear friend, the late Prof. Marc Roubens.

Preface

The reader will find in this monograph a series of tutorials and advanced topics originally written over the last decade as documentation parts for the DIGRAPH3 collection of Python modules. These programming resources—like the `outrankingDigraphs` module—were essentially used both for the computational verification of decision algorithms and for the preparation and illustration of a master's course on algorithmic decision theory taught at the University of Luxembourg from 2010 to 2020. Some resources, like the `randomNumbers` module, served for preparing and illustrating the lectures of a computational statistics course. Curious readers will also discover some resources—the `arithmetics` module—used for preparing and illustrating a first semester course on discrete mathematics.

The DIGRAPH3 Python programming resources are useful in the field of algorithmic decision theory and more specifically for the outranking approach of multiple-criteria decision aiding (MCDA). In this latter scientific field, we address essentially three kinds of usage.

First, we present algorithms and illustrate computing tools for solving either a multiple-criteria first choice selection or a dual last choice rejection problem. We also tackle the problem of how to list a set of items with multiple incommensurable performance criteria either from the best to the worst (ranking problem) or from the worst to the best (ordering problem). Finally, we present order-statistical algorithms for relative or absolute quantiles rating of multiple-criteria performance records.

It is necessary to mention that the DIGRAPH3 resources do not provide a professional Python software library. The collection of Python modules I describe in this book was not built following any professional software development methodology. The design of classes and methods was kept as simple and elementary as was opportune for the author. Sophisticated and cryptic overloading of classes, methods and variables is more or less avoided all over. A simple copy, paste and ad hoc customisation development strategy was generally preferred. As a consequence, the DIGRAPH3 modules keep a large part of independence. Furthermore, the development of the DIGRAPH3 modules being spread over two decades, the programming style did evolve with growing experience and the changes and enhancement coming up with the ongoing new releases of the standard Python3 libraries.

The purpose of this book is to present in a single monograph the methodological and scientific contributions the author made over the past two decades to the multiple-criteria decision aiding field and that are either left unpublished or solely published in very specialised media difficult to access.

This monograph should provide the reader with a self-contained series of tutorials which explain how to solve multiple-criteria selection, as well as ranking or rating decision problems. If successful in this aim, the curious reader will effectively install the DIGRAPH3 programming resources on their laptop and try out and redo for themselves the proposed computations.

The material in this book is valuable for master's students and doctoral candidates in computer science, mathematics, engineering sciences or computational management sciences taking a course on algorithmic decision theory, multiple-criteria decision aiding or decision analysis. Some experience in computer programming, in particular with Python, will assist the reader, but it is not a prerequisite. The many coding examples shown throughout the text are purposely kept elementary from a programming point of view.

Chapters presenting algorithms for ranking multiple-criteria performance records from best to worst—especially when facing big performance tableaux—will be of computational interest for designers of web recommender systems.

Similarly, the relative and absolute quantiles-rating algorithms, discussed and illustrated in several chapters, will be of practical interest for public or private performance auditors.

Finally, the monograph does not provide any mathematical developments or proofs. Those readers interested in the mathematical background of our decision algorithms are invited to consult the references provided at chapter level. Full texts of most of these references may be downloaded from the open access https://orbilu.uni.lu/ repository of the University of Luxembourg.

Acknowledgements

This monograph contains many ideas, methods and tools that are not the author's alone. They have been shared with and enhanced by friends, colleagues and many students. To name the most relevant: *Pascal Bouvry*, *Denis Bouyssou*, *Luis C Dias*, *Claude Lamboray*, *Patrick Meyer*, *Vincent Mousseau*, *Alex Olteanu*, *Marc Pirlot*, the late *Bernard Roy*, *Ulrich Sorger*, *Alexis Tsoukiàs*, *Thomas Veneziano* and, especially, the late *Marc Roubens*.

The UL HPC team, and more specifically *Valentin Plugaru* and *Sébastien Varette*, helped with mastering the multiprocessing experiments on the HPC platforms. The Springer publishing team very kindly assisted me eventually with the manuscript preparation.

The help by everybody is gratefully acknowledged.

Luxembourg City, Luxembourg
Autumn 2021

Raymond Bisdorff

Introduction

The Editing Strategy

In the five parts of this monograph, the reader will find several series of tutorials and advanced topics that present and illustrate computational methods and tools mainly useful in the field of multiple-criteria decision aiding and decision analysis. These methods and tools were designed and implemented first in Python2 and then in Python3 by the author over the last two decades for supporting both the computational verification and validation of decision algorithms as well as the preparation and illustration of a master's course on algorithmic decision theory.

Each chapter illustrates a specific preference modelling aspect, like building a best choice recommendation, ranking or rating a set of potential decision alternatives, or computing the winner of an election. In order to keep parts and chapters more or less self-contained, definitions and explanations of major concepts, like bipolar-valued digraphs, multiple-criteria performance tableaux and outranking situations, may appear several times in the monograph.

Explicit Python programming examples, purposely kept elementary, are shown in numerous terminal session style listings. A complete list of the numbered listings, shown over all the chapters, is printed in the Appendix. These programming examples were all checked against errors with the `doctest` module of the standard Python3 library and should work effectively as such either, in a Python3 interactive terminal console, or for sure in an `ipython` console.[1] Note that the layout of console `print(...)` outcomes has been edited in some listings for easing their reading. Some chapters will rely on a given data file that is made available in the `examples` directory of the DIGRAPH3 resources.

For similarly easing their reading, most chapters do not provide mathematical developments and proofs. Readers interested in such details are invited to consult the references listed separately at the end of each chapter. The author's references

[1] IP[i]: IPython interactive computing, https://ipython.org.

provide full text access to preprints on the open access https://orbilu.uni.lu/ repository of the University of Luxembourg.

Readers interested in the technical aspects of the organisation and implementation of the collection of DIGRAPH3 Python modules are invited to consult the extensive reference manual: https://digraph3.readthedocs.io/en/latest/techDoc.html, assisted by a search page https://digraph3.readthedocs.io/en/latest/search.html covering the whole DIGRAPH3 documentation.

Organisation of the Book

The content of the monograph is divided into five parts.

Part I presents three chapters introducing the DIGRAPH3 programming resources and the main formal objects discussed in this book, namely *bipolar-valued digraphs* and, in particular, *outranking digraphs*.

In Chap. 1, the reader will gain contact with the DIGRAPH3 Python resources. First are given the installation instructions and the list of the main DIGRAPH3 Python modules with their purpose. A Python terminal session using the root `digraphs` module eventually illustrates how to generate, save and inspect a random crisp digraph.

Chapter 2 introduces the bipolar-valued digraph model—the root type of all our digraph models. A randomly bipolar-valued digraph instance is generated. Drawing the digraph, separating its asymmetric and symmetric parts, or its border and inner parts, is illustrated. The initial digraph instance may be reconstructed by epistemic disjunctive fusion from these respective parts. Dual, converse and co-dual transforms as well as symmetric and transitive closures are presented. Complete, empty and indeterminate digraphs are eventually presented.

Chapter 3 presents the bipolar-valued outranking digraph—the main formal object used and discussed in this monograph. We first notice its hybrid type: it is conjointly a multiple-criteria performance tableau and a bipolar-valued relation modelling outranking situations between the given performance records. Pairwise comparisons of decision alternatives and the recording of the digraph characteristic valuation are then illustrated. The co-dual transform of the outranking digraph renders the corresponding strict outranking digraph, i.e. its asymmetric part.

Part II illustrates in eight methodological chapters multiple-criteria performance evaluation models and decision algorithms. These chapters are mostly problem oriented.

Chapter 4 presents the RUBIS best choice recommender system. The approach is illustrated with building a best office-location recommendation. We show how to explore a given performance tableau and compute the corresponding outranking digraph. After presenting the pragmatic principles that govern our best choice recommendation algorithm we solve the best office-location choice problem.

Chapter 5 illustrates a way of creating a new multiple-criteria performance tableau by editing a given template file containing five decision alternatives, three decision objectives and six performance criteria. We discuss in detail how to edit the decision alternatives, the decision objectives, the family of performance criteria, and, finally, the evaluations of the decision alternatives on the performance criteria.

Chapter 6 describes the DIGRAPH3 resources for generating random multiple-criteria performance tableaux. These random performance tableaux instances, mainly meant for illustration and training purposes, were serving the preparation and illustration of the algorithmic decision theory course lectures. The random generators propose several useful models like a cost-benefit tableau, a three objectives—economic, societal and environmental—tableau, and an academic performance tableau.

Chapter 7 is more specifically devoted to handling linear voting profiles and computing the winner of such election results like the simple majority or the instant runoff winner. By following CONDORCET's recipe, we consider pairwise comparisons of election candidates and balance the number of times the first beats the second against the number of times the second beats the first in order to obtain a majority margins digraph, in fact a bipolar-valued digraph. When the voters express contradictory linear voting profiles, one may naturally observe cyclic social preferences without seeing any paradox in this situation. Finally, the chapter presents a more politically realistic generator for random linear voting profiles by taking into account pre-election polls.

Chapter 8 introduces several algorithms for solving multiple-criteria ranking problems via bipolar-valued outranking digraphs. The COPELAND, NETFLOWS, KEMENY, SLATER, KOHLER and the RANKEDPAIRS ranking rules are illustrated with the help of a random outranking digraph. The fitness of their respective ranking result is measured with a bipolar-valued version of KENDALL's ordinal correlation index.

Chapter 9 applies order statistics for sorting a set X of n potential decision alternatives, evaluated on m incommensurable performance criteria, into q quantile equivalence classes. The sorting algorithm is based on pairwise outranking characteristics involving the quantile class limits observed on each criterion. Thus, we may implement a weak ordering algorithm of complexity $O(nmq)$.

Chapter 10 addresses the problem of rating multiple-criteria performance records of a set of potential decision alternatives with respect to performance quantiles learned from similar decision alternatives observed in the past. We show how to incrementally compute performance quantiles from incoming performance tableaux. New performance records may now be rated with respect to such historical quantile norms.

Chapter 11 tackles the ranking of big multiple-criteria performance tableaux with thousands or millions of records. To effectively compute rankings from performance tableaux of these sizes, the chapter proposes a collection of C-compiled and optimised modules that may be run on Linux Debian HPC clusters as available, for instance, at the University of Luxembourg.

Part III delivers three realistic algorithmic decision-making case studies.

Chapter 12 presents a case study concerning the building of a best choice recommendation for Alice, a German student who wants some advice concerning the choice of her future university studies. We present Alice's performance tableau—potential foreign language study programs, her decision objectives, performance criteria and performance evaluations—and build a best choice recommendation for her. A thorough robustness analysis confirms a very best choice.

In Chap. 13 we are resolving with our DIGRAPH3 resources a ranking decision problem based on published data from the Times Higher Education (THE) World University Rankings 2016 by Computer Science (CS) subject. We first have a look into the THE multiple-criteria ranking data with short Python scripts. In a second section, we relax the commensurability hypothesis of the ranking criteria and show how to similarly rank with multiple incommensurable performance criteria of solely ordinal significance. A third section is finally devoted to introduce quality measures for qualifying ranking results.

Chapter 14 presents and discusses how to rate with the help of our DIGRAPH3 resources the apparent student enrolment quality of higher education institutions. The multiple-criteria performance tableau we use is inspired by a 2004 student survey published by DER SPIEGEL magazine and concerning nearly 50,000 students, enrolled in 1 of 15 popular academic subjects, like German studies, life sciences, psychology, law or computer science.

In Chap. 15, we propose a series of decision problems of various difficulties which may serve as exercises and exam questions for an algorithmic decision theory or multiple-criteria decision analysis course. They cover selection, ranking and rating problems.

Part IV presents in five chapters more advanced topics showing some pearls of bipolar-valued epistemic logic.

Starting from a motivating decision problem about how to list, from the best to the worst, a set of movies that are star-rated by journalists and movie critics, Chap. 16 shows that KENDALL's ordinal correlation index tau can be extended to a relational bipolar-valued equivalence measure of bipolar-valued digraphs. This finding gives way, on the one hand, to measure the fitness and fairness of multiple-criteria ranking rules. On the other hand, it provides a tool for illustrating preference divergences between decision objectives and/or performance criteria.

We illustrate in Chap. 17, first, the concept of graph kernel, i.e. maximal independent set of vertices. In non-symmetric digraphs, the kernel concept becomes richer and separates into initial and terminal kernels. In, furthermore, lateralized outranking digraphs, initial and terminal kernels become separate may deliver first, respectively last, choice recommendations. After commenting the tractability of kernel computations, we close the chapter with the solving of bipolar-valued kernel equation systems.

In Chap. 18 we propose to link a qualifying significance majority for outranking situations with a required $\alpha\%$-confidence level. We model therefore the significance weights as random variables following more or less widespread distributions around a mean value that corresponds to the given deterministic significance weights. As the bipolar-valued random credibility of an outranking situation hence results from the simple sum of positive or negative independent random variables, we can apply the central limit theorem (CLT) for computing the bipolar-valued likelihood that the expected significance majority margin will indeed be positive, respectively negative.

In Chap. 19, we study the robustness of the outranking digraph when the criteria significance weights faithfully indicate solely an order of importance. The required cardinal significance weights of the performance criteria represent actually the 'Achilles' heel of the outranking approach. Rarely will indeed a decision maker be cognitively competent for suggesting precise decimal-valued criteria significance weights. This approach leads furthermore to the concept of unopposed or Pareto efficient multi-objective choices.

In a social choice context, where decision objectives would match different political parties, such Pareto efficient choices represent in fact multipartisan social choices. Chapter 20 shows that they may judiciously deliver the primary selection in a two-stage election system. The outranking model is based on bipolar approvals-disapprovals of "*at least as well evaluated as*" statements. A similar approach is put into practice with bipolar approval-disapproval voting systems. When converting such approval-disapproval voting ballots into corresponding performance records, one obtains a $(-1, 0, 1)$-valued evaluative voting system. We eventually show in this chapter that, in such bipolar voting systems, the election winner tends to be among the more or less multipartisan candidates.

Part V illustrates in three chapters computational resources for working with simple undirected graphs.

Chapter 21 introduces bipolar-valued undirected graphs and illustrates several special graph models and algorithms like Q-coloring, maximal independent set (MIS) and clique enumeration, line graphs and maximal matchings, grid graphs, and n-cycle graphs with their non-isomorphic MISs.

Chapter 22 specifically addresses working with tree graphs and graph forests. We illustrate how to generate and recognise random tree graphs and how to compute the centres of a tree and draw a rooted and oriented tree. Finally, algorithms for computing spanning trees and forests are presented.

Chapter 23 eventually presents some famous classes of BERGE graphs, namely comparability, interval, permutation and split graphs. We first present an example of an interval graph which is at the same time a triangulated, a comparability, a split and a permutation graph. The importance of being an interval graph is illustrated with *Claude Berge*'s mystery story. We discuss furthermore the generation of permutation graphs and close with how to recognise that a given graph is in fact a permutation graph.

Highlights

Contrary to what is generally thought, it is the preparation of the multiple-criteria performance tableau that takes most of the decision analysis time, not running any decision algorithms. Designing adequate performance evaluating criteria functions for each decision objective and collecting meaningful and precise evaluations is crucial for the success of the decision-making. This is a very critical and essential step. Chapters 4, 5 and 12 illustrate and discuss in detail coherent multiple-criteria performance tableaux. In order to discover more examples of potential performance tableaux, we provide in Chap. 6 random generators for several common kinds of performance tableaux.

Once the multiple-criteria performance tableau is ready, the thrilling step of discovering the resulting outranking relation starts. Are there many chordless outranking circuits? What is its degree of symmetry? What is its degree of transitivity? If the number of potential decision alternatives is small—less than 30—can one try, in the case of a selection problem, to compute prekernels in order to find potential first- or last-choice decision alternatives? Chapters 4, 12 and 17 are illustrating and discussing this challenging computational problem.

Comparing various ranking rules working on bipolar-valued outranking relations constructed from performance tableaux of various kinds, cost-benefit, 3-objectives and academic a.-o., has made us confident about the fact that convincing criteria for judging the quality of a ranking result may not to be found alone by mathematical properties, like KEMENY optimality or CONDORCET consistency. More useful seems to be the fair balancing of decision objectives and performance criteria. In this respect, it is the NETFLOWS ranking rule which appears to be most effective and often gives fairly balanced multiple-criteria rankings. Chapter 8 on ranking rules, the ranking and rating case studies of Chaps. 13 and 14, and Chap. 16 on bipolar-valued relational equivalence of digraphs illustrate and discuss this important topic.

The bipolar-valued epistemic logic, in which our decision algorithms are computing and expressing their decision solutions, provides effective assistance for coping with missing data and imprecise performance evaluations. Chapters 14 and 16 illustrate this advantage. An efficient robustness analysis becomes furthermore available for handling, on the one side, uncertain criteria significance weights leading in Chap. 18 to $\alpha\%$-confident outranking digraphs. On the other side, Chap. 19 illustrates how to compute robust outranking digraphs and decision solutions when solely ordinal criteria significance weights are given. In Chap. 20, the same kind of robustness analysis proposes strategies for tempering plurality tyranny effects in social choice problems by favouring multipartisan candidates, like two-stage elections with multipartisan primary selection of candidates or bipolar approval-disapproval voting systems.

Noticing the efficiency of the bipolar-valued epistemic logical framework for handling outranking digraphs, we could not resist making in Chaps. 21 and 22 an excursion into the domain of simple undirected graphs and tree graphs. The beautiful book *Algorithmic Graph Theory and Perfect Graphs* by M. Ch. Golumbic gave eventually the opportunity to tackle in the last chapter some famous classes of BERGE graphs.

It is my hope that the reader, by going on, will find the same astonishment and enchantment as I experienced when discovering the simplicity, efficiency and elegance of handling bipolar-valued outranking digraphs and graphs with Python programming resources. Extending the bipolar-valued epistemic logical framework to other computational science domains will prove valuable, I am sure, for many future scientific works.

Contents

Part II Evaluation Models and Decision Algorithms

List of Figures

List of Tables

Listings

Part I
Introduction to the DIGRAPH3 Python Resources

The first part contains three chapters for introducing the DIGRAPH3 software collection of Python programming resources. The first chapter is devoted to the installation of the DIGRAPH3 Python modules and running a first Python terminal session using the DIGRAPH3 3 resources. The second chapter introduces bipolar-valued digraphs, the root type of all available specialised digraphs. The third chapter finally introduces the main formal objects of this book, namely bipolar-valued outranking digraphs.

Chapter 1
Working with the DIGRAPH3 Python Resources

Contents

Abstract The chapter is devoted to a first contact with the DIGRAPH3 Python resources. Following the installation instructions, we list the main Python modules with their purpose and eventually illustrate in a first Python terminal session how to generate, save, and inspect a random crisp digraph.

1.1 Installing the DIGRAPH3 Resources

Using the DIGRAPH3 Python modules is simple.[1] You only need to have installed on your system the Python programming language of version 3 (by default available under Linux and MacOS).

Several download options (easiest under Linux or MacOS) are given; either, by using a `git` client and `clone` a working copy from the `github.com` directory:

```
...$ git clone https://github.com/rbisdorff/Digraph3
```

or from the `sourceforge.net` directory:

```
...$ git clone https://git.code.sf.net/p/digraph3/code Digraph3
```

It is also possible, with a browser access, to download either,

- from the `github.com` link or
- from the `sourceforge.net` link,

[1] See the technical description of the DIGRAPH3 programming resources, Bisdorff (2021).

R. Bisdorff, *Algorithmic Decision Making with Python Resources*,
International Series in Operations Research & Management Science 324,
https://doi.org/10.1007/978-3-030-90928-4_1

the latest distribution `zip` archive and extract it.

On Linux or MacOS, `..$ cd` to the cloned or extracted `Digraph3` directory. The following shell command installs (with *sudo* !) the DIGRAPH3 modules in the system-wide python environment :

```
.../Digraph3$ make install
```

Python-3.8 (or later) environment is recommended (see the `makefile` for adapting the `make install` command to your running python environment).

Whereas the following shell command installs the DIGRAPH3 modules locally in an activated virtual python user environment:

```
.../Digraph3$ make installVenv
```

If the C-compiled modules for Big Data applications are required, it is necessary to previously install the `cython` compiler[2] package in the running Python environment:

```
...$ python3 -m pip install cython
```

It is recommended to run a test suite:

```
.../Digraph3$ make tests
```

Test results are stored in the `Digraph3/test` directory. Notice that the python3 `pytest` package is required:

```
...$ python3 -m pip install pytest
```

A verbose (with `stdout` not captured) `pytest` suite may be run as follows:

```
.../Digraph3$ make verboseTests
```

When the GNU `parallel` shell tool[3] is installed and multiple cores are detected, the tests will be executed in multiprocessing mode:

```
../Digraph3$ make pTests
```

Individual module `pytest` suites are also provided (see the `makefile`), like the one for the `outrankingDigraphs` module (see Chap. 2):

```
../Digraph3$ make outrankingDigraphsTests
```

Dependencies

To be fully functional, the DIGRAPH3 resources mainly need:

- The `graphviz` tools (Gansner and North 2000),[4] and
- The **R** statistics resources[5] to be installed;

---

[2] https://cython.org.

[3] https://www.gnu.org/software/parallel.

[4] https://graphviz.org.

[5] https://www.r-project.org.

- When exploring digraph isomorphisms, the `nauty` isomorphism testing program is required (McKay and Piperno 2013). On Linux you may try: `sudo apt install nauty`. For MacOS, corresponding dmg installers are available for downloading;
- Two specific methods of the `OutrankingDigraph` class for clustering performance criteria or decision alternatives require furthermore the `calmat` matrix computing resource to be installed (see the `calmat` directory in the DIGRAPH3 resources).

1.2 Organisation of the DIGRAPH3 Python Modules

The main data handling modules of the DIGRAPH3 resources are the following:

1. `digraphs`: main part of the DIGRAPH3 source code with the root `Digraph` class.
2. `graphs`: resources for handling undirected graphs with the root `Graph` class and a bridge to the `digraphs` module resources.
3. `perfTabs`: tools for handling multiple-criteria performance tableaux with root `PerformanceTableau` class.
4. `outrankingDigraphs`: root module for handling outranking digraphs with the abstract root `OutrankingDigraph` class and the main `BipolarOutrankingDigraph` class.[6]
5. `votingProfiles`: classes and methods for handling voting ballots and computing election results with main `LinearVotingProfile` class.

Various random generators are provided by the following modules:

1. `randomDigraphs`: various random digraph models like random crisp digraphs (`RandomDigraph` class) or random bipolar-valued digraphs (`RandomValuationDigraph` class).
2. `randomPerfTabs`: various implemented random performance tableau models, like Cost-Benefit tableaux (`RandomCBPerformanceTableau` class) or 3-Objectives tableaux (`Random3ObjectivesPerformanceTableau` class).
3. `randomNumbers`: additional random number generators, not available in the standard Python `random.py` library, like a discrete random variable (`DiscreteRandomVariable` class) or a Cauchy random variable (`CauchyRandomVariable` class)

[6] Notice that the outrankingDigraph class defines a hybrid object type, inheriting conjointly from the `Digraph` class and the `PerformanceTableau` class.

Following modules provide tools for *sorting*, *ranking*, and *rating* problems:

1. `sortingDigraphs`: tools for solving sorting problems with the root `SortingDigraph` class and the main `QuantilesSortingDigraph` class;
2. `linearOrders`: tools for solving linearly ranking or ordering problems with the root `LinearOrder` class;
3. `transitiveDigraphs`: additional tools for handing transitive digraphs with root `TransitiveDigraph` class.

Tools for specifically handling Big Data are eventually provided by the following modules:

1. `performanceQuantiles`: incremental representation of large performance tableaux via binned cumulated density functions per criteria; Depends on the `randomPerfTabs` module.
2. `sparseOutrankingDigraphs`: sparse implementation design for bipolar-valued outranking digraphs of order >500.
3. *Cythonized* modules: C-compiled and optimised Python modules for handling big performance tableaux and bipolar outranking digraphs of order >1000.

Readers interested in technical implementation details are invited to consult the reference manual of the DIGRAPH3 resources, where they will find the documentation and complete source code of all DIGRAPH3 modules, classes, and methods (Bisdorff 2021).

1.3 Starting a DIGRAPH3 Terminal Session

After downloading the DIGRAPH3 resources, one may start an interactive Python3 terminal session in the `Digraph3` directory.

```
$HOME/.../Digraph3$ python3
Python 3.9.6 (v3.9.6:db3ff76da1, Jun 28 2021, 11:49:53)
[Clang 6.0 (clang-600.0.57)] on darwin
Type "help", "copyright", "credits" or
   "license" for more information.
>>>
```

For exploring the classes and methods provided by the DIGRAPH3 modules enter the Python3 commands following the session prompts marked with `>>>` or `...`; the lines without a prompt are output from the Python3 interpreter. Python class names and Boolean parameters start by convention with a capital case; names of other Python objects, like modules, methods, and variables start with a lower case. All Python names and code are shown in a `typewriting` font.

Listing 1.1 Generating a digraph instance

```
>>> from randomDigraphs import RandomDigraph
>>> dg = RandomDigraph(order=5,arcProbability=0.5,\
```

```
3 ...                       seed=101)
4 >>> dg
5   *------- Digraph instance description ------*
6    Instance class        : RandomDigraph
7    Instance name         : randomDigraph
8    Digraph Order         : 5
9    Digraph Size          : 12
10   Valuation domain      : [-1.00; 1.00]
11   Determinateness (%)   : 100.00
12   Attributes          : ['actions','valuationdomain',
13                          'relation','order','name',
14                          'gamma','notGamma']
```

In Listing 1.1 on the facing page we import, for instance, from the `randomDigraphs` module the `RandomDigraph` class in order to generate a random digraph object `dg` of order 5 and arc probability of 50%. The resulting digraph of *order* 5—number of nodes called (decision) *actions*—and *size* 12—number of arcs—is completely determined (see Line 11) .

The content of `dg` may be saved in a file named `tutorialDigraph.py`.

```
>>> dg.save('tutorialDigraph')
 *--- Saving digraph in file: <tutorialDigraph.py> ---*
```

with the following content:

Listing 1.2 A stored digraph instance

```
from decimal import Decimal
from collections import OrderedDict
actions = OrderedDict([
    ('a1', {'shortName': 'a1',
          'name': 'random decision action'}),
    ('a2', {'shortName': 'a2',
          'name': 'random decision action'}),
    ('a3', {'shortName': 'a3',
          'name': 'random decision action'}),
    ('a4', {'shortName': 'a4',
          'name': 'random decision action'}),
    ('a5', {'shortName': 'a5',
          'name': 'random decision action'}),
 ])
valuationdomain = {
     'min': Decimal('-1.0'),
     'med': Decimal('0.0'),
     'max': Decimal('1.0'),
     'hasIntegerValuation': True, # representation format
 }
relation = {
     'a1': {'a1':Decimal('-1.0'), 'a2':Decimal('-1.0'),
              'a3':Decimal('1.0'), 'a4':Decimal('-1.0'),
              'a5':Decimal('-1.0'),},
     'a2': {'a1':Decimal('1.0'), 'a2':Decimal('-1.0'),
              'a3':Decimal('-1.0'), 'a4':Decimal('1.0'),
              'a5':Decimal('1.0'),},
```

```
    'a3': {'a1':Decimal('1.0'), 'a2':Decimal('-1.0'),
             'a3':Decimal('-1.0'), 'a4':Decimal('1.0'),
             'a5':Decimal('-1.0'),},
    'a4': {'a1':Decimal('1.0'), 'a2':Decimal('1.0'),
             'a3':Decimal('1.0'), 'a4':Decimal('-1.0'),
             'a5':Decimal('-1.0'),},
    'a5': {'a1':Decimal('1.0'), 'a2':Decimal('1.0'),
             'a3':Decimal('1.0'), 'a4':Decimal('-1.0'),
             'a5':Decimal('-1.0'),},
}
```

In the DIGRAPH3 resources, all digraph object instances are of root `Digraph` type (see the technical description) and contain at least the following attributes (see Listing 1.1 on page 6 Lines 12–14):

1. A `name` attribute, holding usually the actual name of the stored instance that was used to create the instance;
2. An ordered dictionary of digraph nodes called `actions` (decision alternatives) with at least a `name` attribute;
3. An `order` attribute containing the number of graph nodes (length of the actions dictionary) automatically added by the object constructor;
4. A logical characteristic `valuationdomain` dictionary with three decimal entries: the minimum (-1.0, means certainly false), the median (0.0, means missing information), and the maximum characteristic value ($+1.0$, means certainly true);
5. A double dictionary called `relation` and indexed by an oriented pair of actions (nodes) and carrying a decimal characteristic value in the range of the previous valuation domain;
6. Its associated `gamma` attribute, a dictionary containing the direct successors, respectively, predecessors of each action, automatically added by the object constructor;
7. Its associated `notGamma` attribute, a dictionary containing the actions that are not direct successors, respectively, predecessors of each action, automatically added by the object constructor.

1.4 Inspecting a Digraph Object

Different `show...()` methods, like the `showShort()` method, now reveal us that `dg` is a crisp, irreflexive, and connected digraph of order five (see Listing 1.3 Lines 1, 16, 26, 29).

Listing 1.3 Random crisp digraph object

```
1 >>> dg.showShort()
2  *----- show short -------------*
3  Digraph          : tutorialDigraph
4  Actions          : OrderedDict([
```

```
 ('a1', {'shortName': 'a1', 'name': 'random decision action'}),
 ('a2', {'shortName': 'a2', 'name': 'random decision action'}),
 ('a3', {'shortName': 'a3', 'name': 'random decision action'}),
 ('a4', {'shortName': 'a4', 'name': 'random decision action'}),
 ('a5', {'shortName': 'a5', 'name': 'random decision action'})
 ])
Valuation domain : {
 'min': Decimal('-1.0'),
 'max': Decimal('1.0'),
 'med': Decimal('0.0'), 'hasIntegerValuation': True
 }
>>> dg.showRelationTable()
 * ---- Relation Table -----
   S   |  'a1'  'a2'  'a3'  'a4'  'a5'
 ------|-------------------------------
  'a1' |   -1    -1     1    -1    -1
  'a2' |    1    -1    -1     1     1
  'a3' |    1    -1    -1     1    -1
  'a4' |    1     1     1    -1    -1
  'a5' |    1     1     1    -1    -1
Valuation domain: [-1;+1]
>>> dg.showComponents()
 *--- Connected Components ---*
 1: ['a1', 'a2', 'a3', 'a4', 'a5']
>>> dg.showNeighborhoods()
 Neighborhoods:
   Gamma     :
 'a1': in => {'a2', 'a4', 'a3', 'a5'}, out => {'a3'}
 'a2': in => {'a5', 'a4'}, out => {'a1', 'a4', 'a5'}
 'a3': in => {'a1', 'a4', 'a5'}, out => {'a1', 'a4'}
 'a4': in => {'a2', 'a3'}, out => {'a1', 'a3', 'a2'}
 'a5': in => {'a2'}, out => {'a1', 'a3', 'a2'}
   Not Gamma :
 'a1': in => set(), out => {'a2', 'a4', 'a5'}
 'a2': in => {'a1', 'a3'}, out => {'a3'}
 'a3': in => {'a2'}, out => {'a2', 'a5'}
 'a4': in => {'a1', 'a5'}, out => {'a5'}
 'a5': in => {'a1', 'a4', 'a3'}, out => {'a4'}
```

The `exportGraphViz()` method generates in the current working directory a `tutorialDigraph.dot` file and a `tutorialdigraph.png` picture of the tutorial digraph dg (see Fig. 1.1), if the *graphviz* tools are installed on your system (Gansner and North 2000).

```
>>> dg.exportGraphViz('tutorialDigraph')
 *---- exporting a dot file do GraphViz tools ---------*
 Exporting to tutorialDigraph.dot
 dot -Grankdir=BT -Tpng tutorialDigraph.dot -o
     tutorialDigraph.png
```

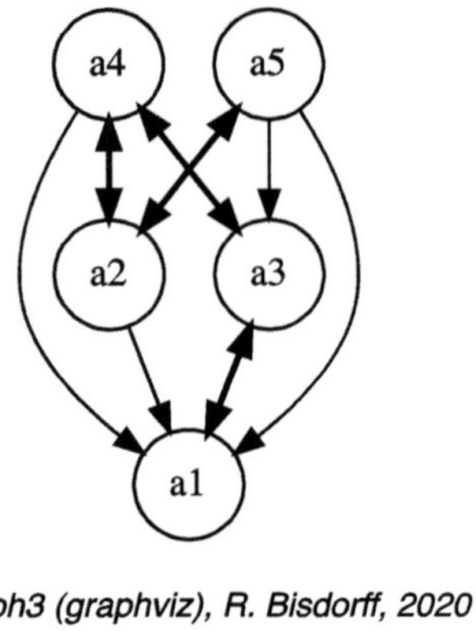

Fig. 1.1 *The tutorial crisp digraph*. The `exportGraphViz()` method is depending on drawing tools from https://graphviz.org. On Linux Ubuntu or Debian you may try '`sudo apt-get install graphviz`' to install them. Under MacOS there are ready `dmg` installers available

Further methods are provided for inspecting this random `Digraph` object `dg`, like the following `showStatistics()` method.

Listing 1.4 Inspecting a `Digraph` object

```
>>> dg.showStatistics()
 *----- general statistics -------------*
 for digraph                : <tutorialDigraph.py>
 order                      : 5 nodes
 size                       : 12 arcs
 undetermined             : 0 arcs
 determinateness (%)        : 100.0
 arc density                : 0.60
 double arc density         : 0.40
 single arc density         : 0.40
 absence density            : 0.20
 strict single arc density: 0.40
 strict absence density     : 0.20
 nbr. of components          : 1
 nbr. of strong components   : 1
 transitivity degree (%)    : 60.0
                            : [0, 1, 2, 3, 4, 5]
 outdegrees distribution    : [0, 1, 1, 3, 0, 0]
 indegrees distribution     : [0, 1, 2, 1, 1, 0]
 mean outdegree             : 2.40
 mean indegree              : 2.40
                            : [0, 1, 2, 3, 4, 5, 6, 7, 8, 9, 10]
 symmetric degrees dist.    : [0, 0, 0, 0, 1, 4, 0, 0, 0, 0, 0]
 mean symmetric degree      : 4.80
 outdegrees concentration index    : 0.1667
 indegrees concentration index     : 0.2333
 symdegrees concentration index    : 0.0333
                                   : [0, 1, 2, 3, 4, 'inf']
 neighbourhood depths distribution: [0, 1, 4, 0, 0, 0]
 mean neighbourhood depth          : 1.80
 digraph diameter                  : 2
 agglomeration distribution        :
```

Relation Map
Ranking rule: Alphabetic

r(x,y)	a1	a2	a3	a4	a5
a1	—	—	+	—	—
a2	+	—	—	+	+
a3	+	—	—	+	—
a4	+	+	+	—	—
a5	+	+	+	—	—

Semantics	
+	certainly valid
+	valid
	indeterminate
−	invalid
—	certainly invalid

Fig. 1.2 *Browsing the relation map of the tutorial digraph.* + Indicates a certainly valid and − indicates a certainly invalid relation. Here we find confirmed again that our random digraph instance, `dg`, is indeed a crisp, i.e. 100% determined irreflexive digraph instance

```
 a1 : 58.33
 a2 : 33.33
 a3 : 33.33
 a4 : 50.00
 a5 : 50.00
 agglomeration coefficient          : 45.00
```

The preceding `show...()` methods usually rely upon corresponding compute methods, like: `computeSize()`, `computeDeterminateness()`, or `computeTransitivityDegree()`.

Listing 1.5 Various `compute...()` methods

```
>>> dg.computeSize()
 12
>>> dg.computeDeterminateness(InPercents=True)
 Decimal('100.00')
>>> dg.computeTransitivityDegree(InPercents=True)
 Decimal('60.00')
```

Mind that `show...()` methods output their results in the Python terminal console. We provide also some `showHTML...()` methods which output their results in a system browser's tab or window (Fig. 1.2).

```
>>> dg.showHTMLRelationMap(relationName='r(x,y)',\
...                          rankingRule=None)
```

Some special types of digraph instances, like the `CirculantDigraph` or the `GridDigraph` classes, are readily available (see Fig. 1.3 on the following page).

Listing 1.6 Circulant digraphs and $n \times m$ grid digraphs

```
>>> from digraphs import CirculantDigraph,GridDigraph
>>> c8 = CirculantDigraph(order=8,circulants=[1,3])
>>> c8.exportGraphViz('c8')
```

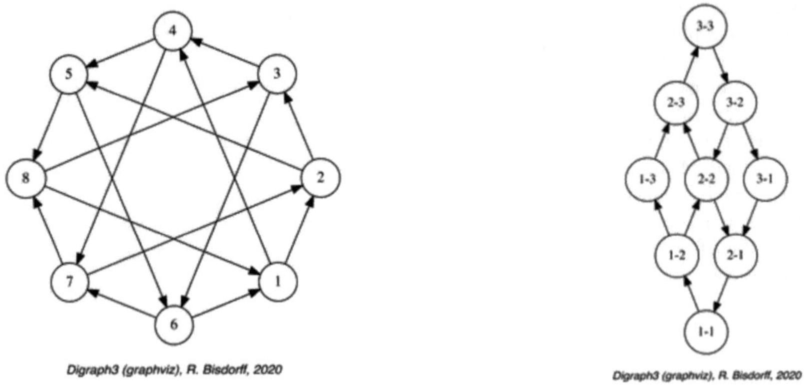

Fig. 1.3 The circulant [1,3] digraph and the 3 grid digraph

```
  *---- exporting a dot file for GraphViz tools ----*
   Exporting to c8.dot
   circo -Tpng c8.dot -o c8.png
>>> grid3 = GridDigraph(n=3,m=3,\
...                     hasMedianSplitOrientation=True)
>>> grid.exportGraphViz('grid3')
  *---- exporting a dot file for GraphViz tools ----*
   Exporting to grid3.dot
   dot -Grankdir=BT -Tpng grid3.dot -o grid3.png
```

The next Chap. 2 introduces the fundamental *bipolar-valued digraph* model which is *root object type* to all the digraph models implemented in the DIGRAPH3 modules (Bisdorff 2021).

References

Bisdorff R (2021) Technical documentation of the Digraph3 collection of Python modules. https://digraph3.readthedocs.io/en/latest/techDoc.html

Gansner E, North S (2000) An open graph visualization system and its applications to software engineering. Softw Pract Exp 30(11):1203–1233. https://graphviz.org/documentation/

McKay B, Piperno A (2013) Practical graph isomorphism, II. J Symb Comput 60:94–112. https://www.cs.sunysb.edu/~algorith/implement/nauty/implement.shtml

Chapter 2
Working with Bipolar-Valued Digraphs

Contents

Abstract The chapter introduces bipolar-valued digraphs, the fundamental root type of all the specialised digraphs implemented in the DIGRAPH3 modules. With the help of a randomly valued digraph, we illustrate some basic digraph manipulation methods, like drawing the digraph, dividing the digraph into its asymmetric and symmetric parts, separating the border from the inner part, computing associated dual, converse and codual digraphs, and operating symmetric and transitive closures.

2.1 Random Bipolar-Valued Digraphs

In Listing 2.1, we generate a uniformly random $[-1.0; +1.0]$-valued digraph of order 7, denoted `rdg` and modelling, for instance, a binary relation $S(x, y)$ defined on the set of nodes of `rdg`. For this purpose, the DIGRAPH3 resources provide in the `randomDigraphs` module a specific `RandomValuationDigraph` class (Bisdorff 2021b).

Listing 2.1 Random bipolar-valued digraph instance

```
>>> from randomDigraphs import RandomValuationDigraph
>>> rdg = RandomValuationDigraph(order=7)
>>> rdg.save('tutRandValDigraph')
```


R. Bisdorff, *Algorithmic Decision Making with Python Resources*,
International Series in Operations Research & Management Science 324,
https://doi.org/10.1007/978-3-030-90928-4_2

```
4  >>> from digraphs import Digraph
5  >>> rdg = Digraph('tutRandValDigraph')
6  >>> rdg
7    *------- Digraph instance description ------*
8     Instance class       : Digraph
9     Instance name        : tutRandValDigraph
10    Digraph Order        : 7
11    Digraph Size         : 22
12    Valuation domain     : [-1.00;1.00]
13    Determinateness (%)  : 75.24
14    Attributes           : ['name','actions','order',
15                             'valuationdomain','relation',
16                             'gamma','notGamma']
```

With the `save()` method (see Listing 2.1 on the previous page Line 3) we keep for future use a backup version of `rdg` which is saved into a file named `tutRandValDigraph.py` in the current working directory. The genuine `Digraph` class constructor can reload the `rdg` object from the stored file (Line 5). One can easily inspect the content of a `Digraph` object with its name `rdg` (Line 6). The digraph size 22 indicates the number of positively valued arcs (Line 11). The valuation domain is uniformly distributed in the interval $[-1.0; 1.0]$, the mean absolute arc valuation being $(0.7524 \times 2) - 1.0 = 0.5048$ (Line 13).

As mentioned in the previous Chap. 1, all objects of `Digraph` type contain at least the list of attributes shown here in Lines 14–16:

- A `name` (string) and a dictionary of `actions` (digraph nodes),
- An `order` (integer) attribute containing the number of nodes,
- A `valuationdomain` dictionary and a double dictionary `relation` representing the adjacency table of the digraph relation,
- A `gamma` and a `notGamma` dictionary containing the direct neighbourhood and not neighbourhood of each node.

The `Digraph` class provides some generic `show...()` methods for exploring the content of a given `Digraph` object, like the `showRelationTable()`, the `showComponents()`, and the `showNeighborhoods()` methods.

Listing 2.2 Example of random valuation digraph

```
>>> rdg.showRelationTable()
 * ---- Relation Table -----
  r(xSy) |  '1'   '2'   '3'   '4'   '5'   '6'   '7'
  -------|------------------------------------------
   '1'   |  0.00 -0.48  0.70  0.86  0.30  0.38  0.44
   '2'   | -0.22  0.00 -0.38  0.50  0.80 -0.54  0.02
   '3'   | -0.42  0.08  0.00  0.70 -0.56  0.84 -1.00
   '4'   |  0.44 -0.40 -0.62  0.00  0.04  0.66  0.76
   '5'   |  0.32 -0.48 -0.46  0.64  0.00 -0.22 -0.52
   '6'   | -0.84  0.00 -0.40 -0.96 -0.18  0.00 -0.22
   '7'   |  0.88  0.72  0.82  0.52 -0.84  0.04  0.00
>>> rdg.showComponents()
```

```
  *--- Connected Components ---*
  1: ['1', '2', '3', '4', '5', '6', '7']
>>> rdg.showNeighborhoods()
  *---- Neighborhoods ------*
     Gamma:
     '1': in => {'5','7','4'}, out => {'5','7','6','3','4'}
     '2': in => {'7','3'},out => {'5','7','4'}
     '3': in => {'7','1'}, out => {'6','2','4'}
     '4': in => {'5','7','1','2','3'}, out => {'5','7','1','6'}
     '5': in => {'1','2','4'}, out => {'1','4'}
     '6': in => {'7','1','3','4'}, out => set()
     '7': in => {'1','2','4'}, out => {'1','2','3','4','6'}
     Not Gamma:
     '1': in => {'6','2','3'}, out => {'2'}
     '2': in => {'5','1','4'}, out => {'1','6','3'}
     '3': in => {'5','6','2','4'}, out => {'5','7','1'}
     '4': in => {'6'}, out => {'2','3'}
     '5': in => {'7','6','3'}, out => {'7','6','2','3'}
     '6': in => {'5','2'}, out => {'5','7','1','3','4'}
     '7': in => {'5','6','3'}, out => {'5'}
```

Positive characteristic values indicate the presence of an arc whereas negative ones indicate their absence. 0.0 values indicate an indeterminate situation: there may be or not be an arc (see Listing 2.2 on the facing page Lines 5–11). Mind that some `Digraph` class methods will ignore the *reflexive* links by considering that they are *indeterminate*, i.e. the characteristic value $r(x\,S\,x)$ for all action x is set to the *median*, i.e. *indeterminate* value 0.0 in this case (Bisdorff 2004).

2.2 Graphviz Drawings

An even better insight into the `Digraph` object `rdg` is given by looking in Fig. 2.1 at its graphviz drawing (Gansner and North 2000).[1]

```
>>> rdg.exportGraphViz('tutRandValDigraph')
 *---- exporting a dot file for GraphViz tools ------*
  Exporting to tutRandValDigraph.dot
  dot -Grankdir=BT -Tpng tutRandValDigraph.dot\
                           -o tutRandValDigraph.png
```

[1] The `exportGraphViz()` method is depending on drawing tools from the graphviz software (https://graphviz.org/). On Linux Ubuntu or Debian you may try to install them with `...$ sudo apt-get install graphviz`. There are ready `dmg` installers for MacOS.

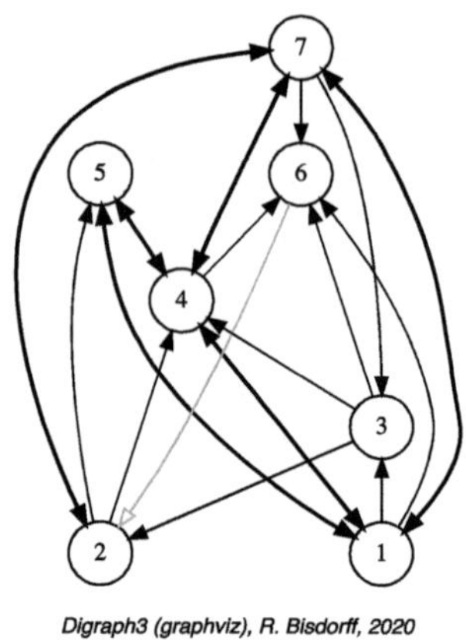

Fig. 2.1 *The tutorial random valuation digraph.* Double links are drawn in bold black with an arrowhead at each end, whereas single asymmetric links are drawn in black with an arrowhead showing the direction of the link. Notice the undetermined relational situation ($r(6\ S\ 2) = 0.00$) observed between nodes `6` and `2`. The corresponding link is marked in grey with an open arrowhead in the drawing

2.3 Asymmetric and Symmetric Parts

Extracting both the *symmetric* as well as the *asymmetric* part of digraph `rdg` may be done with the help of two corresponding classes (see Listing 2.3).

Listing 2.3 Computing asymmetric and symmetric parts

```
>>> from digraphs import AsymmetricPartialDigraph,\
...                      SymmetricPartialDigraph
>>> asymDg = AsymmetricPartialDigraph(rdg)
>>> asymDg.exportGraphViz()
>>> symDg = SymmetricPartialDigraph(rdg)
>>> symDg.exportGraphViz()
```

The relation constructors of the partial objects `asymDg` and `symDg` set to the indeterminate characteristic value 0.0 all non-asymmetric, respectively, non-symmetric links between nodes (see Fig. 2.2 on the next page).

In Listing 2.4 below is shown, for instance, the source code of the `relation` constructor of the `AsymmetricPartialDigraph` class:

Listing 2.4 Computing the asymmetric part of a bipolar-valued relation

```
def _constructRelation(self):
    actions = self.actions
    Min = self.valuationdomain['min']
    Max = self.valuationdomain['max']
    Med = self.valuationdomain['med']
    relationIn = self.relation
    relationOut = {}
    for a in actions:
        relationOut[a] = {}
        for b in actions:
            if a != b:
                if relationIn[a][b] >= Med and relationIn[b][a] <= Med:
                    relationOut[a][b] = relationIn[a][b]
                elif relationIn[a][b] <= Med and relationIn[b][a] >= Med:
                    relationOut[a][b] = relationIn[a][b]
```

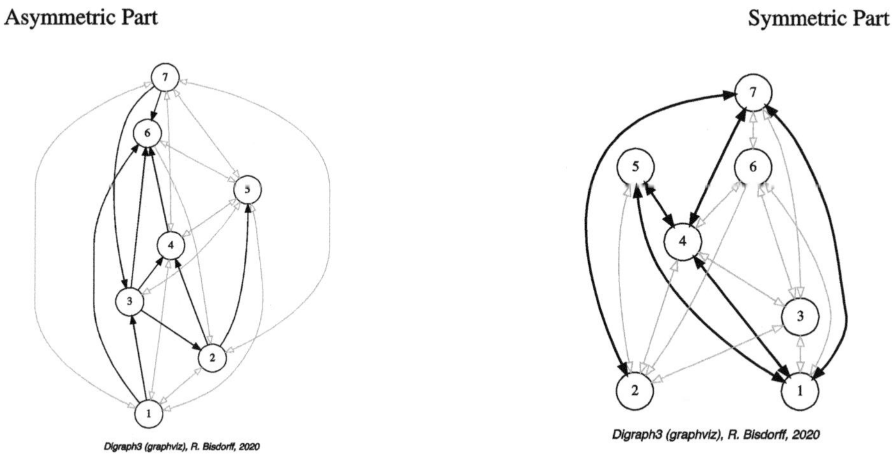

Fig. 2.2 Asymmetric and symmetric part of the tutorial random valuation digraph

```
16                 else:
17                     relationOut[a][b] = Med
18             else: # reflexive links are ignored
19                 relationOut[a][b] = Med
20     return relationOut
```

In Line 17 above all non-asymmetric situations are set to the *indeterminate*—median—characteristic value. Mind by the way that all reflexive relations are similarly set to this indeterminate value (see Line 19).

2.4 Border and Inner Parts

We can also extract the *border*—the part of a digraph induced by the union of its initial and terminal prekernels (see Chap. 17)—as well as, the *inner part*—the complement of the border—with the help of two corresponding classes: `GraphBorder` and `GraphInner` (see Listing 2.5).

Figure 2.3 on the following page shows the border and inner parts of the linear ordering obtained from the tutorial random valuation digraph `rdg` with the NETFLOWS ranking rule (see Sect. 8.3).

Listing 2.5 Border and inner part of a linear order

```
>>> from digraphs import GraphBorder, GraphInner
>>> from linearOrders import NetFlowsOrder
>>> nf = NetFlowsOrder(rdg)
>>> nf.netFlowsOrder
  ['6', '4', '5', '3', '2', '1', '7']
>>> bnf = GraphBorder(nf)
```

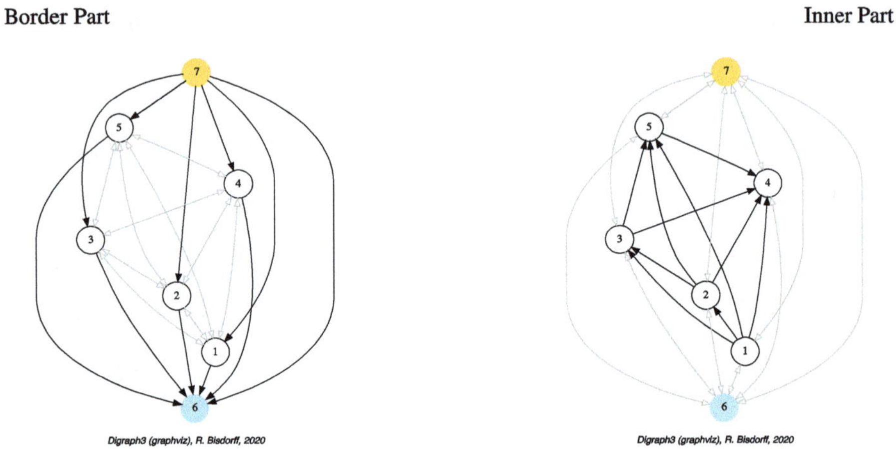

Fig. 2.3 *Border* and *inner* part of a linear order oriented by *terminal* and *initial* kernels

```
7 >>> bnf.exportGraphViz(lastChoice=['6'],firstChoice=['7'])
8 >>> inf = GraphInner(nf)
9 >>> inf.exportGraphViz(lastChoice=['6'],firstChoice=['7'])
```

The drawings in Fig. 2.3 are oriented with the terminal node `6` (`lastChoice` parameter) and initial node `7` (`firstChoice` parameter) (see Listing 2.5 on the previous page Lines 7 and 9).

The class constructors of the partial digraphs `bnf` and `inf` (see Lines 6 and 8) set to the *indeterminate* characteristic value all links not in the *border*, respectively, *not* in the *inner* part (see Fig. 2.3). Being much *denser* than a linear order, the actual inner part of our tutorial random valuation digraph `rdg` is reduced to a single arc between nodes `3` and `4` (see Fig. 2.4 on the facing page right side). Indeed, a complete digraph on the limit has no inner part (privacy!) at all, whereas empty and indeterminate digraphs admit both, an empty border and an empty inner part.

2.5 Fusion by Epistemic Disjunction

The `Digraph` object `rdg` can be restored from both partial objects `asymDg` and `symDg`, or as well from the border `bg` and the inner part `ig`, with a *bipolar disjunctive fusion* operator `o-max`, also called *epistemic disjunction*, available via the `FusionDigraph` class.

Listing 2.6 Epistemic fusion of partial diagraphs

```
>>> from digraphs import FusionDigraph
>>> fusDg = FusionDigraph(asymDg,symDg,operator='o-max')
>>> # fusDg = FusionDigraph(bg,ig,operator='o-max')
>>> fusDg.showRelationTable()
```

Border Part **Inner Part**

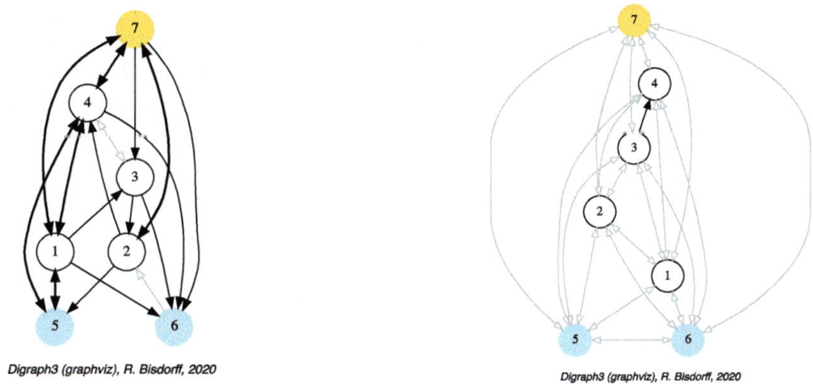

Fig. 2.4 Border and inner part of the tutorial random valuation digraph `rdg`

```
5   * ---- Relation Table -----
6     r(xSy)  |  '1'    '2'    '3'   '4'    '5'    '6'   '7'
7     -------|------------------------------------------------
8      '1'    |  0.00 -0.48  0.70  0.86  0.30  0.38  0.44
9      '2'    | -0.22  0.00 -0.38  0.50  0.80 -0.54  0.02
10     '3'    | -0.42  0.08  0.00  0.70 -0.56  0.84 -1.00
11     '4'    |  0.44 -0.40 -0.62  0.00  0.04  0.66  0.76
12     '5'    |  0.32 -0.48 -0.46  0.64  0.00 -0.22 -0.52
13     '6'    | -0.84  0.00 -0.40 -0.96 -0.18  0.00 -0.22
14     '7'    |  0.88  0.72  0.82  0.52 -0.84  0.04  0.00
```

The epistemic fusion operator `o-max` (see Listing 2.6 Line 2) is defined as follows:

Definition 2.1 (Disjunctive Epistemic Fusion Operator `o-max`) Let r and r' characterise two bipolar-valued epistemic situations:

- `o-max`$(r, r') = \max(r, r')$ when both $r \geqslant 0.0$ and $r' \geqslant 0.0$, i.e. r and r' are positively valid or indeterminate;
- `o-max`$(r, r') = \min(r, r')$ when both $r \leqslant 0.0$ and $r' \leqslant 0.0$, i.e. r and r' are positively invalid or indeterminate;
- otherwise `o-max`$(r, r') = 0.0$, i.e. indeterminate.

Mind that the `o-max` operator, like a mean operator, is *not associative* when more than 2 operands are given. In order to make the `o-max` fusion univocal, the following rule is applied—first, all positive and negative terms are separately aggregated—then the `o-max` fusion is applied to both aggregates.

2.6 Dual, Converse, and Codual Digraphs

The `digraphs` module provides class constructors for computing the *dual*[2] (negated relation), the *converse* (transposed relation), and the *codual* (transposed and negated relation) of the `Digraph` instance `rdg` (see Listing 2.7 Lines 4, 16, 29).

Listing 2.7 Computing associated dual, converse and codual digraphs

```
1  >>> from digraphs import\
2  ...          DualDigraph, ConverseDigraph, CoDualDigraph
3  >>> # dual of rdg
4  >>> ddg = DualDigraph(rdg)
5  >>> ddg.showRelationTable()
6       -r(xSy) |  '1'    '2'    '3'   '4'    '5'    '6'  '7'
7       --------|---------------------------------------------
8       '1 '    |  0.00  0.48 -0.70 -0.86 -0.30 -0.38 -0.44
9       '2'     |  0.22  0.00  0.38 -0.50  0.80  0.54 -0.02
10      '3'     |  0.42  0.08  0.00 -0.70  0.56 -0.84  1.00
11      '4'     | -0.44  0.40  0.62  0.00 -0.04 -0.66 -0.76
12      '5'     | -0.32  0.48  0.46 -0.64  0.00  0.22  0.52
13      '6'     |  0.84  0.00  0.40  0.96  0.18  0.00  0.22
14      '7'     |  0.88 -0.72 -0.82 -0.52  0.84 -0.04  0.00
15 >>> # converse of rdg
16 >>> cdg = ConverseDigraph(rdg)
17 >>> cdg.showRelationTable()
18      * ---- Relation Table -----
19       r(ySx) |  '1'    '2'    '3'    '4'    '5'    '6'    '7'
20      --------|---------------------------------------------
21      '1'     |  0.00 -0.22 -0.42  0.44  0.32 -0.84  0.88
22      '2'     | -0.48  0.00  0.08 -0.40 -0.48  0.00  0.72
23      '3'     |  0.70 -0.38  0.00 -0.62 -0.46 -0.40  0.82
24      '4'     |  0.86  0.50  0.70  0.00  0.64 -0.96  0.52
25      '5'     |  0.30  0.80 -0.56  0.04  0.00 -0.18 -0.84
26      '6'     |  0.38 -0.54  0.84  0.66 -0.22  0.00  0.04
27      '7'     |  0.44  0.02 -1.00  0.76 -0.52 -0.22  0.00
28 >>> # codual of rdg
29 >>> cddg = CoDualDigraph(rdg)
30 >>> cddg.showRelationTable()
31      * ---- Relation Table -----
32      -r(ySx) |  '1'    '2'    '3'    '4'    '5'    '6'    '7'
33      --------|---------------------------------------------
34      '1'     |  0.00  0.22  0.42 -0.44 -0.32  0.84 -0.88
35      '2'     |  0.48  0.00 -0.08  0.40  0.48  0.00 -0.72
36      '3'     | -0.70  0.38  0.00  0.62  0.46  0.40 -0.82
37      '4'     | -0.86 -0.50 -0.70  0.00 -0.64  0.96 -0.52
38      '5'     | -0.30 -0.80  0.56 -0.04  0.00  0.18  0.84
```

[2] Not to be confused with the dual graph of a plane graph g that has a vertex for each face of g. Here we mean the *less than* (strict converse) relation corresponding to a *greater or equal* relation, or the *less than or equal* relation corresponding to a (strict) *better than* relation.

```
      '6'    | -0.38  0.54 -0.84 -0.66  0.22  0.00 -0.04
      '7'    | -0.44 -0.02  1.00 -0.76  0.52  0.22  0.00
```

Computing the *dual*, respectively, the *converse* of a digraph, may also be done with prefixing the `__neg__` (−) or the `__invert__` (∼) operator. The *codual* of a `Digraph` object can, hence, as well be computed with a *composition* (in either order) of both operations (see Line 3 below).

Listing 2.8 Computing the dual, the converse, and the codual of a digraph

```
1  >>> ddg = -rdg    # dual of rdg
2  >>> cdg = ~rdg    # converse of rdg
3  >>> cddg = ~(-rdg) # = -(~(rdg) codual of rdg
4  >>> (-(~rdg)).showRelationTable()
5    * ---- Relation Table -----
6     -r(ySx)  |  '1'    '2'    '3'    '4'    '5'    '6'    '7'
7     ---------|---------------------------------------------
8       '1'    |  0.00   0.22   0.42 -0.44 -0.32   0.84 -0.88
9       '2'    |  0.48   0.00 -0.08   0.40   0.48   0.00 -0.72
10      '3'    | -0.70   0.38   0.00   0.62   0.46   0.40 -0.82
11      '4'    | -0.86 -0.50 -0.70   0.00 -0.64   0.96 -0.52
12      '5'    | -0.30 -0.80   0.56 -0.04   0.00   0.18   0.84
13      '6'    | -0.38   0.54 -0.84 -0.66   0.22   0.00 -0.04
14      '7'    | -0.44 -0.02   1.00 -0.76   0.52   0.22   0.00
```

2.7 Symmetric and Transitive Closures

Symmetric and transitive closures, by default in-site methods, are also available (see Fig. 2.5). Note that it is a good idea, before going ahead with these in-site operations, that irreversibly modifies the original `rdg` object, to previously make a backup version of `rdg`. The simplest storage method, always provided by the generic `Digraph.save()` method, writes out in a named file the Python content of the `Digraph` object in string representation (see Sect. 1.3).

Listing 2.9 Symmetric and transitive closures

```
1 >>> rdg.save('tutRandValDigraph')
2 >>> rdg.closeSymmetric(InSite=True)
3 >>> rdg.closeTransitive(InSite=True)
4 >>> rdg.exportGraphViz('symTransClosure')
```

The `closeSymmetric()` method (see Listing 2.9 Line 2), of complexity $O(n^2)$ where n denotes the digraph's order, changes, on the one hand, all single pairwise links it may detect into double links by operating a disjunction of the pairwise relations. On the other hand, the `closeTransitive()` method (see

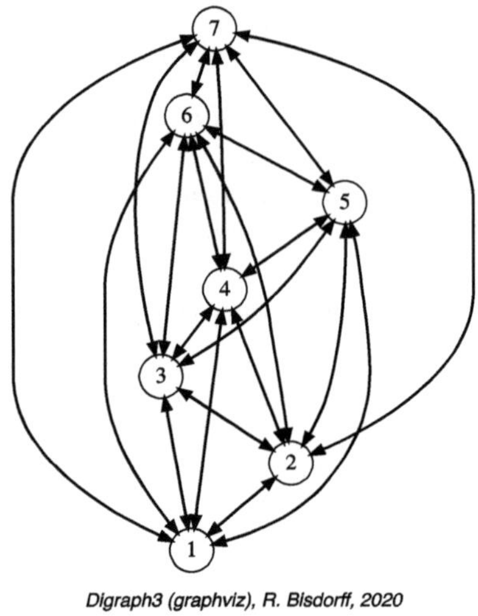

Fig. 2.5 Symmetric and transitive closure of the tutorial random valuation digraph `rdg`

Line 3) implements the *Roy-Warshall* transitive closure algorithm of complexity $O(n^3)$ (Roy 1959; Warshall 1962).

The same `closeTransitive()` with a `Reverse = True` flag may be readily used for eliminating all transitive arcs from a transitive digraph instance. We make usage of this feature when drawing HASSE diagrams of `TransitiveDigraph` objects.

2.8 Strong Components

As the original digraph `rdg` was connected (see above the result of the `showShort()` command), both the symmetric and the transitive closures operated together, will necessarily produce a single strong component, i.e. a **complete** digraph. It is sometimes useful to collapse all strong components in a given digraph and construct the so *collapsed* digraph. Using the `StrongComponentsCollapsedDigraph` constructor here will render a single hyper-node gathering all the original nodes (see Line 7 below).

Listing 2.10 Computing the strong components in a digraph

```
1 >>> from digraphs import StrongComponentsCollapsedDigraph
2 >>> sc = StrongComponentsCollapsedDigraph(rdg)
3 >>> sc.showAll()
4   *----- show detail -----*
5    Digraph              : tutRandValDigraph_Scc
6   *---- Actions ----*
7     ['_7_1_2_6_5_3_4_']
8   *---- Relation Table -----
9       S      |  'Scc_1'
```

```
   -------|---------
    'Scc_1' |  0.00
 *---- strong Components ----*
  short          content
  'Scc_1'        '_7_1_2_6_5_3_4_'
 *---- Neighborhoods ----*
  Gamma        :
  'frozenset({'7','1','2','6','5','3','4'})':
                    in => set(), out => set()
  Not Gamma  :
  'frozenset({'7','1','2','6','5','3','4'})':
                    in => set(), out => set()
```

2.9 CSV Storage

Sometimes it is required to exchange the graph valuation data in CSV format with statistical software like **R**[3] or `gretl`.[4] For this purpose it is possible to export the digraph adjacency table into a CSV file. The characteristic valuation domain is hereby normalised by default to the range $[-1.0, 1.0]$ and the diagonal is put by default to the minimal value -1.0.

```
>>> rdg = Digraph('tutRandValDigraph')
>>> rdg.saveCSV('tutRandValDigraph')
  # content of file tutRandValDigraph.csv
  "d","1","2","3","4","5","6","7"
  "1",-1.0,0.48,-0.7,-0.86,-0.3,-0.38,-0.44
  "2",0.22,-1.0,0.38,-0.5,-0.8,0.54,-0.02
  "3",0.42,-0.08,-1.0,-0.7,0.56,-0.84,1.0
  "4",-0.44,0.4,0.62,-1.0,-0.04,-0.66,-0.76
  "5",-0.32,0.48,0.46,-0.64,-1.0,0.22,0.52
  "6",0.84,0.0,0.4,0.96,0.18,-1.0,0.22
  "7",-0.88,-0.72,-0.82,-0.52,0.84,-0.04,-1.0
```

It is possible to reload a `Digraph` instance from its previously saved CSV file content.

```
>>> from digraphs import CSVDigraph
>>> rdgcsv = CSVDigraph('tutRandValDigraph')
>>> rdgcsv.showRelationTable(ReflexiveTerms=False)
    * ---- Relation Table -----
    r(xSy) |   '1'    '2'    '3'    '4'    '5'    '6'    '7'
    -------|-------------------------------------------------
    '1'    |    -    -0.48   0.70   0.86   0.30   0.38   0.44
    '2'    |  -0.22    -    -0.38   0.50   0.80  -0.54   0.02
    '3'    |  -0.42   0.08    -     0.70  -0.56   0.84  -1.00
    '4'    |   0.44  -0.40  -0.62    -     0.04   0.66   0.76
```

[3] https://www.r-project.org/.

[4] The Gnu Regression, Econometrics and Time-series Library http://gretl.sourceforge.net/.

Tutorial random digraph

r(x S y)	1	2	3	4	5	6	7
1	0.00	-0.48	0.70	0.86	0.30	0.38	0.44
2	-0.22	0.00	-0.38	0.50	0.80	-0.54	0.02
3	-0.42	0.08	0.00	0.70	-0.56	0.84	-1.00
4	0.44	-0.40	-0.62	0.00	0.04	0.66	0.76
5	0.32	-0.48	-0.46	0.64	0.00	-0.22	-0.52
6	-0.84	0.00	-0.40	-0.96	-0.18	0.00	-0.22
7	0.88	0.72	0.82	0.52	-0.84	0.04	0.00

Fig. 2.6 *The valued relation table shown in a browser window*. Positive arcs are shown in green and negative arcs in red. Indeterminate—zero-valued—arcs, like the reflexive diagonal ones or the arc between nodes 6 and 2, are shown in grey

```
   '5'     |  0.32 -0.48 -0.46  0.64   -    -0.22 -0.52
   '6'     | -0.84  0.00 -0.40 -0.96 -0.18   -    -0.22
   '7'     |  0.88  0.72  0.82  0.52 -0.84  0.04   -
```

It is as well possible with the `showHTMLRelationTable()` method to show a coloured version of the valued relation table in a system browser window (see Fig. 2.6).

```
>>> rdgcsv.showHTMLRelationTable(tableTitle="Tutorial random
    digraph")
```

2.10 Complete, Empty, and Indeterminate Digraphs

Let us finally mention some special universal classes of digraphs that are readily available in the `digraphs` module,[5] like:

- the `CompleteDigraph` class,
- the `EmptyDigraph` class, and
- the `IndeterminateDigraph` class,

which set all pairwise characteristic values, respectively, to the *maximum*, the *minimum*, respectively, to the *median*—indeterminate—value.

Listing 2.11 Complete, empty, and indeterminate digraphs

```
>>> from digraphs import CompleteDigraph,EmptyDigraph,\
...                        IndeterminateDigraph
>>> # the empty digraph
>>> e = EmptyDigraph(order=5)
>>> e.showRelationTable()
    * ---- Relation Table -----
      S   |    '1'    '2'    '3'    '4'    '5'
    ---- -|-----------------------------------
    '1'   |  -1.00  -1.00  -1.00  -1.00  -1.00
```

[5] See Bisdorff (2021a).

```
10      '2'    |  -1.00  -1.00  -1.00  -1.00  -1.00
11      '3'    |  -1.00  -1.00  -1.00  -1.00  -1.00
12      '4'    |  -1.00  -1.00  -1.00  -1.00  -1.00
13      '5'    |  -1.00  -1.00  -1.00  -1.00  -1.00
14  >>> e.showNeighborhoods()
15      Neighborhoods:
16        Gamma      :
17      '1': in => set(), out => set()
18      '2': in => set(), out => set()
19      '5': in => set(), out => set()
20      '3': in => set(), out => set()
21      '4': in => set(), out => set()
22        Not Gamma  :
23      '1': in => {'2','4','5','3'}, out => {'2','4','5','3'}
24      '2': in => {'1','4','5','3'}, out => {'1','4','5','3'}
25      '5': in => {'1','2','4','3'}, out => {'1','2','4','3'}
26      '3': in => {'1','2','4','5'}, out => {'1','2','4','5'}
27      '4': in => {'1','2','5','3'}, out => {'1','2','5','3'}
28  >>> # the indeterminate digraph
29  >>> i = IndeterminateDigraph()
30      * ---- Relation Table -----
31        S    |  '1'    '2'    '3'    '4'    '5'
32      ------|------------------------------
33      '1'    |  0.00   0.00   0.00   0.00   0.00
34      '2'    |  0.00   0.00   0.00   0.00   0.00
35      '3'    |  0.00   0.00   0.00   0.00   0.00
36      '4'    |  0.00   0.00   0.00   0.00   0.00
37      '5'    |  0.00   0.00   0.00   0.00   0.00
38  >>> i.showNeighborhoods()
39      Neighborhoods:
40        Gamma      :
41      '1': in => set(), out => set()
42      '2': in => set(), out => set()
43      '5': in => set(), out => set()
44      '3': in => set(), out => set()
45      '4': in => set(), out => set()
46        Not Gamma  :
47      '1': in => set(), out => set()
48      '2': in => set(), out => set()
49      '5': in => set(), out => set()
50      '3': in => set(), out => set()
51      '4': in => set(), out => set()
```

Mind the subtle difference between the neighbourhoods of an *empty* and the neighbourhoods of an *indeterminate* `Digraph` instance. In the first kind, the neighbourhoods are known to be completely *empty* (see Listing 2.11 on the preceding page Lines 23–27) whereas, in the latter, *nothing is known* about the actual neighbourhoods of the nodes (see Lines 47–51). These two cases illustrate why in *bipolar-valued* digraphs, we may sometimes need both a `gamma` **and** a `notGamma` attribute.

Notes

It was *Denis Bouyssou* who first suggested us end of the nineties, when we started to work in Prolog on the computation of fuzzy digraph kernels with finite domain constraint solvers, that the 50% criteria significance majority was a very special value to be carefully taken into account. The converging solution vectors of the fixpoint kernel equations furthermore confirmed this special status of the 50% majority (see Chap. 17). These early insights led to the seminal articles on bipolar-valued epistemic logic where we introduced split truth/falseness semantics for a multi-valued logical processing of fuzzy preference modelling (Bisdorff 2000, 2002). The characteristic valuation domain remained, however, the classical fuzzy [0.0; 1.0] valuation domain.

It is only in 2004, when we succeeded in assessing the stability of the outranking digraph when solely ordinal criteria significance weights are given that it became clear and evident for us that the characteristic valuation domain had to be shifted to a $[-1.0; +1.0]$-valued domain (see Chap. 19 and Bisdorff 2004). In this bipolar valuation, the 50% majority threshold corresponds now to the median 0.0 value, characterising with the correct zero value an epistemic indeterminateness—no knowledge—situation. Furthermore, identifying truth and falseness directly by the sign of the characteristic value revealed itself to be very efficient not only from a computational point of view but also from scientific and semiotic perspectives. A positive (respectively, negative) characteristic value now attest a logically valid (respectively, invalid) statement and a negative affirmation now means a positive refutation and vice versa. Furthermore, the median zero value gives way to efficiently handling partial objects—like the border or the asymmetric part of a digraph—and, even more important from a practical decision making point of view, any missing data.

The bipolar $[-1.0; +1.0]$-valued characteristic domain opened so the way to important new operations and concepts, like the disjunctive epistemic fusion operation seen in Sect. 2.5 on page 18 that confers the outranking digraph a logically sound and epistemically correct definition (Bisdorff 2013). KENDALL's ordinal correlation index could be extended to a bipolar-valued relational equivalence index between digraphs (see Chap. 16 and Bisdorff 2012). Making usage of the bipolar-valued Gaussian error function naturally led to defining a bipolar-valued likelihood function, where a positive, respectively, negative, value gives the likelihood of an affirmation, respectively, a refutation (see Chap. 18 and Bisdorff 2014).

In the following Chap. 3 we introduce the main formal object of this book, namely *bipolar-valued outranking* digraphs.

References

Bisdorff R (2000) Logical foundation of fuzzy preferential systems with application to the electre decision aid methods. Comput Oper Res 27:673–687. http://hdl.handle.net/10993/23724

Bisdorff R (2002) Logical foundation of multicriteria preference aggregation. In: Bouyssou D, Jacquet-Lagrèze E, Perny P, Słowiński R, Vanderpooten D, Vincke P (eds) Aiding decisions with multiple criteria. Kluwer Academic, pp 379–403. http://hdl.handle.net/10993/45313

Bisdorff R (2004) On a natural fuzzification of Boolean logic. In: Klement EP, Pap E (eds) Linz seminar on fuzzy set theory, mathematics of fuzzy systems, Bildungszentrum St. Magdalena, Linz (Austria), pp 20–26. http://hdl.handle.net/10993/23721

Bisdorff R (2012) On measuring and testing the ordinal correlation between bipolar outranking relations. In: Mousseau V, Pirlot M (eds) DAP'2012 from multiple criteria decision aid to preference learning, University of Mons (Belgium), pp 91–100. http://hdl.handle.net/10993/23909

Bisdorff R (2013) On polarizing outranking relations with large performance differences. J Multi-Criteria Decis Anal Wiley 20:3–12. http://hdl.handle.net/10993/245

Bisdorff R (2014) On confident outrankings with multiple criteria of uncertain significance. In: Mousseau V, Pirlot M (eds) DAP'2014 from multiple criteria decision aid to preference learning, Ecole Centrale de Paris, pp 1–6. http://hdl.handle.net/10993/23910

Bisdorff R (2021a) Documentation of the Digraph3 collection of Python modules for algorithmic decision theory. https://digraph3.readthedocs.io/en/latest/

Bisdorff R (2021b) Technical documentation of the Digraph3 collection of Python modules. https://digraph3.readthedocs.io/en/latest/techDoc.html

Gansner E, North S (2000) An open graph visualization system and its applications to software engineering. Softw Pract Exp 30(11):1203–1233. https://graphviz.org/documentation/

Roy B (1959) Transitivité et connexité. C R Acad Sci Paris 249:216–218

Warshall S (1962) A theorem on Boolean matrices. J ACM 9:11–12

Chapter 3
Working with Outranking Digraphs

The rule for the combination of independent concurrent arguments takes a very simple form when expressed in terms of the intensity of belief ... It is this: Take the sum of all the feelings of belief which would be produced separately by all the arguments pro, subtract from that the similar sum for arguments con, and the remainder is the feeling of belief which ought to have the whole. This is a proceeding which men often resort to, under the name of balancing reasons

—C.S. Peirce, The probability of induction (1878)

Contents

Abstract In this chapter, we introduce the main formal object of this book, namely the bipolar-valued outranking digraph. With a randomly generated multiple-criteria performance tableau, we construct the corresponding bipolar-valued outranking relation from pairwise comparisons. The resulting bipolar-valued outranking characteristics may be recoded. Finally, the codual outranking digraph gives us the associated strict outranking relation.

3.1 The Hybrid Outranking Digraph Model

In the `outrankingDigraphs` module, the `BipolarOutrankingDigraph` class provides our standard *outranking digraph* constructor. Such an instance represents a *hybrid* object of both the `PerformanceTableau` type *and* the `Outran-`

R. Bisdorff, *Algorithmic Decision Making with Python Resources*, International Series in Operations Research & Management Science 324,
https://doi.org/10.1007/978-3-030-90928-4_3

kingDigraph type (Bisdorff 2021). A given BipolarOutrankingDigraph object contains hence always at least the following attributes:

1. actions: an ordered dictionary describing the potential decision actions or alternatives with name and comment attributes
2. objectives: a possibly empty ordered dictionary of decision objectives with name and comment attributes, describing the multiple preference dimensions involved in the decision problem
3. criteria: an ordered dictionary of performance criteria, i.e. *preferentially independent* and *non-redundant* decimal-valued evaluation functions used for assessing the performance of each potential decision action with respect to a decision objective
4. evaluation: a double dictionary gathering performance evaluations for each decision alternative on each criterion function
5. valuationdomain: a dictionary with three entries: the minimum (−1.0, *certainly outranked*), the median (0.0, *indeterminate*), and the maximum characteristic value (+1.0, *certainly outranking*)
6. relation: a double dictionary defined on the Cartesian product of the set of decision alternatives capturing the credibility of the pairwise *outranking situation* computed on the basis of the performance differences observed between couples of decision alternatives on the given family of criteria functions

Let us consider, for instance, a random bipolar-valued outranking digraph with seven decision actions denoted a1, a2, ..., a7. We need therefore, first, to generate in Listing 3.1 a corresponding random performance tableau.

Listing 3.1 Generating a random performance tableau

```
>>> from perfTabs import RandomPerformanceTableau
>>> pt = RandomPerformanceTableau(numberOfActions=7,\
...                               seed=100)
>>> pt
  *------- PerformanceTableau instance description ------*
   Instance class     : RandomPerformanceTableau
   Seed               : 100
   Instance name      : randomperftab
   Actions            : 7
   Criteria           : 7
   NaN proportion (%) : 6.1
>>> pt.showActions()
  *----- show digraphs actions --------------*
  key:  a1
   name:        action 1
   comment:     RandomPerformanceTableau() generated.
  key:  a2
   name:        action 2
   comment:     RandomPerformanceTableau() generated.
    ...
    ...
```

```
    key:  a7
     name:          action 7
     comment:       RandomPerformanceTableau() generated.
```

In this example, we consider a family of seven *equi-significant* cardinal *criteria functions* g1, g2, ..., g7, measuring the performance of each alternative on a rational scale from 0.0 (worst) to 100.00 (best). In order to capture the evaluation procedures' potential *uncertainty* and *imprecision*, each criterion function g1 to g7 admits three performance *discrimination thresholds* of 2.5, 5.0, and 80.0 pts for warranting, respectively, any *indifference*, *preference*, or *considerable performance difference* situation.

Listing 3.2 Inspecting the performance criteria

```
1  >>> pt.showCriteria()
2    *----  criteria -----*
3     g1 'RandomPerformanceTableau() instance'
4       Scale = [0.0, 100.0]
5       Weight = 1.0
6       Threshold ind : 2.50 + 0.00x ; percentile: 4.76
7       Threshold pref : 5.00 + 0.00x ; percentile: 9.52
8       Threshold veto : 80.00 + 0.00x ; percentile: 95.24
9     g2 'RandomPerformanceTableau() instance'
10      Scale = [0.0, 100.0]
11      Weight = 1.0
12      Threshold ind : 2.50 + 0.00x ; percentile: 6.67
13      Threshold pref : 5.00 + 0.00x ; percentile: 6.67
14      Threshold veto : 80.00 + 0.00x ; percentile: 100.00
15       ...
16       ...
17    g7 'RandomPerformanceTableau() instance'
18      Scale = [0.0, 100.0]
19      Weight = 1.0
20      Threshold ind : 2.50 + 0.00x ; percentile: 0.00
21      Threshold pref : 5.00 + 0.00x ; percentile: 4.76
22      Threshold veto : 80.00 + 0.00x ; percentile: 100.00
```

On criteria function g1, we observe, for instance, about 5% of *indifference* situations, about 90% of *preference* situations, and about 5% of *considerable* performance difference situations (see Lines 6–8 in Listing 3.2).

The individual *performance evaluation* of all decision alternatives on each criterion is gathered in a *performance table*.

Listing 3.3 Inspecting the performance evaluations

```
>>> pt.showPerformanceTableau()
    *----  performance tableau -----*
    criteria |  'a1'  'a2'  'a3'  'a4'  'a5'  'a6'  'a7'
    ---------|---------------------------------------
      'g1'   |  15.2  44.5  57.9  58.0  24.2  29.1  96.6
      'g2'   |  82.3  43.9   NA   35.8  29.1  34.8  62.2
      'g3'   |  44.2  19.1  27.7  41.5  22.4  21.5  56.9
```

```
8        'g4'       |  46.4  16.2  21.5  51.2  77.0  39.4  32.1
9        'g5'       |  47.7  14.8  79.7  67.5   NA   90.7  80.2
10       'g6'       |  69.6  45.5  22.0  33.8  31.8   NA   48.8
11       'g7'       |  82.9  41.7  12.8  21.9  75.7  15.4   6.0
```

It is noteworthy to mention the three *missing data* (NA) cases: action a3 is missing, for instance, an evaluation on criterion g2 (see Line 6 in Listing 3.3).

3.2 The Bipolar-Valued Outranking Digraph

Given the previous random performance tableau pt, the BipolarOutranking-Digraph class constructor computes the corresponding *bipolar-valued outranking digraph*.

Listing 3.4 Example of random bipolar-valued outranking digraph

```
1  >>> from outrankingDigraphs import \
2  ...                      BipolarOutrankingDigraph
3  >>> odg = BipolarOutrankingDigraph(pt)
4  >>> odg
5    *------- Object instance description ------*
6     Instance class         : BipolarOutrankingDigraph
7     Instance name          : rel_randomperftab
8     Actions                : 7
9     Criteria               : 7
10    Size                   : 20
11    Determinateness (%)    : 63.27
12    Valuation domain       : [-1.00;1.00]
13    Attributes             : [
14         'name', 'actions',
15         'criteria', 'evaluation', 'NA',
16         'valuationdomain', 'relation',
17         'order', 'gamma', 'notGamma', ...
18         ]
```

The resulting digraph contains 20 positive (valid) outranking relations. And, the mean majority criteria significance support of all the pairwise outranking situations is 63.3% (see Lines 10–11 in Listing 3.4).

We can inspect the complete $[-1.0, +1.0]$-valued adjacency table with the showRelationTable() method().

Listing 3.5 Inspecting the valued adjacency table

```
>>> odg.showRelationTable()
  * ---- Relation Table -----
   r(x,y) |  'a1'   'a2'   'a3'   'a4'   'a5'   'a6'   'a7'
   -------|---------------------------------------------
    'a1'  | +1.00  +0.71  +0.29  +0.29  +0.29  +0.29  +0.00
    'a2'  | -0.71  +1.00  -0.29  -0.14  +0.14  +0.29  -0.57
    'a3'  | -0.29  +0.29  +1.00  -0.29  -0.14  +0.00  -0.29
```

```
8      'a4' | +0.00  +0.14  +0.57  +1.00  +0.29  +0.57  -0.43
9      'a5' | -0.29  +0.00  +0.14  +0.00  +1.00  +0.29  -0.29
10     'a6' | -0.29  +0.00  +0.14  -0.29  +0.14  +1.00  +0.00
11     'a7' | +0.00  +0.71  +0.57  +0.43  +0.29  +0.00  +1.00
12   Valuation domain: [-1.0; 1.0]
```

The `BipolarOutrankingDigraph` class constructor computes from the given performance tableau `pt` the characteristic value $r(x \succsim y)$ of a *pairwise outranking* relation $x \succsim y$ in a default *valuation domain* $[-1.0, +1.0]$ with the *median* value 0.0 acting as *indeterminate* characteristic value.

The semantics of $r(x \succsim y)$ are the following:

Definition 3.1 (Semantics of the Outranking Characteristic Function)

a. When $r(x \succsim y) > 0.0$, it is *True* that x *outranks* y, i.e. alternative x is '*at least as well evaluated as*' alternative y on a majority of criteria significance **and** there is no considerable negative performance difference observed in disfavour of x.
b. When $r(x \succsim y) < 0.0$, it is *False* that x *outranks* y, i.e. alternative x is **only** '*evaluated at least as well as*' alternative y on a minority of criteria significance **and** there is no considerable positive performance difference observed in favour of x.
c. When $r(x \succsim y) = 0.0$, it is **indeterminate** whether x outranks y or not.

3.3 Pairwise Comparisons

From above given semantics, we notice in Line 5 in Listing 3.5 on the preceding page that `a1` outranks `a2`: $r(a1 \succsim a2) + 0.71)$, but not `a7`: $r(a1 \succsim a7) = 0.0)$. In order to comprehend the characteristic values shown in the outranking relation table, we can inspect with the `showPairwiseComparison()` method the details of the pairwise multiple-criteria comparison between, for instance, alternatives `a1` and `a2`.

Listing 3.6 Inspecting a pairwise multiple-criteria comparison

```
>>> odg.showPairwiseComparison('a1','a2')
  *------------ pairwise comparison ----*
  Comparing actions : (a1, a2)
  crit. wght. g(a1)  g(a2)    diff   | ind   pref     r()
  -------------------------------      --------------------
   g1   1.00  15.17  44.51  -29.34 | 2.50  5.00    -1.00
   g2   1.00  82.29  43.90  +38.39 | 2.50  5.00    +1.00
   g3   1.00  44.23  19.10  +25.13 | 2.50  5.00    +1.00
   g4   1.00  46.37  16.22  +30.15 | 2.50  5.00    +1.00
   g5   1.00  47.67  14.81  +32.86 | 2.50  5.00    +1.00
   g6   1.00  69.62  45.49  +24.13 | 2.50  5.00    +1.00
   g7   1.00  82.88  41.66  +41.22 | 2.50  5.00    +1.00
   Valuation in range: -7.00 to +7.00;              -------
       r(a1,a2): +5.00/7.00 = +0.71                 +5.00
```

The outranking characteristic value $r(a1 \succsim a2)$ represents the relative *majority margin* resulting from the difference between the significance weights of the criteria in favour and the significance weights of the criteria in disfavour of the statement that alternative `a1` is '*at least as well evaluated as*' alternative `a2`. No considerable performance difference being observed, no polarising situation is triggered in this pairwise comparison.

Such a polarised situation is however observed when we compare the evaluations of alternatives `a1` and `a7` with the `showPairwiseOutrankings()` method.

Listing 3.7 Pairwise comparison with considerable performance difference

```
1  >>> odg.showPairwiseoutrankings('a1','a7')
2   *------------  pairwise comparison ----*
3    Comparing actions : (a1, a7)
4    crit. wght. g(a1)  g(a7)   diff   | ind   pref    r()   |  v     veto
5    -------------------------------    -----------------    -----------
6     g1   1.00  15.17  96.58  -81.41  | 2.50  5.00   -1.00  | 80.00 -1.00
7     g2   1.00  82.29  62.22  +20.07  | 2.50  5.00   +1.00  |
8     g3   1.00  44.23  56.90  -12.67  | 2.50  5.00   -1.00  |
9     g4   1.00  46.37  32.06  +14.31  | 2.50  5.00   +1.00  |
10    g5   1.00  47.67  80.16  -32.49  | 2.50  5.00   -1.00  |
11    g6   1.00  69.62  48.80  +20.82  | 2.50  5.00   +1.00  |
12    g7   1.00  82.88   6.05  +76.83  | 2.50  5.00   +1.00  |
13   ----------------------------------------
14   Valuation in range: -7.00 to +7.00; r(x,y)= +1/7 => 0.0
15  *------------  pairwise comparison ----*
16   Comparing actions : (a1, a7)
17   crit. wght. g(a7)  g(a1)   diff   | ind   pref    r()   |  v     veto
18   -------------------------------    -----------------    -----------
19    g1   1.00  96.58  15.17  +81.41  | 2.50  5.00   +1.00  | 80.00 +1.00
20    g2   1.00  62.22  82.29  -20.07  | 2.50  5.00   -1.00  |
21    g3   1.00  56.90  44.23  +12.67  | 2.50  5.00   +1.00  |
22    g4   1.00  32.06  46.37  -14.31  | 2.50  5.00   -1.00  |
23    g5   1.00  80.16  47.67  +32.49  | 2.50  5.00   +1.00  |
24    g6   1.00  48.80  69.62  -20.82  | 2.50  5.00   -1.00  |
25    g7   1.00   6.05  82.88  -76.83  | 2.50  5.00   -1.00  |
26   ----------------------------------------
27   Valuation in range: -7.00 to +7.00; r(x,y)= -1/7 => 0.0
```

This time, we observe a $(1/7 + 1)/2 = 57.1\%$ majority of criteria significance warranting an '*at least as well evaluated as*' situation between alternative `a1` and alternative `a7`. Yet, we also observe a considerable *negative* performance difference on criterion `g1` (see Line 6 in Listing 3.7). Both contradictory facts trigger eventually in Line 14 an *indeterminate* outranking situation. The inverse polarisation effect appears when considering in Lines 19–25 the converse performance differences between alternative `a7` and alternative `a1`. The considerable better performing situation on criterion `g1` makes doubtful the otherwise "*not at least as well evaluated as*" situation (see Lines 19 and 27).

Notice that the occurrence in a pairwise comparison of conjointly considerable positive and negative performance differences will also trigger an indeterminate outranking situation. When observing at the same time a positive (respectively, negative) "*at least as well evaluated as*" situation and one or more considerable positive (respectively, negative) performance difference, the outranking situation gets validated (respectively, invalidated) for certain (Bisdorff 2013).

3.4 Recoding the Characteristic Valuation Domain

All outranking digraphs, being of root `Digraph` type, inherit the methods available under this latter class. The characteristic valuation domain of a digraph can, for instance, be recoded with the `recodeValuation()` method to the *integer* range $[-7, +7]$, i.e. plus or minus the total significance weights of the family of criteria considered in this example instance `odg`.

Listing 3.8 Recoding the digraph valuation

```
>>> odg.recodeValuation(-7,+7)
>>> odg.valuationdomain['hasIntegerValuation'] = True
>>> Digraph.showRelationTable(odg,ReflexiveTerms=False)
  * ---- Relation Table -----
   r(x,y)   |  'a1'  'a2'  'a3'  'a4'  'a5'  'a6'  'a7'
  ---------|-------------------------------------------
     'a1'   |    -    +5    +2    +2    +2    +2     0
     'a2'   |   -5     -    -1    -1    +1    +2    -4
     'a3'   |   -1    +2     -    -1    -1     0    -1
     'a4'   |    0    +1    +4     -    +2    +4    -3
     'a5'   |   -1     0    +1     0     -    +2    -1
     'a6'   |   -1     0    +1    -1    +1     -     0
     'a7'   |    0    +5    +4    +3    +2     0     -
    Valuation domain: [-7;+7]
```

Notice in Listing 3.8 that the self comparison characteristics $r(x \succsim x)$ may be ignored by setting the `ReflexiveTerms` parameter to `False`. Mind that the trivial reflexive terms of outranking relations are ignored in some of the DIGRAPH3 methods.

3.5 The Strict Outranking Digraph

From theory, we know that a bipolar-valued outranking digraph is *weakly complete*, i.e. if $r(x \succsim y) < 0.0$, then $r(y \succsim x) \geqslant 0.0$. From this property follows that a bipolar-valued outranking digraph verifies the *coduality principle*: the *dual*[1] (negation) of the *converse* (transposition) of the outranking relation $(x \succsim y)$ corresponds to its asymmetric *strict outranking* part $(x \gtrsim y)$ (Bisdorff 2013, 2020).

Visualising the *codual* (*strict*) outranking digraph may be done with a graphviz drawing (Fig. 3.1).[2]

[1] Not to be confused with the dual graph of a plane graph g that has a vertex for each face of g. Here we mean the *less than* (strict converse) relation corresponding to a *greater or equal* relation, or the *less than or equal to* relation corresponding to a (strict) *better than* relation.

[2] The `exportGraphViz()` method is depending on drawing tools from the graphviz software (https://graphviz.org/). On Linux Ubuntu or Debian, you may try `sudo apt-get install graphviz` to install them. There are ready *dmg* installers for Mac OSX.

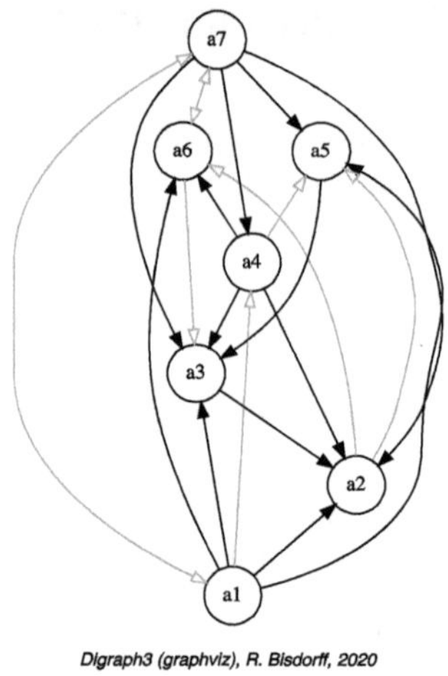

Fig. 3.1 *The strict (codual) outranking digraph.* It becomes readily clear now from the picture that both alternatives `a1` and `a7` are *not outranked* by any other alternatives. Hence, `a1` and `a7` appear as *weak* CONDORCET winners and may be recommended as potential *best* decision alternatives in this illustrative preference modelling example

```
>>> cdodg = -(~odg)
>>> cdodg.exportGraphViz('codualOdg')
 *---- exporting a dot file for GraphViz tools ---------*
  Exporting to codualOdg.dot
  dot -Grankdir=BT -Tpng codualOdg.dot -o codualOdg.png
```

Many more tools for exploiting bipolar-valued outranking digraphs are available in the DIGRAPH3 resources (Bisdorff 2021).

Notes

The seminal work on outranking digraphs goes back to the seventies and eighties when *Bernard Roy* joined the just starting University Paris-Dauphine and founded there the '*Laboratoire d'Analyse et de Modélisation de Systèmes pour l'Aide à la Décision*' (LAMSADE). The LAMSADE became the major site in the development of the outranking approach to multiple-criteria decision aiding (Roy and Bouyssou 1993).

The ongoing success of the original *outranking* concept stems from the fact that it is rooted in a sound pragmatism. The multiple-criteria performance tableau, necessarily associated with a given outranking digraph, is indeed convincingly objective and meaningful (Roy 1991). And, ideas from social choice theory gave initially the insight that a pairwise voting mechanism à la CONDORCET could provide an order-statistical tool for aggregating a set of preference points of view into what *Marc Barbut* called the *central* CONDORCET point of view (de Caritat, Marquis de Condorcet 1784 and Barbut 1980), in fact the median of the multiple preference points of view, at minimal absolute KENDALL's ordinal correlation distance from all individual points of view (see Chap. 16).

Considering thus each performance criterion as a subset of unanimous voters and balancing the votes in favour against considerable counter-performances in disfavour gave eventually rise to the concept of *outranking situation*, a distinctive feature of the Multiple-Criteria Decision Aiding approach (Bisdorff et al. 2015). A modern definition would be an alternative x is said to *outrank* alternative y when—a *significant majority* of criteria confirm that alternative x has to be considered as *at least as well evaluated as* an alternative y (the *concordance* argument), and—no discordant criterion opens to significant doubt the validity of the previous confirmation by revealing a considerable counter-performance of alternative x compared to y (the *discordance* argument).

If the concordance argument was always well received, the discordance argument however, very confused in the beginning (Benayoun et al. 1966), could only be handled in an epistemically correct and logically sound way by using a bipolar-valued epistemic logic (see Definition 3.1 and Bisdorff 2013). The outranking situation had consequently to receive an explicit negative definition: an alternative x is said to *do not outrank* an alternative y when—a *significant majority* of criteria confirm that alternative x has to be considered as *not at least as well evaluated as* alternative y, and—no discordant criterion opens to significant doubt the validity of the previous confirmation by revealing a considerable *better* performance of alternative x compared to y.

Furthermore, the initial conjunctive aggregation of the concordance and discordance arguments had to be replaced by a disjunctive epistemic fusion operation, polarising in a logically sound and epistemically correct way the concordance with the discordance argument. This way, bipolar-valued outranking digraphs gained two very useful properties from a measure theoretical perspective. They are *weakly complete*; incomparability situations are no longer attested by the absence of positive outranking relations, but instead by epistemic indeterminateness. And they verify the *coduality principle*: the negation of the epistemic '*at least as well evaluated as*' situation corresponds formally to the strict converse epistemic '*less well evaluated than*' situation.

In the next methodological Part II, we present and discuss multiple-criteria evaluation models and decision algorithms, like building a best choice recommendation, determining the winner of an election and computing linear rankings or quantile ratings with multiple incommensurable criteria.

References

Barbut M (1980) Médianes, Condorcet et Kendall. Math Sci Hum 69:9–13

Benayoun R, Roy B, Sussmann B (1966) ELECTRE: une méthode pour guider le choix en présence de points de vue multiples. Tech. Rep. 49, Société d'Economie et de Mathématique Appliqué, Direction Scientifique

Bisdorff R (2013) On polarizing outranking relations with large performance differences. J Multi-Criteria Dec Anal Wiley 20:3–12. http://hdl.handle.net/10993/245

Bisdorff R (2020) Lecture 7: best multiple criteria choice: the Rubis outranking method. In: Lectures of the algorithmic decision theory course, University of Luxembourg. http://hdl.handle.net/10993/37933

Bisdorff R (2021) Technical documentation of the Digraph3 collection of Python modules. https://digraph3.readthedocs.io/en/latest/techDoc.html

Bisdorff R, Dias L, Meyer P, Mousseau V, Pirlot M (eds) (2015) Evaluation and decision models with multiple criteria: case studies. International handbooks on information systems. Springer, Berlin. http://hdl.handle.net/10993/23698

de Caritat, Marquis de Condorcet J (1784) Essai sur l'application de l'analyse à la probabilité des décisions rendues à la pluralité des voix. Imprimerie royale, Paris

Roy B (1991) The outranking approach and the foundations of electre methods. Theory Decis 31(1):49–73

Roy B, Bouyssou D (1993) Aide Multicritère à la Décision : Méthodes et Cas. Economica, Paris

Part II
Evaluation Models and Decision Algorithms

The second and methodological part of the book presents in eight chapters multiple-criteria evaluation models and decision algorithms for selecting a best decision alternative, for ranking the potential decision alternative from best to worst, and for relative or absolute quantile rating with the help of bipolar-valued outranking digraphs. We also show how to edit a new multiple-criteria performance tableau and present several models of random performance tableau generators. The last chapter is devoted to HPC ranking of big performance tableaux.

Chapter 4
Building a Best Choice Recommendation

> *... The goal of our research was to design a resolution method ... that is easy to put into practice, that requires as few and reliable hypotheses as possible, and that meets the needs [of the decision maker]...*
>
> *—(Benayoun et al. 1966)*

Contents

Abstract This chapter presents the RUBIS best choice recommender system. Our approach is illustrated with a best office location selection problem. We show how to explore the given performance tableau and compute the corresponding outranking digraph. After presenting the pragmatic principles that govern our best choice recommendation algorithm we solve the best office location choice problem.

4.1 What Office Location to Choose?

A SME, specialised in printing and copy services, has to move into new offices, and its CEO has gathered seven *potential new office locations* (see Table 4.1 on the following page).

Three *decision objectives*, in order of decreasing importance, are guiding the CEO's choice:

1. *maximise* the future turnover of the SME,

R. Bisdorff, *Algorithmic Decision Making with Python Resources*,
International Series in Operations Research & Management Science 324,
https://doi.org/10.1007/978-3-030-90928-4_4

Table 4.1 The potential new office locations

ID	Name	Address	Comment
A	Ave	Avenue de la liberté	High standing city centre
B	Bon	Bonnevoie	Industrial environment
C	Ces	Cessange	Residential suburb location
D	Dom	Dommeldange	Industrial suburb environment
E	Bel	Esch-Belval	New and ambitious urbanisation far from the city
F	Fen	Fentange	Out in the countryside
G	Gar	Avenue de la Gare	Main city shopping street

Table 4.2 The family of performance criteria

Objective	ID	Name	Weight	Comment
Yearly costs	Ct	Costs	45	Annual rent, charges, and cleaning
Future turnover	Pr	Proximity	32	Distance from town centre
Future turnover	V	Visibility	26	Circulation of potential customers
Future turnover	St	Standing	23	Image and presentation
Working conditions	W	Space	10	Working space
Working conditions	Cf	Comfort	6	Quality of office equipment
Working conditions	P	Parking	3	Available parking facilities

2. *minimise* the future yearly costs induced by the moving,
3. *maximise* the new working conditions.

The decision consequences to take into account for evaluating the potential new office locations with respect to each one of the three objectives are modelled by the *family of performance criteria*[1] shown in Table 4.2 below.

In Table 4.2 we notice that the *Costs* criterion admits the highest significance weight (45), followed by the *Future turnover* criteria $(32, 26, 23)$. The *Working conditions* criteria are the less significant $(10, 6, 3)$.[2] It follows that the CEO considers *maximising the future turnover* the most important objective $((32 + 26 + 23) = 81)$, followed by minimising the future yearly costs (45), and less important, *maximising working conditions* $((10 + 6 + 3) = 19)$.

The evaluations of the seven potential new locations on each performance criterion are gathered in a *performance table* shown in Table 4.3 on the next page. All criteria, except the *Costs* Criterion, admit for evaluation a qualitative satisfaction scale from 0% (weakest) to 100% (highest). One may thus notice that location `A` (*Ave*) is the most expensive, but also 100% satisfying the *Proximity* as well as the *Standing* criterion. Whereas location `C` (*Ces*) is the cheapest one; Providing, however, no satisfaction at all on both the *Standing* and the *Working Space* criteria.

[1] See Roy (2000).

[2] These criteria weights were supposedly established with a swing weighing MCDA method (Keeney and Raiffa 1976).

Table 4.3 Performance evaluations of the potential office locations

Criterion	A	B	C	D	E	F	G
Costs	35.0K€	17.8K€	6.7K€	14.1K€	34.8K€	18.6K€	12.0K€
Proximity	100	20	80	70	40	0	60
Visibility	60	80	70	50	60	0	100
Standing	100	10	0	30	90	70	20
Working space	75	30	0	55	100	0	50
Working comfort	0	100	10	30	60	80	50
Parking	90	30	100	90	70	0	80

Concerning yearly costs, we suppose that the CEO is indifferent up to a performance difference of 1000.00€, and he actually prefers a location when there is at least a positive difference of 2500.00€. The evaluations observed on the six qualitative criteria (measured in percentages of satisfaction) are very subjective and rather imprecise. The CEO is hence *indifferent* up to a satisfaction difference of 10%, and he claims a significant *preference* when the satisfaction difference is at least of 20%. Furthermore, a satisfaction difference of 80% represents for him a *considerably large* performance difference, triggering the case given a *polarisation* of the preferential situation (Bisdorff 2013).

In view of Table 4.3, what is now the office location we may recommend to the CEO as **best choice**?

4.2 The Given Performance Tableau

The file `officeChoice.py`, stored in the `examples` directory of the DIGRAPH3 resources, provides a corresponding `PerformanceTableau` object. We can inspect its actual content with the computing resources provided by the `perfTabs` module .

Listing 4.1 Inspecting the `officeChoice` performance tableau

```
>>> from perfTabs import PerformanceTableau
>>> pt = PerformanceTableau('officeChoice')
>>> pt
 *------- PerformanceTableau instance description ------*
   Instance class      : PerformanceTableau
   Instance name       : officeChoice
   Actions             : 7
   Objectives          : 3
   Criteria            : 7
   NaN proportion (%)  : 0.0
   Attributes          : ['name', 'actions', 'objectives',
                          'criteria', 'weightPreorder',
                          'NA', 'evaluation']
>>> pt.showPerformanceTableau()
```

```
*----  performance tableau -----*
 Criteria|  'Ct'         'Cf'    'P'    'Pr'    'St'    'V'     'W'
 Weights |  45.00        6.00   3.00  32.00   23.00   26.00   10.00
 --------|---------------------------------------------------------
  'Ave'  | -35000.00    0.00  90.00 100.00  100.00   60.00   75.00
  'Bon'  | -17800.00  100.00  30.00  20.00   10.00   80.00   30.00
  'Ces'  |  -6700.00   10.00 100.00  80.00    0.00   70.00    0.00
  'Dom'  | -14100.00   30.00  90.00  70.00   30.00   50.00   55.00
  'Bel'  | -34800.00   60.00  70.00  40.00   90.00   60.00  100.00
  'Fen'  | -18600.00   80.00   0.00   0.00   70.00    0.00    0.00
  'Gar'  | -12000.00   50.00  80.00  60.00   20.00  100.00   50.00
```

We thus recover all the input data shown in Sect. 4.1. Notice that the *negative* evaluations of the *Costs* criterion indicate a negative preference direction: the *lower* the costs, the *better* it is.

The `showCriteria()` method evaluates the actual preference discrimination we observe on each performance criterion.

Listing 4.2 Inspecting the performance criteria

```
1  >>> pt.showCriteria(IntegerWeights=True)
2   *----  criteria -----*
3    Ct 'Costs'
4     Preference direction: min
5     Scale = (0.00, 50000.00)
6     Weight = 45
7     Threshold ind : 1000.00 + 0.00x ;  percentile:  9.52
8     Threshold pref : 2500.00 + 0.00x ; percentile: 14.29
9    Cf 'Comfort'
10    Preference direction: max
11    Scale = (0.00, 100.00)
12    Weight = 6
13    Threshold ind : 10.00 + 0.00x ;  percentile:   9.52
14    Threshold pref : 20.00 + 0.00x ; percentile:  28.57
15    Threshold veto : 80.00 + 0.00x ; percentile:  90.48
16    ...
17    ...
```

On the *Costs* criterion, 9.5% of the performance differences are considered insignificant and 14.3% below the preference discrimination threshold (see Lines 7–8 in Listing 4.2). On the qualitative *Comfort* criterion, we observe again 9.5% of insignificant performance differences (line 13). Due to the imprecision in the subjective evaluations, we notice here 28.6% of performance differences below the preference discrimination threshold (Line 14). Furthermore, $100.0 - 90.5 = 9.5\%$ of the performance differences are judged *considerably large* (Line 15) and will trigger hence a polarisation of the concerned outranking situations (Bisdorff 2013). Same information is available for all the other criteria. A colourful comparison of all the evaluations is shown in Fig. 4.1 on the facing page by the *heatmap* statistics, illustrating the respective quantile class of each evaluation. As the set of potential alternatives is tiny, we choose here a classification into performance quintiles.

Heatmap of Performance Tableau 'officeChoice'

criteria	Ct	Pr	V	St	W	Cf	P
weights	+45.00	+32.00	+26.00	+23.00	+10.00	+6.00	+3.00
Ave	-35000	100	60	100	75	0	90
Bon	-17800	20	80	10	30	100	30
Ces	-6700	80	70	0	0	10	100
Dom	-14100	70	50	30	55	30	90
Bel	-34800	40	60	90	100	60	70
Fen	-18600	0	0	70	0	80	0
Gar	-12000	60	100	20	50	50	80

Color legend:

quantile	20.00%	40.00%	60.00%	80.00%	100.00%

Fig. 4.1 Unranked heatmap of the office choice performance tableau

```
>>> pt.showHTMLPerformanceHeatmap(colorLevels=5,\
...                                rankingRule=None)
```

Location `Ave` shows extreme and contradictory evaluations: highest *Costs* and no *Working Comfort* on the one hand, and total satisfaction with respect to *Standing*, *Proximity*, and *Parking facilities* on the other hand. Similar, but opposite, situation is given for location `Ces`: unsatisfactory *Working Space*, no *Standing*, and no *Working Comfort* on the one hand, and lowest *Costs*, best *Proximity*, and *Parking facilities* on the other hand. Contrary to these contradictory alternatives, we observe two appealing compromise alternatives: locations `Dom` and `Gar`. Finally, location `Fen` is clearly the less satisfactory alternative of all. To help now the CEO choosing the best office location, we are going to compute pairwise outranking situations on the set of potential decision alternatives (see Bisdorff 2013).

4.3 Computing the Outranking Digraph

Definition 4.1 (Outranking Situation) For two potential decision alternatives x and y:

- 'Alternative x *outranks* alternative y', denoted $(x \succsim y)$, is given when:
  1. A *majority* of criteria significance warrants that alternative x is *at least as well evaluated as* alternative y, and
  2. *No considerable* negative performance difference is observed on any criterion.
- 'Alternative x *does not outrank* y', denoted $(x \not\succsim y)$, is given when:
  1. Only a *minority* of criteria significance warrants that alternative x is *at least as well evaluated as* alternative y, and
  2. *No considerable* positive performance difference is observed on any criterion.

Fig. 4.2 *Bipolar-valued adjacency matrix*. In the resulting outranking relation we may notice, on the one hand, that Alternative D is *positively outranking* all other potential office locations. On the other hand, alternatives A (the most expensive) and C (the cheapest) are *not outranked* by any other location

Valued Adjacency Matrix

r(x S y)	A	B	C	D	E	F	G
A	-	0.00	1.00	0.30	0.78	0.00	0.00
B	0.00	-	0.00	-0.56	0.00	1.00	-0.60
C	0.00	0.00	-	0.46	0.00	1.00	0.10
D	0.10	0.56	0.02	-	0.46	1.00	0.25
E	0.52	0.00	0.00	-0.10	-	1.00	-0.42
F	0.00	-1.00	-1.00	-1.00	-1.00	-	-1.00
G	0.00	0.92	-0.10	1.00	0.54	1.00	-

Valuation domain: [-1.00; +1.00]

- Otherwise, the outranking situation between alternatives x and y is considered to be *indeterminate*.

The credibility of each pairwise outranking situation, denoted $r(x \succsim y)$, is measured in a bipolar significance valuation $[-1.0, 1.0]$, where *positive* terms $r(x \succsim y) > 0.0$ indicate a *validated outranking*, and *negative* terms $r(x \succsim y) < 0.0$ indicate an *invalidated outranking*, i.e. a *validated outranked* situation. The *median* value $r(x \succsim y) = 0.0$ represents an *indeterminate* situation (see Bisdorff 2004, 2013).

For computing such a bipolar-valued binary outranking relation from the given performance tableau `pt`, we use the `BipolarOutrankingDigraph` class from the `outrankingDigraphs` module. The corresponding `showHTMLRelationTable()` method shows here the resulting bipolar-valued adjacency matrix in a system browser window (see Fig. 4.2).

Listing 4.3 Computing a bipolar-valued outranking digraph

```
>>> from outrankingDigraphs import BipolarOutrankingDigraph
>>> bod = BipolarOutrankingDigraph(pt)
>>> bod.showHTMLRelationTable()
```

Alternative D gives a CONDORCET winner,[3] whereas alternatives A (the most expensive) and C (the cheapest) give in fact *weak* CONDORCET winners.

```
>>> bod.computeCondorcetWinners()
 ['D']
>>> bod.computeWeakCondorcetWinners()
 ['A', 'C', 'D']
```

From theory, we know that outranking digraphs are *weakly complete*, i.e. for all x and y in X, $r(x \succsim y) < 0.0$ implies that $r(y \succsim x) \geqslant 0.0$. And, they verify the *coduality principle*: $r(x \not\succsim y) = r(y \precnsim x)$ (Bisdorff 2013).[4]

[3] See Chap. 7 on computing the winner of an election.

[4] Not to be confused with the dual graph of a plane graph `g` that has a vertex for each face of `g`. Here we mean the *less than* (strict converse) relation corresponding to a *greater or equal* relation, or the *less than or equal* relation corresponding to a (strict) *better than* relation.

Definition 4.2 (Strict Outranking Situation) For two potential decision alternatives x and y, 'x *strictly outranks* alternative y', denoted $(x \succnsim y)$, when 'x *outranks* alternative y' $((x \succsim y) > 0.0)$ and "y *does not outrank* alternative x" $((y \succsim x) < 0.0)$.

Following from the coduality principle, the strict outranking digraph `bodcd` is built from the outranking digraph `bod` with a codual transform (see Sect. 2.6).

```
>>> bodcd = ~(-bod)   # codual tranform
>>> bodcd
*------- Object instance description ------*
Instance class          : BipolarOutrankingDigraph
Instance name           : converse-dual-rel_officeChoice
Actions                 : 7
Criteria                : 7
Size                    : 10
Determinateness (%)     : 72.38
Valuation domain        : [-1.00;1.00]
```

4.4 Designing a Best Choice Recommender System

Solving a best choice problem consists traditionally in finding *the* unique best decision alternative. We adopt here instead a modern recommender system's approach which shows a non-empty subset of decision alternatives which contains by construction the potential best alternative(s).

The five *pragmatic principles* for computing such a *best choice recommendation* (BCR) are the following:

`P1`: *Elimination for well motivated reasons*; each eliminated alternative has to be strictly outranked by at least one alternative in the BCR.

`P2`: *Minimal size*; the BCR must be as limited in cardinality as possible.

`P3`: *Efficient and informative*; the BCR must not contain a self-contained sub-recommendation.

`P4`: *Effectively better*; the BCR must not be ambiguous in the sense that it may not be both a first choice as well as a last choice recommendation.

`P5`: *Maximally determined*; the BCR is, of all potential best choice recommendations, the most determined one in the sense of the epistemic characteristics of the bipolar-valued outranking relation.

Let X be the set of potential decision alternatives. Let Y be a non-empty subset of X, called a *choice* in the strict outranking digraph $G(X, r(\succnsim)$. We can now qualify a BCR Y in following terms:

- Y is called strictly *outranking* (respectively, *outranked*) when for all not selected alternative x there exists an alternative $y \in X$ retained such that $r(y \succnsim x) > 0.0$ (respectively, $r(y \precnsim x) > 0.0$). Such a choice verifies principle `P1`.

- Y is called *weakly independent* when for all $x \neq y$ in Y we observe $r(x \gnsim y) \leq 0.0$. Such a choice verifies principles P3 (*internal stability*).
- Y is conjointly a strictly *outranking* (respectively, *outranked*) **and** *weakly independent* choice. Such a choice is called an *initial* (respectively, *terminal*) *prekernel*.[5] The initial prekernel now verifies principles P1, P2, P3, and P4.
- To finally verify principle P5, we recommend among all potential initial prekernels, a *most determined* one, i.e. a strictly *outranking* and *weakly independent* choice supported by the highest criteria significance. And in this most determined initial prekernel we eventually retain the alternative(s) that are included with highest criteria significance.[6]

Mind that a given strict outranking digraph may not always admit prekernels. This is the case when the digraph contains chordless circuits of odd length (see Chap. 17). Luckily, our strict outranking digraph bodcd here does not show any chordless outranking circuits; a fact we can check with the computeChordlessCircuits() method followed by the showChordlessCircuits() method.[7]

```
>>> bodcd.computeChordlessCircuits()
  []
>>> bodcd.showChordlessCircuits()
 No circuits observed in this digraph.
```

When observing chordless odd outranking circuits, we need to break them open with the BrokenCocsDigraph class at their weakest link, before enumerating the prekernels (Bisdorff 2021).

We are ready now for building a best choice recommendation.

4.5 Computing the RUBIS Best Choice Recommendation

The showBestChoiceRecommendation() method computes the RUBIS best choice recommendation directly from the outranking digraph *bod*. By default this method is operating on the *codual* (strict) outranking digraph where chordless odd circuits have been broken up (see the CoDual and BrokenCocs parameters in Listing 4.4 Line 2):

Listing 4.4 Computing the best choice recommendation

```
1 >>> bod.showBestChoiceRecommendation(\
2 ...                CoDual=True, BrokenCocs=True # default
    settings\
```

[5] See Chap. 17 on computing kernels in digraphs.

[6] See Sect. 17.6.

[7] The computeChordlessCircuits() and showChordlessCircuits() methods are separate because there are various methods available for enumerating the chordless circuits in a digraph (Bisdorff 2010).

```
3  ...                  ChoiceVector = True)
4   * --- First and last choice recommendation(s) ---*
5    (in decreasing order of determinateness)
6    Credibility domain: [-1.00,1.00]
7    === >> potential first choice(s)
8    * choice                 : ['A', 'C', 'D']
9      independence           : 0.00
10     dominance              : 0.10
11     absorbency             : 0.00
12     covering (%)           : 41.67
13     determinateness (%)    : 50.59
14     - characteristic vector = {
15       'D': +0.02, 'G': 0.00, 'C': 0.00, 'A': 0.00,
16       'F': -0.02, 'E': -0.02, 'B': -0.02, }
17   === >> potential last choice(s)
18   * choice                 : ['A', 'F']
19     independence           : 0.00
20     dominance              : -0.52
21     absorbency             : 1.00
22     covered (%)            : 50.00
23     determinateness (%)    : 50.00
24     - characteristic vector = {
25       'G': 0.00, 'F': 0.00, 'E': 0.00, 'D': 0.00,
26       'C': 0.00, 'B': 0.00, 'A': 0.00, }
```

It is interesting to notice in Line 8 above that the RUBIS *first choice recommendation* consists actually in the previously mentioned set of weak CONDORCET winners: A, C, and D (see Fig. 4.2 on page 46). In the corresponding prekernel characteristic vector (see Lines 3, 15 and Sect. 17.6), representing the bipolar credibility degree with which each alternative may indeed be included in, or excluded from this recommendation, we find that alternative D is the only positively validated one, whereas both extreme alternatives—A (the most expensive) and C (the cheapest)—stay in an indeterminate situation (see Bisdorff 2006; Bisdorff et al. 2006). They may *be or not be* potential best choices. Notice furthermore that compromise alternative G, while not actually being included in this strictly outranking prekernel, shows as well an indeterminate situation with respect to *being or not being* recommended as potential best choice. Alternatives B, E, and F are all negatively included, i.e. positively excluded, from this best choice recommendation (see Line 16).

To inspect why alternative D is the only positive best choice recommendation, we shall compare now the evaluations of alternatives D and G in a pairwise perspective.

Listing 4.5 Inspecting pairwise comparison between alternatives G and D

```
>>> bod.showPairwiseComparison('G','D')
*------------ pairwise comparison ----*
 Comparing actions : ('G', 'D')
 crit.  wght.    g(x)      g(y)      diff.   |  ind.     pref.   concord.
 ====================================================================
  Costs 45.00 -12000.00 -14100.00 +2100.00 | 1000.00 2500.00  +45.00
  Comf.  6.00     50.00     30.00   +20.00 |   10.00   20.00   +6.00
  Park.  3.00     80.00     90.00   -10.00 |   10.00   20.00   +3.00
  Prox. 32.00     60.00     70.00   -10.00 |   10.00   20.00  +32.00
```

```
 Stdg. 23.00      20.00      30.00    -10.00 |   10.00   20.00  +23.00
 Visi. 26.00     100.00      50.00    +50.00 |   10.00   20.00  +26.00
 Spac. 10.00      50.00      55.00     -5.00 |   10.00   20.00  +10.00
 ======================================================================
  Valuation in range: -145.00 to +145.00; global concordance: +145.00
```

In Listing 4.5 on the preceding page, we notice that, with the given preference discrimination thresholds, alternative `G` is actually *certainly at least as well evaluated as* alternative `D`: $r(G \succsim D) = +145/145 = +1.0$.

Yet, we must as well acknowledge in Listing 4.6 that the cheapest alternative `C` is in fact *strictly outranking* alternative `G`: $r(C \succsim G) = +15/145 > 0.0$, and $r(G \succsim C) = -15/145 < 0.0$.

Listing 4.6 Inspecting pairwise comparison between alternatives `C` and `G`

```
>>> bod.showPairwiseComparison('C','G')
 *------------  pairwise comparison ----*
  Comparing actions : (C,G)/(G,C)
  crit. wght.   g(x)       g(y)      diff.  |   ind.   pref.       (C,G)/(G,C)
 ==========================================================================
  'C'   45.00 -6700.00 -12000.00 +5300.00 | 1000.00 2500.00    +45.00/-45.00
  'Cf'   6.00    10.00     50.00   -40.00 |   10.00   20.00     -6.00/ +6.00
  'P'    3.00   100.00     80.00   +20.00 |   10.00   20.00     +3.00/ -3.00
  'Pr'  32.00    80.00     60.00   +20.00 |   10.00   20.00    +32.00/-32.00
  'St'  23.00     0.00     20.00   -20.00 |   10.00   20.00    -23.00/+23.00
  'V'   26.00    70.00    100.00   -30.00 |   10.00   20.00    -26.00/+26.00
  'W'   10.00     0.00     50.00   -50.00 |   10.00   20.00    -10.00/+10.00
                                                              --------------
   Valuation in range: -145 to +145;     r(C >= G)/r(G >= c): +15.00/-15.00
```

Let us finally notice in Listing 4.4 on page 48 Line 18 that both alternatives `A` and `F` are reported as potential last choice recommendation. Yet, this last choice recommendation appears to be globally indeterminate (Lines 25–26). This confirms the *incomparability* status of alternative `A` (see Fig. 4.3 on the next page).

```
>>> bodcd.exportGraphViz(fileName='bestOfficeChoice',\
...                      firstChoice=['C','D'],\
...                      lastChoice=['F'])
 *---- exporting a dot file for GraphViz tools ---------*
  Exporting to bestOfficeChoice.dot
  dot -Grankdir=BT -Tpng bestOfficeChoice.dot \
                      -o bestOfficeChoice.png
```

4.6 Weakly Ordering the Outranking Digraph

To get a global insight in the overall strict outranking situations, we may use the `RankingByChoosingDigraph` class imported from the `transitiveDigraphs` module for computing a *ranking-by-choosing* result from the codual, i.e. the strict outranking digraph instance `bodcd` (see above). If the computing node supports multiple processor cores, *first* and *last* choosing iterations may be run in parallel (see Line 4 in Listing 4.7 on the facing page).

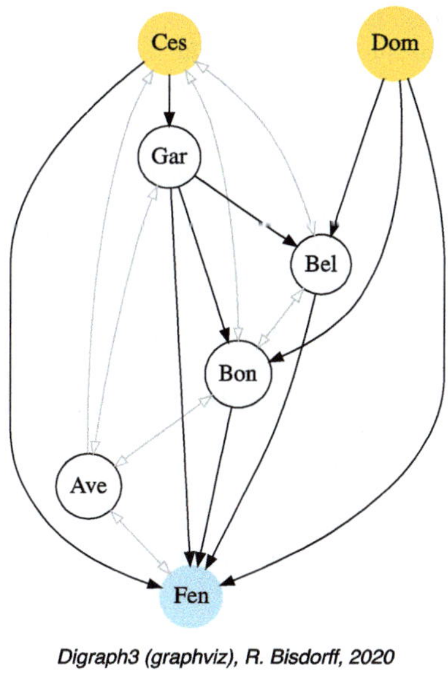

Fig. 4.3 *Best office choice recommendation from strict outranking digraph.* Notice that location A (*Ave*) (the most expensive) is appearing *incomparable* to all the other alternatives

Listing 4.7 Ranking-by-choosing the outranking digraph

```
>>> from transitiveDigraphs import\
...                     RankingByChoosingDigraph
>>> rbc = RankingByChoosingDigraph(bodcd)
 Threading ... # multiprocessing if 2 cores are available
 Exiting computing threads
>>> rbc.showRankingByChoosing()
 Ranking by Choosing and Rejecting
    1st ranked ['D']
       2nd ranked ['C', 'G']
       2nd last ranked ['B', 'C', 'E']
    1st last ranked ['A', 'F']
>>> rbc.exportGraphViz('officeChoiceRanking')
 *---- exporting a dot file for GraphViz tools ---------*
  Exporting to officeChoiceRanking.dot
  dot -Grankdir=TB -Tpng officeChoiceRanking.dot\
                    -o officeChoiceRanking.png
```

The best choice recommendation hence depends on the very importance the CEO is attaching to each of his decision objectives. In the given setting here, where he considers that *maximising the future turnover* is the most important objective followed by *minimising the Costs* and, less important, *maximising the working conditions*, location D represents actually the *best compromise*. However, if *Costs* do not play much a role, it would be perhaps better to decide to move to the most advantageous location A; or if, on the contrary, *Costs* do matter a lot, moving to the cheapest alternative C could definitely represent a more convincing recommendation (Fig. 4.4).

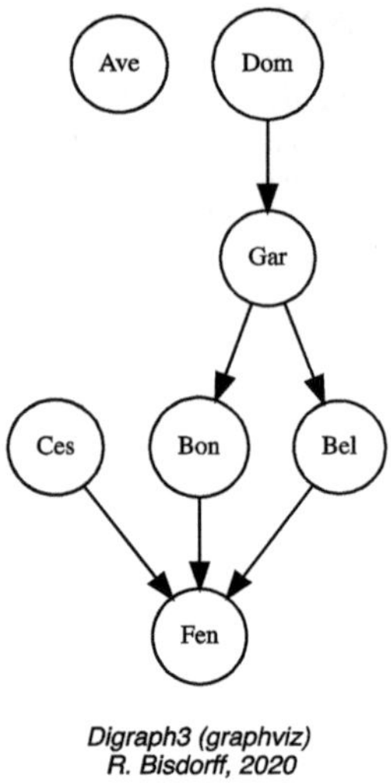

Fig. 4.4 *Ranking-by-choosing the potential office locations.* In this *ranking-by-choosing* method, where we operate the *epistemic fusion* of iterated (strict) first and last choices, compromise alternative `Dom` is now ranked before compromise alternative `Gar`. The overall partial ordering result shows again the important fact that the most expensive location, `Ave`, and the cheapest location, `Ces`, due to their contradictory performances appear both *incomparable* with most of the other alternatives

It might be worth editing the criteria significance weights in the `officeChoice.py` data file in such a way that:

- All three decision objectives are considered *equally important*, and
- All criteria under each objective are considered *equi-significant*.

What will become the best choice recommendation under this working hypothesis?[8]

In the next Chap. 5 we precisely show how to edit a new performance tableau from a given template file.

Notes

Following a seminar presentation in 2005 at the LAMSADE,[9] where the author promoted the use of kernels of the outranking digraph as suitable candidates for delivering best choice recommendations (Bisdorff 2005), a critical discussion started about the methodological requirement for a convincing best choice recommendation to be internally stable (pragmatic principle `P3`). *Denis Bouyssou* illustrated his doubts with the potential outranking digraph shown in Fig. 4.5 on the next page.

---

[8] See also the notes of Lecture 7 from the MICS Algorithmic Decision Theory course (Bisdorff 2020).

[9] Laboratoires d'Analyse et de Modélisation de Systèmes d'Aide à la Décision, Université Paris-Dauphine, UMR 7243 CNRS.

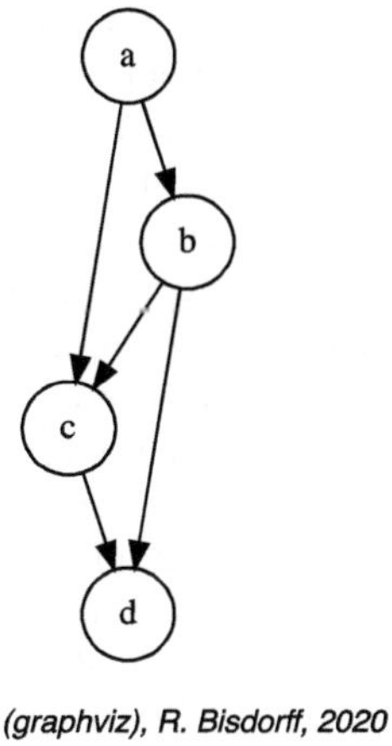

Fig. 4.5 *The internal stability of a best choice recommendation in question*
The only kernel of this digraph is the pair {a, d}; yet, it is an ambiguous recommendation, as {a, d} is conjointly an outranking and outranked choice. If the instability of the best choice recommendation is, however, not considered a problem, then the choice {a, b} shows the most convincing strict outranking quality and could be considered in priority for a potential best choice recommendation

His commentary was the following: Adding alternative d to the set of potential best choice candidates is not convincing as there exists in the given digraph the node b, which is better evaluated than d. The argument that the incomparability between a and d should favour d as potential best choice is interesting but another hypothesis could be that b perhaps outranks a. In this latter case, it seems clear that the actual best choice recommendation should be reduced to node b, unless one disposes of other information, like a performance tableau and/or the actual computation method of the outranking situations. In any case, one has to be very clear about the available information when judging a best choice procedure.

It became thereafter obvious for us all that both the lack of a specific performance tableau as well as the lack of a precisely defined algorithm for computing valid outranking situations do not allow to judge if a given digraph does indeed model a potential outranking relation. In our present bipolar-valued epistemic approach, a valid outranking digraph instance, following from a given performance tableau and the disjunctive epistemic fusion construction of the outranking relation (see Chap. 3), will necessarily verify the weak completeness condition and the coduality principle. As a consequence, incomparability situations are now modelled by epistemic indeterminateness not by the actual absence of a reciprocal outranking relation.

The digraph put forward by *D. Bouyssou* in the October 2005 discussion is not weakly complete—node a is not outranking node d and vice versa—and does hence not represent, in our present sense, a valid outranking digraph instance. Yet, it may be a partial tournament and as such it could be a strict outranking digraph, i.e. the asymmetric part—the codual—of a valid outranking digraph. In this case, nodes a and d—the kernel of the strict outranking digraph—would actually for sure outrank each other and, hence, represent both indifferently the natural best choice recommendation. However, in this not strict codual digraph, node a becomes also the unique CONDORCET winner—outranking for sure all other nodes—and gives hence the evident unique best choice recommendation.

Only after 2013, when the weak completeness and the coduality properties of the outranking digraph were discovered, it became obvious that the initial prekernels of the strict outranking digraph, coupled with the solution of the corresponding kernel equation system, were in fact delivering the most convincing best choice recommendations (see Chap. 17, Sect. 20.4 and Bisdorff 2013). It stays an interesting open mathematical problem to show (or not) that both necessary conditions—weak completeness and coduality—are also sufficient for qualifying any bipolar-valued digraph as potential instance of an outranking digraph.

References

Benayoun R, Roy B, Sussmann B (1966) ELECTRE: une méthode pour guider le choix en présence de points de vue multiples. Tech. Rep. 49, Société d'Economie et de Mathématique Appliqué, Direction Scientifique

Bisdorff R (2004) On a natural fuzzification of Boolean logic. In: Klement EP, Pap E (eds) Linz seminar on fuzzy set theory, mathematics of fuzzy systems, Bildungszentrum St. Magdalena, Linz (Austria), pp 20–26. http://hdl.handle.net/10993/23721

Bisdorff R (2005) Exploitation en problématique du choix d'une relation de surclassement valuée. http://hdl.handle.net/10993/47659

Bisdorff R (2006) On enumerating the kernels in a bipolar-valued digraph. Ann Lamsade 6:1–38. http://hdl.handle.net/10993/38741

Bisdorff R (2010) Enumerating chordless circuits in directed graphs. In: ORBEL24-2010, 24th annual conference of the Belgian Operational Research Society (ORBEL aka Sogesci-B.V.W.B.), January 28–29, Liège (BE), Université de Liège (BE), pp 1–12. http://hdl.handle.net/10993/23926

Bisdorff R (2013) On polarizing outranking relations with large performance differences. J Multi-Criteria Decis Anal Wiley 20:3–12. http://hdl.handle.net/10993/245

Bisdorff R (2020) Lecture 7: Best multiple criteria choice: the Rubis outranking method. In: Lectures of the algorithmic decision theory course, University of Luxembourg, http://hdl.handle.net/10993/37933

Bisdorff R (2021) Technical documentation of the Digraph3 collection of Python modules. https://digraph3.readthedocs.io/en/latest/techDoc.html

Bisdorff R, Pirlot M, Roubens M (2006) Choices and kernels from bipolar valued digraphs. Eur J Oper Res 175:155–170. http://hdl.handle.net/10993/23720

Keeney R, Raiffa H (1976) Decisions with multiple objectives: preferences and value tradeoffs. Cambridge University Press

Roy B (2000) A French-English decision aiding glossary. Newsl Eur Working Group Multicriteria Aid Decis 3(Suppl 1):1–10

Chapter 5
How to Create a New Multiple-Criteria Performance Tableau

Contents

Abstract The chapter illustrates a way of creating a new `PerformanceTableau` instance by editing a given template with five decision alternatives, three decision objectives, and six performance criteria. We discuss in detail editing the decision alternatives, the decision objectives, the family of performance criteria, and finally, the evaluations of the decision alternatives on the performance criteria.

5.1 Editing a Template File

For easing the editing of a new multiple-criteria performance tableau, we provide the following `perfTab_Template.py` file in the `examples` directory of the DIGRAPH3 resources.

Listing 5.1 `PerformanceTableau` object template

```
################################################
# Digraph3 documentation
# Template for creating a new PerformanceTableau instance
# (C) R. Bisdorff Mar 2021
# Digraph3/examples/perfTab_Template.py
################################################
from decimal import Decimal
from collections import OrderedDict
#####
# edit the decision actions
# avoid special characters, like '_', '/' or ':',
# in action identifiers and short names
actions = OrderedDict([
```


R. Bisdorff, *Algorithmic Decision Making with Python Resources*,
International Series in Operations Research & Management Science 324,
https://doi.org/10.1007/978-3-030-90928-4_5

```
    ('a1', {
     'shortName': 'action1',
     'name': 'decision alternative a1',
     'comment': 'some specific features of this alternative',
      }),
     ...
     ...
  ])
  #####
  # edit the decision objectives
  # adjust the list of performance criteria
  # and the total weight (sum of the criteria weights)
  # per objective
  objectives = OrderedDict([
   ('obj1', {
    'name': 'decision objective obj1',
    'comment': "some specific features of this objective",
    'criteria': ['g1', 'g2'],
    'weight': Decimal('6'),
    }),
    ...
    ...
   ])
  #####
  # edit the performance criteria
  # adjust the objective reference
  # Left Decimal of a threshold = constant part and
  #  right Decimal = proportional part of the threshold
  criteria = OrderedDict([
   ('g1', {
    'shortName': 'crit1',
    'name': "performance criteria 1",
    'objective': 'obj1',
    'preferenceDirection': 'max',
    'comment': 'measurement scale type and unit',
    'scale': (Decimal('0.0'), Decimal('100.0'),
    'thresholds': {'ind':  (Decimal('2.50'), Decimal('0.0')),
                   'pref': (Decimal('5.00'), Decimal('0.0')),
                   'veto': (Decimal('60.00'), Decimal('0.0'))
                  },
    'weight': Decimal('3'),
    }),
    ...
    ...
   ])
  #####
  # default missing data symbol = -999
  NA = Decimal('-999')
  #####
  # edit the performance evaluations
  # criteria to be minimized take negative evaluations
  evaluation = {
   'g1': {
      'a1':Decimal("41.0"),
      'a2':Decimal("100.0"),
      'a3':Decimal("63.0"),
      'a4':Decimal('23.0'),
      'a5': NA,
     },
   # g2 is of ordinal type and scale 0-10
   'g2': {
      'a1':Decimal("4"),
      'a2':Decimal("10"),
      'a3':Decimal("6"),
      'a4':Decimal('2'),
      'a5':Decimal('9'),
     },
```

```
81     # g3 has preferenceDirection = 'min'
82     'g3': {
83        'a1':Decimal("-52.2"),
84        'a2':NA,
85        'a3':Decimal("-47.3"),
86        'a4':Decimal('-35.7'),
87        'a5':Decimal('-68.00'),
88       },
89     ...
90     ...
91     }
92   ####################
```

The template file, shown in Listing 5.1 on page 55, contains first the instructions to import the required standard Python `Decimal` and `OrderedDict` classes (see Lines 7–8). Four main sections are following: the potential decision *actions*, the decision *objectives*, the performance *criteria*, and finally the performance *evaluations*.

5.2 Editing the Decision Alternatives

Decision alternatives are stored in attribute `actions` under the `OrderedDict` format. The `OrderedDict` object keeps this initial order when iterating over the decision alternatives.[1]

Required attributes of each decision alternative, besides the object identifier,[2] are: `shortName`, `name`, and `comment` (see Lines 15–17 in Listing 5.1 on page 55). The `shortName` attribute is essentially used when showing the performance tableau or the performance heatmap in a browser view.

The random performance tableau models, introduced in Chap. 6, use the `actions` attribute for storing special features of the decision alternatives. The *Cost-Benefit* model, for instance, uses a `type` attribute for distinguishing between *advantageous*, *neutral*, and *cheap* alternatives (see Sect. 6.3). The *3-Objectives* model keeps a detailed record of the performance profile per decision objective and the corresponding random generators per performance criteria as shown in Listing 5.2.

Listing 5.2 Example of decision alternative description

```
>>> from randomPerfTabs import \
...              Random3ObjectivesPerformanceTableau
>>> t = Random3ObjectivesPerformanceTableau()
>>> t.actions
  OrderedDict([
```

[1] See the `OrderedDict` description in the Python documentation (Python Software Foundation 2021).

[2] Mind that *graphviz* drawings require node identifier strings without any special characters like '_' or '/'.

```
    ('p01', {'shortName': 'p01',
             'name': 'action p01 Eco~ Soc- Env+',
             'comment': 'random public policy',
             'Eco': 'fair',
             'Soc': 'weak',
             'Env': 'good',
             'profile': {'Eco':'fair',
                         'Soc':'weak',
                         'Env':'good'}
             'generators': {'ec01': ('triangular', 50.0, 0.5),
                            'so02': ('triangular', 30.0, 0.5),
                            'en03': ('triangular', 70.0, 0.5),
                            ...
                           },
             }
     ),
    ...
    ])
```

The second section of the template file concerns the decision *objectives*.

5.3 Editing the Decision Objectives

The minimal required attributes of the ordered decision *objectives* dictionary, besides the individual objective identifiers, are `name`, `comment`, `criteria`: the list of significant performance criteria, and `weight`: the importance of the decision objective. The latter attribute contains the sum of the *significance* weights of the objective's criteria list (see Lines 27–33 in Listing 5.1 on page 55).

The `objectives` attribute is methodologically useful for specifying the performance criteria significance in building decision recommendations. Mostly, we assume indeed that decision objectives are all equally important and the performance criteria are equi-significant per objective. This is, for instance, the default setting in the random *3-Objectives* performance tableau model.

Listing 5.3 Example of decision objectives' description

```
>>> # t = Random3ObjectivesPerformanceTableau()
>>> t.objectives
 OrderedDict([
 ('Eco',
  {'name': 'Economical aspect',
   'comment': 'Random3ObjectivesPerformanceTableau generated',
   'criteria': ['ec01', 'ec06', 'ec09'],
   'weight': Decimal('48')}),
  ('Soc',
   {'name': 'Societal aspect',
    'comment': 'Random3ObjectivesPerformanceTableau generated',
    'criteria': ['so02', 'so12'],
    'weight': Decimal('48')}),
```

```
14   ('Env',
15    {'name': 'Environmental aspect',
16     'comment': 'Random3ObjectivesPerformanceTableau generated',
17     'criteria': ['en03', 'en04', 'en05', 'en07',
18                  'en08', 'en10', 'en11', 'en13'],
19     'weight': Decimal('48')})
20 ])
```

The importance weight sums up to 48 for each one of the three example decision objectives shown in Listing 5.3 on the facing page so that the significance weight of each one of the 3 economic criteria is set to 16, of both societal criteria is set to 24, and of each one of the 6 environmental criteria is set to 8 (see Lines 8, 13 and 19).

Mind that the `objectives` attribute is always present in a `PerformanceTableau` object instance, even when empty. In this case, we consider that each performance criterion canonically represents its own decision objective. The criterion significance weight equals in this case the corresponding decision objective's importance weight.

The third section of the template file concerns now the *performance criteria*.

5.4 Editing the Family of Performance Criteria

In order to assess how well each potential decision alternative is satisfying a given decision objective, we need *performance criteria*, i.e. decimal-valued evaluation functions gathered in an ordered `criteria` dictionary. The required attributes (see Listing 5.4), besides the criteria identifiers, are the usual `shortName`, `name`, and `comment`. Specific for a criterion is furthermore the `objective` reference, the significance `weight`, the evaluation `scale` (minimum and maximum performance values), the `preferenceDirection` ('`max`' or '`min`'), and the performance discrimination `thresholds` attributes.

Listing 5.4 Example of performance criteria description

```
criteria = OrderedDict([
  ('g1', {
   'shortName': 'crit1',
   'name': "performance criteria 1",
   'comment': 'measurement scale type and unit',
   'objective': 'obj1',
   'weight': Decimal('3'),
   'scale': (Decimal('0.0'), Decimal('100.0'),
   'preferenceDirection': 'max',
   'thresholds': {'ind':  (Decimal('2.50'), Decimal('0.0')),
                  'pref': (Decimal('5.00'), Decimal('0.0')),
                  'veto': (Decimal('60.00'), Decimal('0.0'))
                 },
  }),
  ...
  ...])
```

In our bipolar-valued outranking approach, all performance criteria implement *decimal-valued* evaluation functions, where preferences are either *increasing* or *decreasing* with measured performances. In order to model a *coherent* performance tableau, the family of decision criteria must satisfy two methodological requirements:

1. *Independence*: Each decision criterion implements an evaluation that is *functionally independent* of the evaluation of the other decision criteria, i.e. the performance evaluated on one of the criteria does not *constrain* in any sense the performance evaluated on any other criterion.
2. *Non-redundancy*: Each performance criterion is only *significant* for a *single* decision objective.

For taking into account any, usually *unavoidable*, *imprecision* of the performance evaluation procedures, we may specify three performance *discrimination thresholds*: an *indifference* (`ind`), a *preference* (`pref`), and a *considerable performance difference* (`veto`) threshold (see Listing 5.4 on the preceding page Lines 10–12). The left decimal number of a threshold description tuple indicates a *constant part*, whereas the right decimal number indicates a *proportional* part.

On the template performance criterion `g1`, shown in Listing 5.4 on the previous page, we observe, for instance, an evaluation scale from 0.0 to 100.0 with a constant *indifference* threshold of 2.5, a constant *preference* threshold of 5.0, and a constant *considerable performance difference* threshold of 60.0. The latter threshold will trigger, the case given, a *polarisation* of the outranking statement (Bisdorff 2013).

In a random *Cost-Benefit* performance tableau model we may obtain by default the following description of a cardinal *Costs* criterion:

Listing 5.5 Example of cardinal *Costs* criterion

```
>>> from randomPerfTabs import \
...             RandomCBPerformanceTableau
>>> tcb = RandomCBPerformanceTableau()
>>> tcb.showObjectives()
 *------ decision objectives -------*
 C: Costs
   c1 random cardinal cost criterion 6
   Total weight: 6.00 (1 criteria)
   ...
   ...
>>> tcb.criteria
 OrderedDict([
  ('c1', {'preferenceDirection': 'min',
          'scaleType': 'cardinal',
          'objective': 'C',
          'shortName': 'c1',
          'name': 'random cardinal cost criterion',
          'scale': (0.0, 100.0),
          'weight': Decimal('6'),
          'randomMode': ['triangular', 50.0, 0.5],
          'comment': 'Evaluation generator: triangular law ...',
```

```
22             'thresholds': OrderedDict([
23                ('ind', (Decimal('1.49'), Decimal('0'))),
24                ('pref', (Decimal('3.7'), Decimal('0'))),
25                ('veto', (Decimal('67.71'), Decimal('0')))
26                ])
27             }),
28    ...
29    ...
30  ])
```

Criterion `c1` appears here to be a cardinal criterion to be minimised and significant for the *Costs* (`C`) decision objective (see Line 13 in Listing 5.5 on the facing page). The `showCriteria()` prints out the corresponding performance discrimination thresholds.

```
1  >>> tcb.showCriteria(IntegerWeights=True)
2   *----  criteria -----*
3     c1 'Costs/random cardinal cost criterion'
4      Preference direction: min
5      Scale = (0.0, 100.0)
6      Weight = 6
7      Threshold ind : 1.49 + 0.00x ; percentile: 5.13
8      Threshold pref : 3.70 + 0.00x ; percentile: 10.26
9      Threshold veto : 67.71 + 0.00x ; percentile: 96.15
10     ...
11     ...
```

The *indifference* threshold on this criterion amounts to a constant value of 1.49 (Line 6 above). More or less 5% of the observed performance differences on this criterion appear hence to be *insignificant*. Similarly, with a preference threshold of 3.70, about 90% of the observed performance differences are preferentially *significant* (Line 7). Furthermore, $100.0 - 96.15 = 3.85\%$ of the observed performance differences appear to be *considerable* (Line 8) and will trigger, the case given, a *polarisation* of the corresponding outranking situations.

After the performance criteria descriptions, we are ready for recording the actual *performance evaluations*.

5.5 Editing the Performance Evaluations

The individual evaluations of each decision alternative on each decision criterion are recorded in a double *criterion* × *action* dictionary called `evaluation` (see Listing 5.5). As we may encounter cases of missing data, we previously define a *missing data* symbol `NA` which is set to a value disjoint from all the measurement scales, by default `Decimal('-999')` (see Line 2 in Listing 5.6 on the next page).

Listing 5.6 Editing performance evaluations

```
1  #----------
2  NA = Decimal('-999')
3  #----------
4  evaluation = {
5    'g1': {
6       'a1':Decimal("41.0"),
7       'a2':Decimal("100.0"),
8       'a3':Decimal("63.0"),
9       'a4':Decimal('23.0'),
10      'a5': NA,  # missing data
11   },
12   ...
13   ...
14   # g3 has preferenceDirection = 'min'
15   'g3': {
16      'a1':Decimal("-52.2"), # negative grades
17      'a2':NA,
18      'a3':Decimal("-47.3"),
19      'a4':Decimal('-35.7'),
20      'a5':Decimal('-68.00'),
21   },
22   ...
23   ...
24 }
```

Notice in Listing 5.6 that on a criterion with `preferenceDirection = 'min'` all performance evaluations are recorded as *negative* values (Lines 16 and following).

We can now inspect below the eventually edited complete template performance tableau `perfTab_Template.py` with the `showPerformanceTableau()` method.

```
>>> from perfTabs import PerformanceTableau
>>> pt = PerformanceTableau('perfTab_Template')
>>> pt.showPerformanceTableau(ndigits=1)
 *----  performance tableau -----*
  Criteria   |  'g1'    'g2'   'g3'   'g4'    'g5'    'g6'
  Actions    |    3       3      6      2       2       2
  ---------|----------------------------------------
  'action1'  |  41.0    4.0  -52.2   71.0   63.0   22.5
  'action2'  | 100.0   10.0     NA   89.0   30.7   75.0
  'action3'  |  63.0    6.0  -47.3   55.4   63.5     NA
  'action4'  |  23.0    2.0  -35.7   83.5   37.5   54.9
  'action5'  |    NA    9.0  -68.0   10.0   88.0   75.0
```

Computing below the associated outranking digraph `bod` allows checking the potential presence of any polarised outranking situations.

```
>>> from outrankingDigraphs import BipolarOutrankingDigraph
>>> bod = BipolarOutrankingDigraph(pt)
>>> bod.showPolarisations()
 *----  Negative polarisations ----*
```

```
 number of negative polarisations : 1
 1: r(a4 >= a2)  = -0.44
    criterion: g1
    Considerable performance difference : -77.00
    Veto discrimination threshold       : -60.00
    Polarisation: r(a4 >= a2) = -0.44 ==> -1.00
*----  Positive polarisations ----*
 number of positive polarisations : 1
 1: r(a2 >= a4)  = 0.56
    criterion: g1
    Considerable performance difference : +77.00
    Counter-veto threshold              : 60.00
    Polarisation: r(a2 >= a4) = 0.56 ==> +1.00
```

Indeed, due to the considerable positive performance difference (+77.00) observed on performance criterion `g1`, alternative `a2` "*for sure outranks*" alternative `a4`, respectively, `a4` "*does for sure not outrank*" `a2`.

5.6 Inspecting the Template Outranking Relation

In Listing 5.7, the `showRelationTable()` method prints out the outranking relation table.

Listing 5.7 The template outranking relation

```
1 >>> bod.showRelationTable()
2  * ---- Relation Table -----
3    r   |  'a1'   'a2'   'a3'   'a4'   'a5'
4  -----|-----------------------------------
5  'a1' | +1.00  -0.44  -0.22  -0.11  +0.06
6  'a2' | +0.44  +1.00  +0.33  +1.00  +0.28
7  'a3' | +0.67  -0.33  +1.00  +0.00  +0.17
8  'a4' | +0.11  -1.00  +0.00  +1.00  +0.06
9  'a5' | -0.06  -0.06  -0.17  -0.06  +1.00
```

One may notice in this outranking relation table above that decision alternative `a2` positively *outranks* all the other four alternatives (Line 6). Similarly, alternative `a5` is positively *outranked* by all the other alternatives (see Line 9). We can orient this way the *graphviz* drawing of the template outranking digraph.

```
>>> bod.exportGraphViz(fileName= 'template',\
...                    firstChoice =['a2'],\
...                    lastChoice=['a5'])
  *---- exporting a dot file for GraphViz tools ------*
   Exporting to template.dot
   dot -Grankdir=BT -Tpng template.dot -o template.png
```

In Fig. 5.1 on the following page, alternatives *action3* (`a3`) and *action4* (`a4`) appear actually *incomparable*. In Listing 5.7, their pairwise outranking characteristics show indeed the *indeterminate* value 0.00 (Lines 7–8).

Checking their pairwise comparison may be done with the `showPairwiseComparison()` method:

```
>>> bod.showPairwiseComparison('a3','a4')
 *------------  pairwise comparison ----*
  Comparing actions : ('a3','a4')
  crit. wght.  g(a3)  g(a4)    diff | ind   pref   r()   |
  -------------------------------   ----------------
  'g1'  3.00  63.00   23.00  +40.00 | 2.50  5.00  +3.00 |
  'g2'  3.00   6.00    2.00   +4.00 | 0.00  1.00  +3.00 |
  'g3'  6.00 -47.30  -35.70  -11.60 | 0.00 10.00  -6.00 |
  'g4'  2.00  55.40   83.50  -28.10 | 2.09  4.18  -2.00 |
  'g5'  2.00  63.50   37.50  +26.00 | 0.00 10.00  +2.00 |
  'g6'   NA   54.90
  Outranking characteristic value:   r(a3 >= a4) = +0.00
  Valuation in range: -18.00 to +18.00
```

The incomparability situation between `a3` and `a4` results in fact from a perfect balancing of positive (+8) and negative (−8) criteria significance weights.

The five decision alternatives can finally be ranked with a heatmap browser view following the COPELAND ranking rule (see Sect. 8.2), a rule which consistently reproduces the partial outranking order shown in Fig. 5.1.

```
>>> bod.showHTMLPerformanceHeatmap(ndigits=1,\
...     colorLevels=5, Correlations=True,\
...     rankingRule='Copeland',\
...     pageTitle=\
...      'Heatmap of the template performance tableau')
```

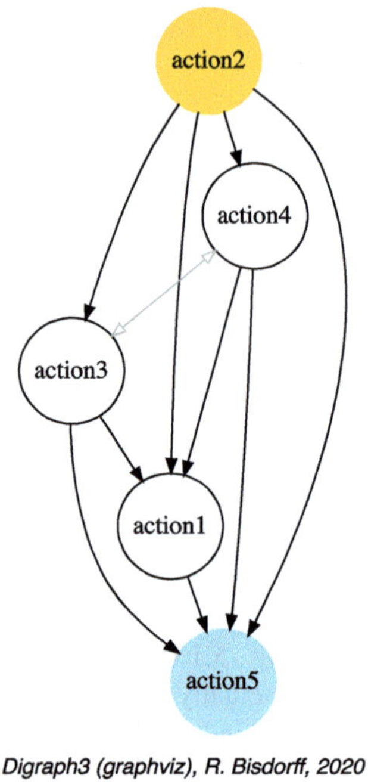

Fig. 5.1 *The template outranking digraph*. The digraph `bod` models in fact a *partial order* on the five potential decision alternatives

criteria	crit4	crit1	crit3	crit2	crit6	crit5
weights	+2.00	+3.00	+6.00	+3.00	+2.00	+2.00
tau(*)	+0.60	+0.40	+0.35	+0.20	+0.10	-0.60
action2	89.0	100.0	NA	10.0	75.0	30.7
action3	55.4	63.0	-47.3	6.0	NA	63.5
action4	83.5	23.0	-35.7	2.0	54.9	37.5
action1	71.0	41.0	-52.2	4.0	22.5	63.0
action5	10.0	NA	-68.0	9.0	75.0	88.0

Color legend:

quantile	20.00%	40.00%	60.00%	80.00%	100.00%

(*) tau: *Ordinal (Kendall) correlation between marginal criterion and global ranking relation*
Outranking model: **standard**, *Ranking rule:* **Copeland**
Ordinal (Kendall) correlation between global ranking and global outranking relation: **+1.000**
Mean marginal correlation (a) : **+0.228**
Standard marginal correlation deviation (b) : **+0.322**
Ranking fairness (a) - (b) : **-0.094**

Fig. 5.2 COPELAND ranked heatmap of the template performance tableau

criteria	crit2	crit6	crit1	crit4	crit3	crit5
weights	+3.00	+2.00	+3.00	+2.00	+6.00	+2.00
tau(*)	+0.60	+0.50	+0.40	+0.20	-0.05	-0.20
action2	10.0	75.0	100.0	89.0	NA	30.7
action3	6.0	NA	63.0	55.4	-47.3	63.5
action5	9.0	75.0	NA	10.0	-68.0	88.0
action4	2.0	54.9	23.0	83.5	-35.7	37.5
action1	4.0	22.5	41.0	71.0	-52.2	63.0

Color legend:

quantile	20.00%	40.00%	60.00%	80.00%	100.00%

(*) tau: *Ordinal (Kendall) correlation between marginal criterion and global ranking relation*
Outranking model: **standard**, *Ranking rule:* **NetFlows**
Ordinal (Kendall) correlation between global ranking and global outranking relation: **+0.920**
Mean marginal correlation (a) : **+0.206**
Standard marginal correlation deviation (b) : **+0.286**
Ranking fairness (a) - (b) : **-0.081**

Fig. 5.3 NETFLOWS ranked heatmap of the template performance tableau

Due to an 8 against 7 *plurality tyranny* effect, the COPELAND ranking rule, essentially based on crisp majority outranking counts, puts here alternative *action5* (a5) last, despite its excellent evaluations observed on criteria g2, g5 and g6 (Fig. 5.2).

A slightly *fairer* ranking result may be obtained in Fig. 5.3 with the NETFLOWS ranking rule (see Sect. 8.3).

```
>>> bod.showHTMLPerformanceHeatmap(ndigits=1,\
...     colorLevels=5, Correlations=True,\
...     rankingRule='NetFlows',\
...     pageTitle=\
...       'Heatmap of the template performance tableau')
```

It might be opportune to furthermore study the robustness of the apparent outranking situations when assuming *uncertain* or solely *ordinal* criteria significance weights (see Chaps. 18 and 19).

The next Chap. 6 describes DIGRAPH3 resources for generating random multiple-criteria performance tableaux of various kinds like Cost-Benefit, three-objectives, or academic tableaux.

References

Bisdorff R (2013) On polarizing outranking relations with large performance differences. J Multi-Criteria Decis Anal Wiley 20:3–12. http://hdl.handle.net/10993/245

Python Software Foundation (2021) Python documentation. https://docs.python.org/3/

Chapter 6
Generating Random Performance Tableaux

Contents

Abstract The chapter describes the DIGRAPH3 resources for generating random multiple-criteria performance tableaux like a *Cost-Benefit* tableau, a three Objectives—*economic*, *societal*, and *environmental*—tableau, and an *academic* performance tableau.

6.1 Introduction

The `randomPerfTabs` module provides several classes for generating random performance tableaux models of different kind, mainly for the purpose of testing implemented methods and tools presented and discussed in the Algorithmic Decision Theory course lectures at the University of Luxembourg. This chapter introduces several of these performance tableau models.

The simplest class, called `RandomPerformanceTableau`, generates a set of n decision actions, a family of m real-valued performance criteria, ranging by default from 0.0 to 100.0, associated with default discrimination thresholds: 2.5 (`ind`), 5.0 (`pref`), and 60.0 (`veto`). The generated random evaluations are by default `Beta(2,2)` distributed on each measurement scale.

One of the most useful models involving two decision objectives, named *Costs* (to be minimised), respectively, *Benefits* (to be maximised), is provided by the `RandomCBPerformanceTableau` class; its purpose being to generate more or less contradictory performances on these two, usually conflicting, objectives. *Low costs* will randomly be correlated with *low benefits*, whereas *high costs* will randomly be correlated with *high benefits*.

R. Bisdorff, *Algorithmic Decision Making with Python Resources*,
International Series in Operations Research & Management Science 324,
https://doi.org/10.1007/978-3-030-90928-4_6

Many public policy decision problems often involve three more or less conflicting decision objectives taking into account *economical*, *societal* as well as *environmental* aspects. For this type of performance tableau model, the `randomPerfTabs` module provides a specific class, called `Random3ObjectivesPerformanceTableau`.

Based on their grades obtained in a number of examinations, deciding which students validate or not their academic studies is the genuine decision practice of universities and academies. To thoroughly study these kind of decision problems, the `randomPerfTabs` module provides a corresponding performance tableau model, called `RandomAcademicPerformanceTableau`, which gathers grades obtained by a given number of students in a given number of weighted courses.

6.2 Random Standard Performance Tableaux

The `RandomPerformanceTableau` class, the simplest of the kind, specialises the generic `PerformanceTableau` class, and takes the following parameters:

- `numberOfActions` := number of decision actions.
- `numberOfCriteria` := number of performance criteria.
- `weightDistribution` := `'random'` (default) | `'fixed'` | `'equisignificant'`:
  - If `'random'`, weights are uniformly selected randomly from the given weight scale;
  - If `'fixed'`, the weightScale must provide a corresponding weights distribution;
  - If `'equi-significant'`, all criterion weights are put to unity.
- `weightScale` := `[Min,Max]` (default =(1, `numberOfCriteria`).
- `IntegerWeights` := `True` (default) | `False` (normalised to proportions of 1.0).
- `commonScale` := `[a,b]`; common performance measuring scales (default = [0.0, 100.0]).
- `commonThresholds` := [(`q0`, `q1`), (`p0`, `p1`), (`v0`, `v1`)]; indifference(`q`), preference (`p`) and considerable performance difference (`v`) discrimination thresholds. For each threshold type $x \in \{\mathtt{q}, \mathtt{p}, \mathtt{v}\}$, the float $x0$ value represents a *constant percentage* of the common scale and the float $x1$ value a *proportional value* of the actual performance measure. Default values are [(2.5.0, 0.0), (5.0, 0.0), (60.0, 0, 0)].

- `commonMode` := common distribution of random performance measurements:[1]
    - (`'beta'`,`None` (default setting), (α, β)), a beta generator with default $\alpha = 2$ and $\beta = 2$ parameters.
    - (`'uniform'`,`None`, `None`), uniformly distributed float values on the given common scales' range [`Min, Max`];
    - (`'normal'`, μ, σ), truncated Gaussian distribution, by default $\mu = (b-a)/2$ and $\sigma = (b-a)/4$;
    - (`'triangular'`, *mode*, *repartition*), generalised triangular distribution with a probability repartition parameter specifying the probability mass accumulated until the mode value. By default, *mode* $= (b - a)/2$ and `repartition`$=0.5$.[2]
- `valueDigits` := `integer`, precision of performance measurements (2 decimal digits by default).
- `missingDataProbability`$:= 0.0 \leq$ `float` ≤ 1.0; probability of missing performance evaluation on a criterion for an alternative (default 0.025).
- `NA` := `Decimal` (default = −999); missing data symbol.

Code Example

Listing 6.1 Generating a random performance tableau

```
>>> from randomPerfTabs import RandomPerformanceTableau
>>> t = RandomPerformanceTableau(numberOfActions=21,\
...                            numberOfCriteria=13,seed=100)
>>> t.actions
  {'a01': {
    'comment': 'RandomPerformanceTableau() generated.',
    'name': 'random decision action'
    },
   'a02': { ... },
    ...
   }
>>> t.criteria
  {'g01': {
    'thresholds': {
      'ind' : (Decimal('10.0'), Decimal('0.0')),
      'veto': (Decimal('80.0'), Decimal('0.0')),
      'pref': (Decimal('20.0'), Decimal('0.0'))},
    'scale': [0.0, 100.0],
    'weight': Decimal('1'),
    'name': 'RandomPerformanceTableau() instance',
    'comment': "Arguments: weightDistribution=random;
                           weightScale=(1, 1);
```

[1] See Lecture 3 of the Computational Statistic Course (Bisdorff 2020).

[2] The `randomNumbers` module provides for this purpose the `ExtendedTriangularRandomVariable` class.

```
23                                  commonMode=None"
24       },
25     'g02':  { ... },
26     ...
27   }
28 >>> t.NA
29   Decimal('-999')
30 >>> t.evaluation
31   {'g01': {'a01': Decimal('15.17'),
32            'a02': Decimal('44.51'),
33            'a03': Decimal('-999'), # missing evaluation
34         ...  },
35     ...
36   }
37 >>> t.showHTMLPerformanceTableau()
```

Highest and lowest evaluations on each criterion are marked in *light green*, respectively, in *light red*. Notice that missing (`NA`) evaluations are recorded in a performance tableau by default as `Decimal('-999')` value (see Listing 6.1 on the preceding page Lines 28–29; Fig. 6.1).

Performance table randomperftab

criteria	g01	g02	g03	g04	g05	g06	g07	g08	g09	g10	g11	g12	g13
weight	1.00	1.00	1.00	1.00	1.00	1.00	1.00	1.00	1.00	1.00	1.00	1.00	1.00
a01	15.17	46.37	82.88	41.14	59.94	41.19	58.68	44.73	22.19	64.64	34.93	42.36	17.55
a02	44.51	16.22	41.66	53.58	31.39	65.22	71.96	57.84	78.08	77.37	8.30	63.41	61.55
a03	NA	21.53	12.82	56.93	26.80	48.03	54.35	62.42	94.27	73.57	71.11	21.81	56.90
a04	58.00	51.16	21.92	65.57	59.02	44.77	37.49	58.39	80.79	55.39	46.44	19.57	39.22
a05	24.22	77.01	75.74	83.87	40.85	8.55	85.44	67.34	57.40	39.08	64.83	29.37	96.39
a06	29.10	39.35	15.45	34.99	49.12	11.49	28.44	52.89	64.24	62.92	58.28	32.02	10.25
a07	96.58	32.06	6.05	49.56	NA	66.06	41.64	13.08	38.31	24.82	48.39	57.03	42.91
a08	82.29	47.67	9.96	79.43	29.45	84.17	31.99	90.88	39.58	50.78	61.88	44.40	48.26
a09	43.90	14.81	60.55	42.37	6.72	56.14	34.20	51.54	21.79	79.13	50.95	93.16	81.89
a10	38.75	79.70	27.88	42.39	71.88	66.09	58.33	58.88	17.10	44.25	48.73	30.63	52.73
a11	35.84	67.48	38.81	33.75	26.87	64.10	71.95	62.72	NA	85.80	58.37	49.33	NA
a12	29.12	13.97	67.45	38.60	48.30	11.87	NA	57.76	74.86	26.57	48.80	43.57	7.68
a13	34.79	90.72	38.93	57.38	64.14	97.86	91.16	43.80	33.68	38.98	28.87	63.36	60.03
a14	62.22	80.16	19.26	62.34	60.96	24.72	73.63	71.21	56.43	46.12	26.09	51.43	12.86
a15	44.23	69.62	94.95	34.95	63.46	52.97	98.84	78.74	36.64	65.12	22.46	55.52	68.79
a16	19.10	45.49	65.63	64.96	50.57	55.91	10.02	34.70	29.31	50.15	70.68	62.57	71.09
a17	27.73	22.03	48.00	79.38	23.35	74.03	58.74	59.42	50.95	82.27	49.20	43.27	38.61
a18	41.46	33.83	7.97	75.11	49.00	55.70	64.99	38.47	49.86	17.45	28.08	35.21	67.81
a19	22.41	NA	34.86	49.30	65.18	39.84	81.16	NA	55.99	66.55	55.38	43.08	29.72
a20	21.52	69.98	71.81	43.74	24.53	55.39	52.67	13.67	66.80	57.46	70.81	5.41	76.05
a21	56.90	48.80	31.66	15.31	40.57	58.14	70.19	67.23	61.10	31.04	60.72	22.39	70.38

Fig. 6.1 Browser view on random performance tableau instance

6.3 Random Cost-Benefit Performance Tableaux

The `randomPerfTabs` module provides a `RandomCBPerformanceTableau` class for generating *Costs* versus *Benefits* organised performance tableaux. The random generator is following by default the directives below:

- Three types of decision actions are distinguished: *cheap*, *neutral*, and *expensive* ones with an equal proportion of 1/3. Two types of weighted criteria are also distinguished: *Costs* criteria to be *minimised*, and *Benefits* criteria to be *maximised*, in the proportions 1/3, respectively, 2/3;
- Random performances on each type of criteria are drawn, either from an ordinal scale [0; 10], or from a cardinal scale [0.0; 100.0], following a parametric triangular law of mode: 30% performance for cheap, 50% for neutral, and 70% performance for expensive decision actions, with constant probability repartition 0.5 on each side of the respective mode;
- Costs criteria use mostly cardinal scales (3/4), whereas Benefits criteria use mostly ordinal scales (2/3);
- The sum of weights of the Costs criteria equals by default the sum of weights of the Benefits criteria: `weighDistribution = 'equiobjectives'`;
- On cardinal criteria, both of cost or of benefit type, following constant performance discrimination quantiles are instantiated: 5% indifferent situations, 90% preference situations, and 5% considerable performance difference situations.

Parameters

- If `numberOfActions == None`, a uniform random number between 10 and 31 of *cheap*, *neutral* or *advantageous* actions (equal 1/3 probability each type) actions is instantiated. Minimal number of decision actions required is 3;
- If `numberOfCriteria == None`, a uniform random number between 5 and 21 of cost or benefit criteria (1/3, respectively, 2/3 probability) is instantiated;
- `weightDistribution` := `'equisignificant'` (default) | `'equiobjectives'` | `'fixed'` | `'random'`;
- default `weightScale` for `'random'` weight distribution is 1 - `numberOfCriteria`;
- All *cardinal* criteria are evaluated with decimals between 0.0 and 100.0 whereas *ordinal* criteria are evaluated with integers between 0 and 10.
- `commonThresholds` is obsolete. Preference discrimination is specified as *percentiles* of concerned performance differences (see below).
- `commonPercentiles` := {`'ind':5, 'pref':10, 'veto':95`} are expressed in percents (reversed for vetoes), and only concern cardinal criteria.
- `missingDataProbability` := 0.0 ≤ `float` ≤ 1.0 ; probability of missing performance evaluation on a criterion for an alternative (default 0.025).
- `NA` := `Decimal` (default = −999); missing data symbol.

Example Python Session

Listing 6.2 Generating a random Cost-Benefit performance tableau

```
1 >>> from randomPerfTabs import\
2 ...                 RandomCBPerformanceTableau
3 >>> t = RandomCBPerformanceTableau(
4 ...         numberOfActions=7,\
5 ...         numberOfCriteria=5,\
6 ...         weightDistribution='equiobjectives',\
7 ...         commonPercentiles={'ind':0.05,
8 ...                            'pref':0.10,\
9 ...                            'veto':0.95},\
10 ...          seed=100)
11 >>> t.showActions()
12   *----- show decision action --------------*
13     key:  a1
14       short name: a1c
15       name:  random cheap decision action
16     key:  a2
17       short name: a2n
18       name:  random neutral decision action
19     ...
20     key:  a7
21       short name: a7a
22       name:  random advantageous decision action
23 >>> t.showCriteria()
24   *----  criteria -----*
25    b1 'random ordinal benefit criterion'
26     Preference direction: max
27     Scale = (0, 10)
28     Weight = 3
29     ...
30    c1 'random cardinal cost criterion'
31     Preference direction: min
32     Scale = (0.0, 100.0)
33     Weight = 2
34     Threshold ind  :  1.76 + 0.00x ; percentile:  9.5
35     Threshold pref :  2.16 + 0.00x ; percentile: 14.3
36     Threshold veto : 73.19 + 0.00x ; percentile: 95.2
37     ...}
```

In Listing 6.2 one may notice the three types of decision actions (see Lines 12–22), as well as the two types of criteria with either an *ordinal* or a *cardinal* performance measuring scale (Lines 24–36). In the latter case, by default about 5% of the random performance differences will be below the *indifference* and 10% below the *preference* discriminating threshold. About 5% will be *considerably large*. More statistics about the generated performance evaluations can be inspected with the `showStatistics()` method.

```
>>> t.showStatistics()
    *-------- Performance tableau summary statistics -------*
    Instance name       : randomCBperftab
    Actions             : 7
    Criteria            : 5
     Criterion name        : b1
       Criterion weight     : 3
       criterion scale     : 0.00 - 10.00
       mean evaluation     : 5.14
       standard deviation : 2.64
       maximal evaluation : 8.00
       quantile Q3 (x_75) : 8.00
       median evaluation  : 6.50
       quantile Q1 (x_25) : 3.50
       minimal evaluation : 1.00
       mean absolute difference      : 2.94
       standard difference deviation : 3.74
      ...
     Criterion name        : c1
       Criterion weight     : 2
       criterion scale     : -100.00 - 0.00
       mean evaluation     : -49.32
       standard deviation : 27.59
       maximal evaluation : 0.00
       quantile Q3 (x_75) : -27.51
       median evaluation  : -35.98
       quantile Q1 (x_25) : -54.02
       minimal evaluation : -91.87
       mean absolute difference      : 28.72
       standard difference deviation : 39.02
     ...
```

A heatmap view with five colour levels gives the result shown in Fig. 6.2 on the following page.

```
>>> t.showHTMLPerformanceHeatmap(colorLevels=5,\
...         rankingRule=None,\
...         pageTitle='Random Cost-Benefit Performance Tableau')
```

Such a performance tableau may be stored with the `save()` method and reloaded with the `PerformanceTableau` class:

```
>>> t.save('temp')
    *----- saving performance tableau in XMCDA 2.0 format
       -------------*
    File: temp.py saved !
>>> from perfTabs import PerformanceTableau
>>> t = PerformanceTableau('temp')
```

criteria	b1	b2	c1	c2	c3
weights	+3.00	+3.00	+2.00	+2.00	+2.00
a1c	2.00	20.66	-28.65	-79.26	-3.19
a2n	3.00	47.34	-15.47	-52.85	-44.47
a3n	3.00	57.44	-81.69	-77.77	-66.35
a4c	2.00	69.06	-77.23	-57.77	-34.80
a5a	7.00	35.23	-57.33	-39.65	-43.45
a6n	9.00	52.77	-39.33	-58.08	-71.82
a7a	5.00	92.47	-79.57	-34.87	-87.54

Color legend:

quantile	20.00%	40.00%	60.00%	80.00%	100.00%

Fig. 6.2 Unordered heatmap of a random Cost-Benefit performance tableau

6.4 Random Three Objectives Performance Tableaux

The `randomPerfTabs` module provides a `Random3ObjectivesPerformanceTableau` class for generating random tableaux concerning public policies evaluated with respect to three decision objectives taking, respectively, into account *economical*, *societal* as well as *environmental* aspects. Each potential public policy is qualified randomly as performing *weakly* (−), *fairly* (∼) or *well* (+) with respect to each one of the three objectives.

Generator Directives Are the Following

- `numberOfActions` = 20 (default), minimal number required is 3;
- `numberOfCriteria` = 13 (default),
- `weightDistribution` = `'equiobjectives'` (default) | `'random'` | `'equisignificant'`,
- `weightScale` = (1,`numberOfCriteria`): only used when random criterion weights are requested,
- `integerWeights` = `True` (default): `False` gives normalised rational weights,
- `commonScale` = (0.0,100.0),
- `commonThresholds` = [(5.0, 0.0), (10.0, 0.0), (60.0, 0.0)]: Performance discrimination thresholds may be set for `'ind'`, `'pref'` and `'veto'` thresholds,
- `commonMode` = [`'triangular'`,`'variable'`,0.5]: random number generators of various other types (*'uniform'*,*'beta'*) are available. If the mode of the `'triangular'` distribution is set to `'variable'`, three modes at 0.3(−), 0.5(∼), respectively, 0.7(+) of the common scale span are set at random for each coalition and action.
- `valueDigits` = 2 (default): evaluations are encoded as decimals,

- `missingDataProbability` = 0.05 (default): random insertion of missing values with given probability,
- `NA` := `Decimal` (default = `Decimal('-999')`; missing data symbol,
- `seed` = `None` (default).

Example Python Session

Listing 6.3 Generating a random three objectives performance tableau

```
1  >>> from randomPerfTabs import\
2  ...            Random3ObjectivesPerformanceTableau
3  >>> t = Random3ObjectivesPerformanceTableau(\
4  ...           numberOfActions=7,\
5  ...           numberOfCriteria=13,\
6  ...           weightDistribution='equiobjectives',\
7  ...           seed=120)
8  >>> t
9    *------- PerformanceTableau instance description ------*
10    Instance class      : Random3ObjectivesPerformanceTableau
11    Seed                : 120
12    Instance name       : random3ObjectivesPerfTab
13    Actions             : 7
14    Objectives          : 3
15    Criteria            : 13
16    NaN proportion (%)  : 2.2
17    Attributes          : ['name', 'valueDigits', 'BigData',
18       'OrdinalScales', 'missingDataProbability',
19       'negativeWeightProbability', 'randomSeed',
20       'sumWeights', 'valuationPrecision', 'commonScale',
21       'objectiveSupportingTypes', 'actions', 'objectives',
22       'criteriaWeightMode', 'criteria',
23       'weightPreorder', 'evaluation', 'NA']
```

The default missing data probability is 5%. In our random instance here we observe 2.2% of missing evaluations (see Line 16). The 'equiobjectives' directive (see Line 6) results hence in a balanced total weight (72.00) for each decision objective as is made apparent below with the `showObjectives()` method.

Listing 6.4 Inspecting the three objectives

```
>>> t.showObjectives()
 *------ show objectives -------"
  Eco: Economical aspect
   ec01 criterion of objective Eco 18
   ec05 criterion of objective Eco 18
   ec06 criterion of objective Eco 18
   ec12 criterion of objective Eco 18
   Total weight: 72.00 (4 criteria)
  Soc: Societal aspect
   so02 criterion of objective Soc 24
   so11 criterion of objective Soc 24
   so13 criterion of objective Soc 24
```

```
13    Total weight: 72.00 (3 criteria)
14   Env: Environmental aspect
15    en03 criterion of objective Env 12
16    en04 criterion of objective Env 12
17    en07 criterion of objective Env 12
18    en08 criterion of objective Env 12
19    en09 criterion of objective Env 12
20    en10 criterion of objective Env 12
21    Total weight: 72.00 (6 criteria)
```

In Listing 6.4 on the previous page, we notice that four *equisignificant* criteria (`ec01`, `ec05`, `ec06`, and `ec12`) of significance weight 18 assess, for instance, the performance of the public policies from an *economic* point of view (Lines 3–8). Three *equisignificant* criteria of weight 24 do the same from a *societal* (Lines 9–13), and six of weight 12 from an *environmental* point of view (Lines 14–21).

Variable *triangular* modes: 0.3, 0.5, or 0.7 of the span of the measure scale, give a different performance status for each public policy with respect to the three decision objectives.

```
>>> t.showActions()
  key:  p1
   short name:  p1
   name:        action p1 Eco- Soc+ Env+
   profile:     {'Eco':'weak', 'Soc':'good', 'Env':'good'}
  key:  p2
   short name:  p2
   name:        action p2 Eco- Soc- Env-
   profile:     {'Eco':'weak', 'Soc':'weak', 'Env':'weak'}
   ...
  key:  p7
   short name:  p7
   name:        action p7 Eco+ Soc- Env-
   profile:     {'Eco':'good', 'Soc':'weak', 'Env':'weak'}
```

Policy `p1`, for instance, will probably show *good* performances with respect to the *societal* and *environmental* aspects, and *weak* performances with respect to the *economical* aspect, whereas policy `p2`, being weak with respect to all the three objectives, will probably appear among the last recommended policies.

We may inspect in Fig. 6.3 on the facing page the given random three-objectives performance tableau with the `showHTMPerformanceTableau()` method.

```
>>> t.showHTMLPerformanceTableau()
```

Light green cells show the highest and light red the lowest evaluations. Policy `p1` obtains thus the lowest evaluation (17.09/100.00) on the environmental criterion `en09`, whereas policy `p3` obtains the highest evaluations on four out of the six *environmental* criteria.

No trivial best choice becomes apparent when looking at the performance tableau shown in Fig. 6.3 on the next page. Let us therefore compute a Rubis best choice recommendation (see Chap. 4).

Performance table random3ObjectivesPerfTab

criteria	ec01	ec05	ec06	ec12	en03	en04	en07	en08	en09	en10	so02	so11	so13
weight	18.00	18.00	18.00	18.00	12.00	12.00	12.00	12.00	12.00	12.00	24.00	24.00	24.00
p1	31.05	26.95	35.91	32.85	68.88	70.52	22.94	50.04	17.06	32.52	20.90	NA	62.37
p2	12.54	11.31	29.63	9.77	43.09	36.04	32.96	54.35	82.25	22.29	75.76	51.42	27.96
p3	30.70	58.45	38.31	24.89	72.36	87.02	30.57	80.17	90.38	86.60	54.51	43.21	6.58
p4	16.18	61.70	60.27	15.10	26.63	66.28	12.84	58.78	86.68	57.92	NA	71.91	70.18
p5	86.00	89.43	44.41	18.32	31.10	74.55	29.89	40.26	26.86	41.91	23.83	57.41	22.41
p6	77.65	94.24	24.75	65.64	3.85	36.15	47.36	81.22	72.54	80.54	27.25	37.04	34.11
p7	72.26	44.62	70.55	29.82	52.69	37.80	50.06	28.27	28.53	29.50	29.81	41.38	26.31

Fig. 6.3 Browser view on the given random three-objectives performance tableau

Listing 6.5 What is the public policy to recommend as best choice?

```
>>> from outrankingDigraphs import\
...         BipolarOutrankingDigraph
>>> g = BipolarOutrankingDigraph(t)
>>> g.showBestChoiceRecommendation()
 ***********************
 Rubis best choice recommendation(s) (BCR)
 (in decreasing order of determinateness)
 Credibility domain: [-1.00,1.00]
 === >> potential first choice(s)
 * choice              : ['p3', 'p4', 'p5', 'p6']
  independence        : 0.00
  dominance           : 0.17
  absorbency          : -1.00
  covering (%)        : 41.67
  determinateness (%) : 52.98
  - most credible action(s) = { 'p3': 0.17, }
 === >> potential last choice(s)
 * choice              : ['p1', 'p2', 'p5', 'p7']
  independence        : 0.00
  dominance           : -0.44
  absorbency          : 0.19
  covered (%)         : 41.67
  determinateness (%) : 50.79
  - most credible action(s) = { 'p7': 0.06, }
```

Policy p3 gives the most credible first choice recommendation with the support of a 57.5% majority of significance, whereas policy p7 gives a credible last choice recommendation.

Policy p5 represents an ambiguous first as well as last choice candidate. A drawing of the strict outranking digraph oriented by first and last recommendations shows in Fig. 6.4 on the following page the complete preferential picture.

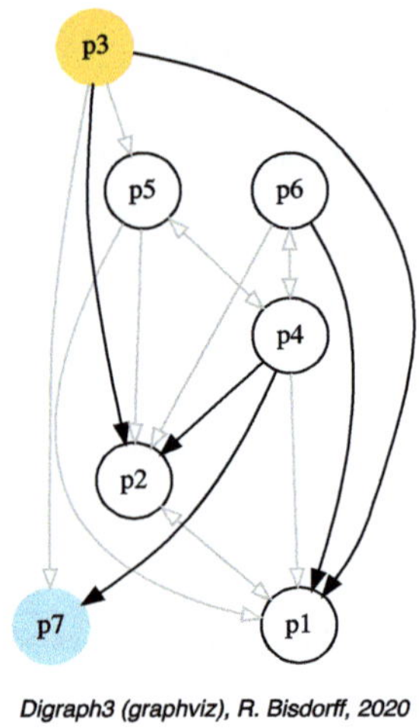

Fig. 6.4 *The strict outranking digraph oriented by first and last choice recommendations.* Policy p5 appears incomparable—in a strict outranking sense—to all the other six policies, whereas policies p1, p2, and p7 appear strictly outranked

```
>>> (~(-g)).exportGraphViz(\
...                 fileName='3ObjPerfTabBestChoice',\
...                 firstChoice=['p3'],lastChoice=['p7'])
  *---- exporting a dot file for GraphViz tools ---------*
   Exporting to 3ObjPerfTabBestChoice.dot
   dot -Grankdir=BT -Tpng 3ObjPerfTabBestChoice.dot\
                      -o 3ObjPerfTabBestChoice.png
```

A heatmap view in Fig. 6.5 on the next page on the COPELAND ranked performance tableau confirms the best choice recommendation (see Sect. 8.2).

```
>>> t.showHTMLPerformanceHeatmap(\
...                           Correlations=True,\
...                           colorLevels=5,ndigits=1,\
...                           rankingRule='Copeland')
```

The COPELAND ranking, based on the bipolar outranking digraph, confirms policy p3 as first-ranked. Notice also that the three strict outranked policies do effectively appear in the last positions.

Heatmap of Performance Tableau 'random3ObjectivesPerfTab'

criteria	en10	ec05	en09	en04	en08	ec01	ec12	ec06	so11	so02	en07	so13	en03
weights	+12.00	+18.00	+12.00	+12.00	+12.00	+18.00	+18.00	+18.00	+24.00	+24.00	+12.00	+24.00	+12.00
tau(*)	+0.79	+0.52	+0.45	+0.40	+0.38	+0.14	+0.12	+0.10	+0.02	-0.02	-0.10	-0.10	-0.17
p3	86.6	58.5	90.4	87.0	80.2	30.7	24.9	38.3	43.2	54.5	30.6	6.6	72.4
p4	57.9	61.7	86.7	66.3	58.8	16.2	15.1	60.3	71.9	NA	12.8	70.2	26.6
p6	80.5	94.2	72.5	36.1	81.2	77.7	65.6	24.8	37.0	27.2	47.4	34.1	3.9
p5	41.9	89.4	26.9	74.5	40.3	86.0	18.3	44.4	57.4	23.8	29.9	22.4	31.1
p7	29.5	44.6	28.5	37.8	28.3	72.3	29.8	70.5	41.4	29.8	50.1	26.3	52.7
p1	32.5	26.9	17.1	70.5	50.0	31.1	32.9	35.9	NA	20.9	22.9	62.4	68.9
p2	22.3	11.3	82.2	36.0	54.4	12.5	9.8	29.6	51.4	75.8	33.0	28.0	43.1

Color legend:

quantile	20.00%	40.00%	60.00%	80.00%	100.00%

(*) **tau:** *Ordinal (Kendall) correlation between marginal criterion and global ranking relation*
Outranking model: **standard**, *Ranking rule*: **Copeland**
Ordinal (Kendall) correlation between global ranking and global outranking relation: **+0.896**
Mean marginal correlation (a) : **+0.161**
Standard marginal correlation deviation (b) : **+0.258**
Ranking fairness (a) - (b) : **-0.098**

Fig. 6.5 Browser view on the COPELAND ranked performance tableau

6.5 Random Academic Performance Tableaux

The `RandomAcademicPerformanceTableau` class generates performance tableaux with random grades for a given number of students in different courses.
Generator Directives

- `numberOfStudents` := `Integer` (default 10),
- `numberOfCourses` := `Integer` (default 5),
- `weightDistribution` := '*equisignificant*' | '*random*' (default),
- `weightScale` := 1, 1 - `numberOfCourses` (default when random)),
- `IntegerWeights` := `Boolean` (True = default),
- `commonScale` := `(Integer,integer)` (0, 20) (default),
- `ndigits` := `Integer` (default 0),
- `WithTypes` := `Boolean` (default False),
- `commonMode` := ('`triangular`',xm=14,r=0.25) (default),
- `commonThresholds` := `'ind'`:(0,0), `'pref'`:(1,0) (default),
- `missingDataProbability` := 0.0 (default),
- `NA` := `Decimal` (default = −999); missing data symbol.

When Parameter `WithTypes` is set to `True`, the students are randomly allocated to one of the following four categories—`weak` (1/6), `fair` (1/3), `good` (1/3), and `excellent` (1/3)—in the bracketed proportions. In a default 0–20 grading range, the random range of a weak student is 0–10, of a fair student 4–16, of a good student 8–20, and of an excellent student 12–20. The random grading generator follows in this case a double triangular probability law with *mode* (xm) equal to the middle of the random range and median repartition ($r = 0.5$) of probability each side of the mode.

Listing 6.6 Generating a random academic performance tableau

```
1  >>> from randomPerfTabs import RandomAcademicPerformanceTableau
2  >>> t = RandomAcademicPerformanceTableau(\
3  ...         numberOfStudents=11,\
4  ...         numberOfCourses=7, missingDataProbability=0.03,\
5  ...         WithTypes=True, seed=100)
6  >>> t
7    *------- PerformanceTableau instance description ------*
8     Instance class    : RandomAcademicPerformanceTableau
9     Seed              : 100
10    Instance name     : randstudPerf
11    Actions           : 11
12    Criteria          : 7
13    NA proportion(%)  : 5.2
14    Attributes        : ['randomSeed', 'name', 'actions',
15              'criteria', 'evaluation', 'weightPreorder']
16 >>> t.showPerformanceTableau()
17   *---- performance tableau -----*
18    Courses |   'm1'  'm2'  'm3'  'm4'  'm5'  'm6'  'm7'
19      ECTS  |    2     1     3     4     1     1     5
20   ---------|---------------------------------------
21     's01f' |   12    13    15    08    16    06    15
22     's02g' |   10    15    20    11    14    15    18
23     's03g' |   14    12    19    11    15    13    11
24     's04f' |   13    15    12    13    13    10    06
25     's05e' |   12    14    13    16    15    12    16
26     's06g' |   17    13    10    14    NA    15    13
27     's07e' |   12    12    12    18    NA    13    17
28     's08f' |   14    12    09    13    13    15    12
29     's09g' |   19    14    15    13    09    13    16
30     's10g' |   10    12    14    17    12    16    09
31     's11w' |   10    10    NA    10    10    NA    08
32 >>> t.weightPreorder
33   [['m2', 'm5', 'm6'], ['m1'], ['m3'], ['m4'], ['m7']]
```

The example random tableau, generated, for instance, above with `missingDataProbability` = 0.03, `WithTypes` = `True` and `seed` = 100 (see Listing 6.6 Lines 2–5), results in a set of two excellent (`s05e` and `s07e`), five good (`s02g`, `s03g`, `s06g`, `s09g` and `s10g`), three fair (`s01f`, `s04f` and `s08f`) and one weak (`s11w`) student. Notice that six students get a grade below the course validating threshold of 10 and we observe four missing grades (`NA`), two in course `m5` and, one in courses `m3` and `m6` (see Lines 21–31).

A statistical summary of the students' grades obtained in the highest weighted course, namely `m7`, is shown with the `showCourseStatistics()` method.

Listing 6.7 Student performance summary statistics per course

```
>>> t.showCourseStatistics('m7')
  *----- Summary performance statistics ------*
   Course name     : m7
   Course weight   : 5
   Students        : 11
```

```
Grading scale   : 0.00 - 20.00
Missing evaluations : 0
Mean evaluation        : 12.82
Standard deviation     : 3.79
Maximal evaluation     : 18.00
Quantile Q3 (x_75)     : 16.25
Median evaluation      : 14.00
Quantile Q1 (x_25)     : 10.50
Minimal evaluation     : 6.00
Mean absolute difference      : 4.30
Standard difference deviation : 5.35
```

In Listing 6.7 on the preceding page, the Course m7 evaluation statistics show a mean grade of 12.82 and a median grade of 14. Maximal (respectively, minimal) grade is 18 (respectively, 6).

With a ranked heatmap view on all the grades we now get in Fig. 6.6 a global pictures of the performance of all the 11 students. The ranking shown here, produced with the NETFLOWS ranking rule (see Sect. 8.3), reports a high correlation of +0.887 with the corresponding bipolar-valued outranking digraph (see Chap. 16).

The NETFLOWS ranking represents also a rather *fair weighted consensus* between the individual courses' marginal rankings as is made apparent with the `showRankingConsensusQuality()` method in Listing 6.8 on the next page.

criteria	m7	m4	m2	m3	m1	m6	m5
weights	+5.00	+4.00	+1.00	+3.00	+2.00	+1.00	+1.00
tau(*)	+0.73	+0.31	+0.29	+0.20	+0.11	+0.09	+0.07
s07e	17.00	18.00	12.00	12.00	12.00	13.00	NA
s02g	18.00	11.00	15.00	20.00	10.00	15.00	14.00
s09g	16.00	13.00	14.00	15.00	19.00	13.00	9.00
s05e	16.00	16.00	14.00	13.00	12.00	12.00	15.00
s06g	13.00	14.00	13.00	10.00	17.00	15.00	NA
s03g	11.00	11.00	12.00	19.00	14.00	13.00	15.00
s10g	9.00	17.00	12.00	14.00	10.00	16.00	12.00
s01f	15.00	8.00	13.00	15.00	12.00	6.00	16.00
s08f	12.00	13.00	12.00	9.00	14.00	15.00	13.00
s04f	6.00	13.00	15.00	12.00	13.00	10.00	13.00
s11w	8.00	10.00	10.00	NA	10.00	NA	10.00

Color legend:

quantile	20.00%	40.00%	60.00%	80.00%	100.00%

(*) tau: *Ordinal (Kendall) correlation between marginal criterion and global ranking relation*
Outranking model: **standard**, *Ranking rule*: **NetFlows**
Ordinal (Kendall) correlation between global ranking and global outranking relation: **+0.887**

Fig. 6.6 Ranking the students in a performance heatmap view

Listing 6.8 Consensus quality of the students' ranking

```
>>> from outrankingDigraphs import\
...                 BipolarOutrankingDigraph
>>> g = BipolarOutrankingDigraph(t)
>>> t.showRankingConsensusQuality(\
...                g.computeNetFlowsRanking())
  Consensus quality of ranking:
  ['s07', 's02', 's09', 's05', 's06', 's03', 's10',
   's01', 's08', 's04', 's11']
  Criterion (weight): correlation
  -----------------------------
     'm7'  (5): +0.727
     'm4'  (4): +0.309
     'm2'  (1): +0.291
     'm3'  (3): +0.200
     'm1'  (2): +0.109
     'm6'  (1): +0.091
     'm5'  (1): +0.073
  Summary:
   Weighted mean marginal correlation (a): +0.361
   Standard deviation (b)               : +0.248
   Ranking fairness (a)-(b)             : +0.113
```

The correlation with the marginal course rankings follows, except for course m2, the order of the given ECTS weights. Course `m7`, with the highest relative ECTS (5), is most present (+0.727) in the global NETFLOWS ranking and no marginal course ranking appears negatively correlated with this global ranking. The mean weighted correlation is +0.361.

Useful multiple-criteria ranking rules are presented and discussed in detail in Chap. 8. The next Chap. 7 is concerned with yet another kind of performance evaluation models which is discussed in social choice theory, namely *linear voting profiles*.

Reference

Bisdorff R (2020) Lecture 3: continuous random variables. In: Lectures of the computational statistics course, University of Luxembourg. http://hdl.handle.net/10993/37870

Chapter 7
Who Wins the Election?

Contents

Abstract This chapter is more specifically devoted to handling linear voting profiles and computing the winner of such election results. By following CONDORCET's recipe, we consider pairwise comparisons of election candidates and balance the number of times the first beats the second against the number of times the second beats the first. Thus we obtain the majority margins digraph, in fact a bipolar-valued digraph. When the voters express contradictory linear voting profiles one naturally observes cyclic social preferences without seeing any paradox in this situation. Finally we present a more politically realistic generator for random linear voting profiles which takes into account pre-election polls.

7.1 Linear Voting Profiles

The `votingProfiles` module provides resources like the `LinearVotingProfile` class for handling election results (Bisdorff 2020a). To illustrates these resources let us consider elections involving a set of eligible candidates and a set of weighted voters, who express their voting preferences in a complete linear ranking (without ties) of the eligible candidates. The module provides a `RandomLinearVotingProfile` class for generating random instances of such `LinearVotingProfile` type. In the interactive Python session shown in Listing 7.1, a random linear voting profile is generated, for instance, for the election of three candidates by five voters:

R. Bisdorff, *Algorithmic Decision Making with Python Resources*,
International Series in Operations Research & Management Science 324,
https://doi.org/10.1007/978-3-030-90928-4_7

Listing 7.1 Example of random linear voting profile

```
1 >>> from votingProfiles import\
2 ...                          RandomLinearVotingProfile
3 >>> lvp = RandomLinearVotingProfile(numberOfVoters=5,\
4 ...                               numberOfCandidates=3,\
5 ...                               RandomWeights=True)
6 >>> lvp.candidates
7   OrderedDict([ ('c1',{'name':'Candidate 1}),
8                 ('c2',{'name':'Candidate 2'}),
9                 ('c3',{'name':'Candidate 3'}) ])
10 >>> lvp.voters
11   OrderedDict([('v1',{'weight': 2}),
12                ('v2':{'weight': 3}),
13                ('v3',{'weight': 1}),
14                ('v4':{'weight': 5}),
15                ('v5',{'weight': 4})])
16 >>> lvp.linearBallot
17     {'v1': ['c2', 'c1', 'c3',],
18      'v2': ['c3', 'c1', 'c2',],
19      'v3': ['c1', 'c3', 'c2',],
20      'v4': ['c1', 'c2', 'c3',],
21      'v5': ['c3', 'c1', 'c2',]}
```

The `LinearVotingProfile` data concerning the eligible candidates and the voters is internally stored in two ordered dictionaries: attribute `candidates` (Line 6) and attribute `voters` (Line 10). Notice that in this random example, the five voters are weighted. The linear voting ballots are stored in a standard dictionary: attribute `linearBallot` (Line 16). These ballots can be inspected with the `showLinearBallots()` method.

Listing 7.2 Showing linear voting profiles

```
>>> lvp.showLinearBallots()
    voters(weight)       candidates rankings
    v1(2):          ['c2', 'c1', 'c3']
    v2(3):          ['c3', 'c1', 'c2']
    v3(1):          ['c1', 'c3', 'c2']
    v4(5):          ['c1', 'c2', 'c3']
    v5(4):          ['c3', 'c1', 'c2']
    nbr. of voters: 15
```

Editing of this linear voting profile may be done by saving first the data into a file, then edit this file, and reload it again.

```
>>> lvp.save(fileName='tutorialLinearVotingProfile1')
  *--- Saving linear profile in file:
                <tutorialLinearVotingProfile1.py> ---*
>>> from votingProfiles import LinearVotingProfile
>>> lpv = LinearVotingProfile('tutorialLinearVotingProfile1')
```

7.2 Computing the Winner

The computeUninominalVotes() and the computeSimpleMajority-Winner() methods compute *uninominal votes*, i.e. how many times a candidate was ranked first, and who is consequently the *simple majority* (plurality) winner(s) in this election.

```
>>> lvp.computeUninominalVotes()
  {'c1': 6, 'c2': 2, 'c3': 7}
>>> lvp.computeSimpleMajorityWinner()
  ['c3']
```

As we observe no absolute majority (8/15) of votes for any one of the three candidates, one may look, with the computeInstantRunoffWinner() method, for the *instant runoff* winner instead (Bisdorff 2020a).

Listing 7.3 Example instant run off winner

```
>>> lvp.computeInstantRunoffWinner(Comments=True)
   Half of the Votes =  7.50
     ==> stage =  1
         remaining candidates ['c1', 'c2', 'c3']
         uninominal votes {'c1': 6, 'c2': 2, 'c3': 7}
         minimal number of votes =  2
         maximal number of votes =  7
         candidate to remove =  c2
         remaining candidates =  ['c1', 'c3']
     ==> stage =  2
         remaining candidates ['c1', 'c3']
         uninominal votes {'c1': 8, 'c3': 7}
         minimal number of votes =  7
         maximal number of votes =  8
         candidate a1 obtains an absolute majority
    Instant run off winner: ['c1']
```

In Listing 7.3 no candidate obtains at stage 1 an absolute majority of votes. Candidate c2 obtains the minimal number of votes (2/15) and is, hence, eliminated. At stage 2, candidate c1 eventually obtains an absolute majority of the votes (8/15) and is hence elected.

One may also follow the *Chevalier de Borda*'s advice and, after a *rank analysis* of the linear ballots, compute the BORDA *score*—the average rank—of each candidate and hence determine the BORDA *winner(s)* (de Borda 1781).

Listing 7.4 Example of BORDA rank scores

```
>>> lvp.computeRankAnalysis()
 {'c1': [6, 9, 0], 'c2': [2, 5, 8], 'c3': [7, 1, 7]}
>>> v.computeBordaScores()
 OrderedDict([
    ('c1', {'BordaScore': 24, 'averageBordaScore': 1.6}),
    ('c3', {'BordaScore': 30, 'averageBordaScore': 2.0}),
    ('c2', {'BordaScore': 36, 'averageBordaScore': 2.4}) ])
>>> lvp.computeBordaWinners()
 ['c1']
```

In Listing 7.4, Candidate `c1` obtains the minimal BORDA score, followed by candidate `c3` and finally candidate `c2`. The corresponding BORDA *rank analysis table* may be printed out with the `showRankAnalysisTable()` method.

Listing 7.5 Rank analysis example with BORDA scores

```
>>> lvp.showRankAnalysisTable()
 *----  Borda rank analysis tableau -----*
  candi- | alternative-to-rank |      Borda
  dates  |   1     2     3     | score   average
  -------|-------------------------------------
    'c1' |   6     9     0     | 24/15    1.60
    'c3' |   7     1     7     | 30/15    2.00
    'c2' |   2     5     8     | 36/15    2.40
```

In our randomly generated election results, we are lucky: Candidate `c1` is both the instant runoff winner and the BORDA winner (see Listings 7.3 on the previous page and 7.4). However, one could also follow the *Marquis de Condorcet*'s advice, and compute the *majority margins* obtained by voting for each individual pair of candidates (de Caritat, Marquis de Condorcet 1784).

7.3 The Majority Margins Digraph

Candidate `c1`, for instance, is ranked four times before and once behind candidate `c2` (see Listing 7.2 on page 84). Hence, due to the voters' weights, the corresponding *majority margin* M(`c1`, `c2`) amounts to $(3+1+5+4) - (2) = +11$. These pairwise *majority margins* define on the set of candidates what we call a *majority margins digraph*. The `MajorityMarginsDigraph` class is handling such kind of digraphs.

Listing 7.6 Example of *Majority Margins* digraph

```
>>> from votingProfiles import MajorityMarginsDigraph
>>> mmdg = MajorityMarginsDigraph(lvp,\
...                              IntegerValuation=True)
>>> mmdg
 *------- Digraph instance description ------*
  Instance class      : MajorityMarginsDigraph
  Instance name       : rel_randomLinearVotingProfile1
  Digraph Order       : 3
  Digraph Size        : 3
  Valuation domain    : [-15.00;15.00]
  Determinateness (%) : 55.56
  Attributes          : ['name', 'actions', 'voters',
                         'ballot', 'valuationdomain',
                         'relation', 'order',
                         'gamma', 'notGamma']
>>> mmdg.showAll()
 *----- show detail -------------*
```

```
18 Digraph          : rel_randLinearVotingProfile1
19 *---- Candidates ----*
20  ['c1', 'c2', 'c3']
21 *---- Characteristic valuation domain ----*
22 {'max': Decimal('15.0'),
23  'med': Decimal('0'),
24  'min': Decimal('-15.0'),
25  'hasIntegerValuation': True}
26 *---- majority margins -----*
27     M(x,y)  |  'c1'   'c2'   'c3'
28    --------|-------------------
29     'c1'   |    0     11      1
30     'c2'   |  -11      0     -1
31     'c3'   |   -1      1      0
32   Valuation domain: [-15;+15]
```

Notice in Listing 7.6 Lines 29–31 that, in the case of linear voting profiles, majority margins always verify a *zero-sum property*: $M(x, y) + M(y, x) = 0$ for all candidates x and y. This is not true in general for arbitrary voting profiles. The *majority margins* digraph of linear voting profiles defines in fact a *weak tournament* and belongs, hence, to the class of *self-codual*[1] bipolar-valued digraphs (see Sect. 2.6).

A candidate x, showing a positive majority margin $M(x, y)$, is beating candidate y with an absolute majority in a pairwise voting. Hence, a candidate showing only positive terms in their row in the *majority margins* digraph relation table, beats all other candidates with absolute majority of votes. CONDORCET recommends to declare this candidate (is always unique, why?) the winner of the election. Here we are lucky, it is again candidate `c1` who is the CONDORCET winner (see Line 29 in Listing 7.6 on the preceding page). This result is confirmed below by the `computeCondorcetWinners()` method.

```
>>> mmdg.computeCondorcetWinners()
 ['c1']
```

By seeing the majority margins like a *bipolar-valued characteristic function* of a global voters' preference relation defined on the set of eligible candidates, we may reuse all operational resources of the generic `Digraph` class,[2] and especially its `exportGraphViz()` method for visualising in Fig. 7.1 on the following page the election result.

---

[1] The class of self-codual bipolar-valued digraphs consists of all weakly asymmetric digraphs, i.e. digraphs containing only asymmetric and/or indeterminate links. Limit cases consist of, on the one side, full tournaments with indeterminate reflexive links, and, on the other side, fully indeterminate digraphs. In this class, the converse (inverse $\sim$) operator is indeed identical to the dual (negation $-$) one.

[2] See Chap. 2.

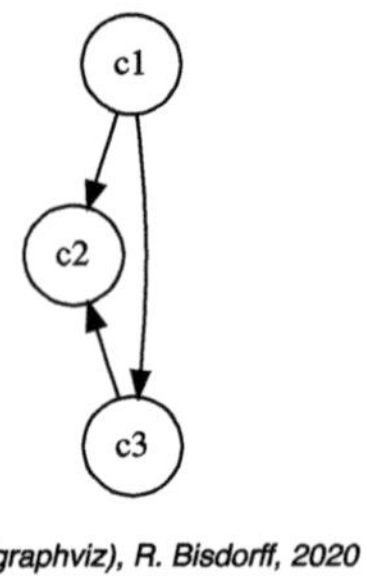

Fig. 7.1 *Visualising an election result.* In the Figure we notice that the *majority margins* digraph from our example linear voting profile models in fact a linear ranking of the candidates: `c1` > `c3` > `c2`, the same actually as modelled by the BORDA scores (see Listing 7.4 on page 85)

```
>>> mmdg.exportGraphViz(\
...             fileName='tutorialLinearBallots',\
...             graphType='pdf')
 *---- exporting a dot file for GraphViz tools ---------*
  Exporting to tutorialLinearBallots.dot
  dot -Grankdir=BT -Tpng tutorialLinearBallots.dot \
                 -o tutorialLinearBallots.pdf
```

That a majority margins digraph models a linear ranking of the eligible candidates, as shown in Fig. 7.1, represents a very unlikely event. Usually, when aggregating linear ballots, there appear cyclic social preferences.

7.4 Cyclic Social Preferences

Let us consider, for instance, in Listing 7.7 the following linear voting profile `v` and construct the corresponding majority margins digraph.

Listing 7.7 Example of cyclic social preferences

```
>>> v.showLinearBallots()
   voters(weight): candidates rankings
    v1(1): ['c1', 'c3', 'c5', 'c2', 'c4']
    v2(1): ['c1', 'c2', 'c4', 'c3', 'c5']
    v3(1): ['c5', 'c2', 'c4', 'c3', 'c1']
    v4(1): ['c3', 'c4', 'c1', 'c5', 'c2']
    v5(1): ['c4', 'c2', 'c3', 'c5', 'c1']
    v6(1): ['c2', 'c4', 'c5', 'c1', 'c3']
    v7(1): ['c5', 'c4', 'c3', 'c1', 'c2']
    v8(1): ['c2', 'c4', 'c5', 'c1', 'c3']
    v9(1): ['c5', 'c3', 'c4', 'c1', 'c2']
>>> mmdg = MajorityMarginsDigraph(v)
>>> mmdg.showRelationTable(ReflexiveTerms=False)
  * ---- Relation Table -----*
```

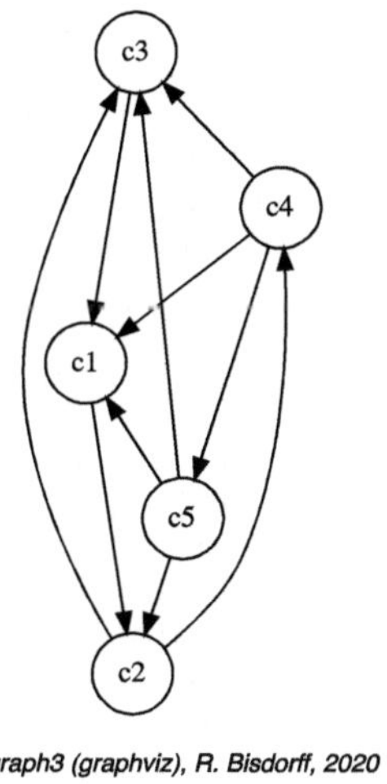

Fig. 7.2 *Cyclic social preferences*. c1 > c2 > c3 > c1, for instance

```
15     M(x,y)  | 'c1'  'c2'  'c3'  'c4'  'c5'
16     -------|------------------------------
17      'c1'  |   -     1    -1    -5    -3
18      'c2'  |  -1     -     1     1    -1
19      'c3'  |   1    -1     -    -3    -1
20      'c4'  |   5    -1     3     -     1
21      'c5'  |   3     1     1    -1     -
22     Valuation domain: [-9;+9]
```

Now there does not exist anymore a completely positive row in the relation table (see Lines 17–21 in Listing 7.7). No one of the five candidates is beating all the others with an absolute majority of votes. There is no CONDORCET winner anymore. In fact, when looking in Fig. 7.2 at a `graphviz` drawing of this *majority margins* digraph, we may observe *cyclic* voters' preferences, like c1 > c2 > c3 > c1.

```
>>> mmdg.exportGraphViz('cycles',graphType='pdf')
 *---- exporting a dot file for GraphViz tools ----*
  Exporting to cycles.dot
  dot -Grankdir=BT -Tpng cycles.dot -o cycles.pdf
```

There may be many circuits appearing in a *majority margins* digraph. The `computeChordlessCircuits` method detects and enumerates all minimal chordless circuits in a `Digraph` instance.

```
>>> mmdg.computeChordlessCircuits()
 [(['c2', 'c3', 'c1'], frozenset({'c2', 'c3', 'c1'})),
  (['c2', 'c4', 'c5'], frozenset({'c2', 'c5', 'c4'})),
  (['c2', 'c4', 'c1'], frozenset({'c2', 'c1', 'c4'}))]
```

In our example voting profile v, we actually obtain three chordless social preference circuits and determining the winner of this election result becomes non-trivial. There is, for instance, no more any instant runoff winner.

```
>>> v.computeInstantRunoffWinner(Comments=True)
Total number of votes =  9.000
Half of the Votes =  4.50
 ==> stage =  1
    remaining candidates ['c2', 'c5', 'c1', 'c4', 'c3']
    uninominal votes {'c2': 2.0, 'c5': 3.0, 'c1': 2.0, 'c4':
        1.0, 'c3': 1.0}
    minimal number of votes =  1.0
    maximal number of votes =  3.0
    candidate to remove =  c3
    candidate to remove =  c4
    remaining candidates =  ['c2', 'c5', 'c1']
 ==> stage =  2
    remaining candidates ['c2', 'c5', 'c1']
    uninominal votes {'c2': 3.0, 'c5': 3.0, 'c1': 3.0}
    minimal number of votes =  3.0
    maximal number of votes =  3.0
    candidate to remove =  c1
    candidate to remove =  c5
    candidate to remove =  c2
    remaining candidates =  []
[]
```

CONDORCET's approach for determining the winner of an election is *not decisive* in this example voting profile. One needs therefore, the case given, exploiting more sophisticated approaches for finding the winner of the election on the basis of the given linear ballots (see Chap. 8 and Bisdorff et al. 2008).

The NETFLOWS ranked heatmap view on voting profiles with the `showHTMLVotingHeatmap()` method is one such tool for showing convincing voting results (see Sect. 8.3).

Listing 7.8 NETFLOWS ranked heatmap view on a voting profile

```
>>> v.showHTMLVotingHeatmap(rankingRule='Netflows',\
...                         colorLevels=3,Correlations=True)
```

It is worthwhile noticing in Fig. 7.3 on the facing page that the compromise NETFLOWS ranking `c4` > `c5` > `c2` > `c3` > `c1`, shown in this heatmap, results in the highest possible *ordinal correlation* index of +0.778 with the majority margins digraph (see Chap. 16). This NETFLOWS ranking result corresponds also to the BORDA scores ranking.[3]

Listing 7.9 Rank analysis table with BORDA scores

```
>>> v.showRankAnalysisTable()
  *----  Borda rank analysis tableau -----*
   candi- | alternative-to-rank         |       Borda
   dates  |  1    2    3    4    5      | score   average
```

[3] Mind that BORDA scores require the unrealistic working hypothesis that one knows how to precisely code in numbers the marginal linear ranks per voter.

Voting Heatmap

criteria	weight	tau*	c4	c5	c2	c3	c1
v5	-1.00	+0.60	1	4	2	3	5
v3	-1.00	+0.60	3	1	2	4	5
v8	-1.00	+0.40	2	3	1	5	4
v7	-1.00	+0.40	2	1	5	3	4
v6	-1.00	+0.40	2	3	1	5	4
v9	-1.00	+0.20	3	1	5	2	4
v4	-1.00	+0.00	2	4	5	1	3
v2	-1.00	-0.40	3	5	2	4	1
v1	-1.00	-0.80	5	3	4	2	1

Color legend:

quantile	33.33%	66.66%	100.00%

(*) **tau:** *Ordinal (Kendall) correlation between marginal criterion and global ranking relation*
Outranking model: **robust**, *Ranking rule*: **NetFlows**
Ordinal (Kendall) correlation between global ranking and global outranking relation: **+0.778**

Fig. 7.3 *Visualising a linear voting profile in a* NETFLOWS *ranked heatmap.* As the number of voters is usually much larger than the number of eligible candidates, the voting heatmap is by default transposed into a voters × candidates view. Notice that the importance weights of the voters are *negative*, which means that the preference direction of the criteria (in this case the individual voters) is *decreasing* (`min`), i.e. goes from lowest (best) rank to highest (worst) rank

```
    -------|---------------------------------------------------
      'c4' |  1    4    3    0    1    |  23     2.56
      'c5' |  3    0    3    2    1    |  25     2.78
      'c2' |  2    3    0    1    3    |  27     3.00
      'c3' |  1    2    2    2    2    |  29     3.22
      'c1' |  2    0    1    4    2    |  31     3.44
```

Let us eventually notice in the rank analysis table, shown in Listing 7.9 above, that the uninominal plurality winner would be, with three votes, candidate `c5`, whereas the best NETFLOWS and BORDA ranked candidate `c1` obtains with candidate `c3` only one vote.

But do represent our example linear voting profiles here realistic election outcomes?

7.5 On Generating Realistic Random Linear Voting Profiles

By default, the `RandomLinearVotingProfile` class generates random linear voting profiles where every candidate has the same uniform probability to be ranked at a certain position by all the voters. Each voter's random linear ballot is indeed generated via a uniform shuffling of the list of candidates.

In reality, political election data are quite different. There usually will be different favourite and marginal candidates for each political party. To simulate these aspects with our random generator, we are using two random exponentially distributed polls of the candidates and consider a bipartisan political landscape with a certain random balance (default theoretical party repartition = 0.50) between the two sets of potential party supporters. A certain theoretical proportion (default = 0.1) will not support any party.

Let us generate in Listing 7.10 such a linear voting profile for an election with 1000 voters and 15 candidates.

Listing 7.10 Generating a linear voting profile with random polls

```
>>> from votingProfiles import\
...                RandomLinearVotingProfile
>>> lvp = RandomLinearVotingProfile(\
...             numberOfCandidates=15,\
...             numberOfVoters=1000,\
...             WithPolls=True,\
...             partyRepartition=0.5,\
...             other=0.1,\
...             seed=0.9189670954954139)
>>> lvp
 *------- VotingProfile instance description ------*
  Instance class : RandomLinearVotingProfile
  Instance name  : randLinearProfile
  Candidates     : 15
  Voters         : 1000
  Attributes     : ['name', 'seed', 'candidates',
                    'voters', 'RandomWeights',
                    'sumWeights', 'poll1', 'poll2',
                    'bipartisan', 'linearBallot',
                    'ballot']
>>> lvp.showRandomPolls()
  Random repartition of voters
   Party-1 supporters : 460 (46.0%)
   Party-2 supporters : 436 (43.6%)
   Other voters       : 104 (10.4%)
  *---------------- random polls ---------------*
   Party-1(46.0%) | Party-2(43.6%)|  expected
  ---------------------------------------------
   c06 : 19.91%   | c11 : 22.94%  | c06 : 15.00%
   c07 : 14.27%   | c08 : 15.65%  | c11 : 13.08%
   c03 : 10.02%   | c04 : 15.07%  | c08 : 09.01%
   c13 : 08.39%   | c06 : 13.40%  | c07 : 08.79%
   c15 : 08.39%   | c03 : 06.49%  | c03 : 07.44%
   c11 : 06.70%   | c09 : 05.63%  | c04 : 07.11%
   c01 : 06.17%   | c07 : 05.10%  | c01 : 05.06%
   c12 : 04.81%   | c01 : 05.09%  | c13 : 05.04%
   c08 : 04.75%   | c12 : 03.43%  | c15 : 04.23%
   c10 : 04.66%   | c13 : 02.71%  | c12 : 03.71%
   c14 : 04.42%   | c14 : 02.70%  | c14 : 03.21%
   c05 : 04.01%   | c15 : 00.86%  | c09 : 03.10%
   c09 : 01.40%   | c10 : 00.44%  | c10 : 02.34%
   c04 : 01.18%   | c05 : 00.29%  | c05 : 01.97%
   c02 : 00.90%   | c02 : 00.21%  | c02 : 00.51%
```

In this example (see above Lines 21 and following), we obtain 460 Party-1 supporters (46%), 436 Party-2 supporters (43.6%), and 104 other voters (10.4%). Favourite candidates of Party-1 supporters, with more than 10%, appear to be `c06` (19.91%), *c*07 (14.27%), and `c03` (10.02%). Whereas for Party-2 supporters, favourite candidates appear to be `c11` (22.94%), followed by `c08` (15.65%), *c*04 (15.07%) and `c06` (13.4%). Being *first* choice for Party-1 supporters and *fourth* choice for Party-2 supporters, this candidate `c06` is a natural candidate for clearly winning this election game (see Listing 7.11).

Listing 7.11 The uninominal and BORDA election winner

```
>>> lvp.computeSimpleMajorityWinner()
 ['c06']
>>> lvp.computeInstantRunoffWinner()
 ['c06']
>>> lvp.computeBordaWinners()
 ['c06']
```

Is candidate `c06` also a CONDORCET winner? To verify, we start by creating the corresponding *majority margins* digraph *mmdg* with the help of the `MajorityMarginsDigraph` class. The created digraph instance contains 15 *actions*—the candidates—and 104 *oriented arcs*—the *positive* majority margins—(see Listing 7.12 Lines 7–8).

Listing 7.12 A majority margins digraph constructed from a linear voting profile

```
1 >>> from votingProfiles import MajorityMarginsDigraph
2 >>> mmdg = MajorityMarginsDigraph(lvp)
3 >>> mmdg
4  *------- Digraph instance description ------*
5   Instance class       : MajorityMarginsDigraph
6   Instance name        : rel_randLinearProfile
7   Digraph Order        : 15
8   Digraph Size         : 104
9   Valuation domain     : [-1000.00;1000.00]
10  Determinateness (%)  : 67.08
11  Attributes           : ['name', 'actions', 'voters',
12                           'ballot', 'valuationdomain',
13                           'relation', 'order',
14                           'gamma', 'notGamma']
```

The `showHTMLRelationTable()` method visualises the resulting pairwise majority margins by showing the HTML formatted version of the `mmdg` relation table in a browser view.

```
>>> mmdg.showHTMLRelationTable(\
...          tableTitle='Pairwise majority margins',
...          relationName='M(x,y)')
```

A complete light green *row* in Fig. 7.4 on the following page reveals a CONDORCET winner, whereas a complete light green *column* reveals a CONDORCET

Fig. 7.4 *Browsing the majority margins. Light green* cells contain the positive majority margins, whereas *light red* cells contain the negative majority margins

Pairwise majority margins

M(x,y)	c01	c02	c03	c04	c05	c06	c07	c08	c09	c10	c11	c12	c13	c14	c15
c01	–	768	-138	108	478	-436	-198	-140	238	440	-268	148	50	202	218
c02	-768	–	-796	-484	-368	-858	-828	-772	-546	-496	-800	-722	-768	-696	-658
c03	138	796	–	160	590	-286	-80	-8	372	522	-158	280	210	360	338
c04	-108	484	-160	–	184	-370	-180	-288	160	136	-420	16	-62	56	30
c05	-478	368	-590	-184	–	-730	-640	-472	-234	-116	-550	-442	-522	-376	-386
c06	436	858	286	370	730	–	248	234	574	692	102	556	482	566	520
c07	198	828	80	180	640	-248	–	0	358	602	-94	304	266	384	420
c08	140	772	8	288	472	-234	0	–	436	396	-176	276	134	298	244
c09	-238	546	-372	-160	234	-574	-358	-436	–	116	-594	-126	-194	-90	-14
c10	-440	496	-522	-136	116	-692	-602	-396	-116	–	-510	-310	-442	-304	-266
c11	268	800	158	420	550	-102	94	176	594	510	–	388	268	474	292
c12	-148	722	-280	-16	442	-556	-304	-276	126	310	-388	–	-92	100	148
c13	-50	768	-210	62	522	-482	-266	-134	194	442	-268	92	–	158	186
c14	-202	696	-360	-56	376	-566	-384	-298	90	304	-474	-100	-158	–	68
c15	-218	658	-338	-30	386	-520	-420	-244	14	266	-292	-148	-186	-68	–

Valuation domain: [-1000; +1000]

loser. We recover again candidate `c06` as CONDORCET winner,[4] whereas the obvious CONDORCET loser is here candidate `c02`, the candidate with the lowest support in both parties (see Listing 7.10 on page 92 Line 43).

With a same bipolar *first-ranked* and *last ranked* selection procedure, we may weakly rank the candidates (with possible ties) by iterating these *first-ranked* and *last ranked* choices among the remaining candidates (Bisdorff 1999).

Before showing the *ranking-by-choosing* result, we have to, first, compute the iterated bipolar selection procedure (see Listing 7.13).

Listing 7.13 Ranking by iterating choosing the *first* and *last* remaining candidates

```
>>> cdg.showRankingByChoosing()
 Ranking by Choosing and Rejecting
   1st best ranked ['c06']
     2nd best ranked ['c11']
       3rd  best ranked ['c07','c08']
         4th best ranked ['c03']
           5th best ranked ['c01']
             6th best ranked ['c13']
               7th first ranked ['c04']
               7th last ranked ['c12']
             6th last ranked ['c14']
           5th last ranked ['c15']
         4th last ranked ['c09']
       3rd last ranked ['c10']
     2nd last ranked ['c05']
   1st last ranked ['c02']
```

[4] The concept of CONDORCET winner—a generalisation of absolute majority winners—proposed by CONDORCET in 1784, is an early historical example of initial digraph kernel (see Chap. 17).

The first selection concerns `c06` (first) and `c02` (last), followed by `c11` (first) opposed to `c05` (last), and so on, until there remains at iteration step 7 a last pair of candidates, namely `[c04, c12]` (see Lines 9–10).

Notice furthermore the 3rd-best ranked candidates at iteration step 3 (see Line 5), namely the pair (`c07, c08`). Both candidates represent indeed conjointly the *3rd best ranked* choice. We obtain hence a *weak ranking*, i.e. a ranking with a tie.

Let us mention that the *instant runoff* procedure, we used before when operated with a `Comments=True` parameter setting, will deliver a more or less similar *reversed* linear *ordering-by-rejecting* result, namely `[c02, c10, c14, c05, c09, c13, c12, c15, c04, c01, c08, c03, c07, c11, c06]`, ordered from the *last* to the *first* choice.

Remarkable about both these *ranking-by-choosing* or *ordering-by-rejecting* results is the fact that the random voting behaviour, simulated here with the help of two discrete random variables,[5] Defined, respectively, by the two random party polls, is rendering a ranking that is more or less in accordance with the simulated balance of the polls: Party-1 supporters : 460; Party-2 supporters: 436 (see Listing 7.10 on page 92 Lines 29–43 third column). Despite a random voting behaviour per voter, the given polls apparently show a *very strong incidence* on the eventual election result. In order to avoid any manipulation of the election outcome, public media are therefore in some countries not allowed to publish polls during the last weeks before a general election.

Mind that the specific *ranking-by-choosing* procedure, we use here on the majority margins digraph, operates the selection procedure by extracting at each step *initial* and *terminal* prekernels, i.e. difficult operational problems (see Chap. 17 and Bisdorff 1999); A technique that does not allow in general to tackle voting profiles with much more than 30 candidates.

Next Chap. 8 on multiple incommensurable criteria ranking methods presents more methods and tools for ranking from pairwise majority margins when a larger number of potential candidates or decision alternatives is given.

References

Bisdorff R (1999) Bipolar ranking from pairwise fuzzy outrankings. Belg J Oper Res Stat Comput Sci 37(4):379–387. http://hdl.handle.net/10993/38738

Bisdorff R (2020a) Lecture 2: Who wins the election? In: Lectures of the algorithmic decision theory course, University of Luxembourg. http://hdl.handle.net/10993/37933

Bisdorff R (2020b) Lecture 4: Discrete random variables. In: Lectures of the computational statistics course, University of Luxembourg. http://hdl.handle.net/10993/37870

[5] Discrete random variables with a given empirical probability law (here the polls) are provided in the `randomNumbers` module by the `DiscreteRandomVariable` class (Bisdorff 2020b).

Bisdorff R, Meyer P, Roubens M (2008) Rubis: a bipolar-valued outranking method for the best choice decision problem. Q J Oper Res 6(2):143–165. http://hdl.handle.net/10993/23716
de Borda J (1781) Mémoire sur les élections au scrutin. Mémoires de l'Académie royale des sciences, Paris
de Caritat, Marquis de Condorcet J (1784) Essai sur l'application de l'analyse à la probabilité des décisions rendues à la pluralité des voix. Imprimerie royale, Paris

Chapter 8
Ranking with Multiple Incommensurable Criteria

... whether we are deciding between buying different commodity baskets, or making choices about what to do on a holiday, or deciding for whom to vote for in an election, we are inescapably involved in evaluating alternatives with non–commensurable aspects.

—Amartya Sen, Idea of Justice (Sen 2009)

Contents

Abstract The DIGRAPH3 python resources provide several algorithms for solving the multiple incommensurable criteria ranking problem via bipolar-valued outranking digraphs. The COPELAND, NETFLOWS, KEMENY, SLATER, KOHLER, and the RANKEDPAIRS ranking rules are introduced and illustrated with a random outranking digraph.

8.1 The Ranking Problem

We need to rank without ties a set X of items (usually decision alternatives) that are evaluated on multiple incommensurable performance criteria; yet, for which we know their pairwise bipolar-valued *strict outranking* characteristics, i.e. $r(x \gtrsim\!\!\!\!\!/\; y)$ for all $x, y \in X$ (see Sect. 3.5 and Bisdorff 2013).

R. Bisdorff, *Algorithmic Decision Making with Python Resources*,
International Series in Operations Research & Management Science 324,
https://doi.org/10.1007/978-3-030-90928-4_8

Let us consider in Listing 8.1 a didactic outranking digraph `g` generated from a random *Cost-Benefit* performance tableau (see Sect. 6.3) concerning 9 decision alternatives evaluated on 13 performance criteria. The `BipolarOutrankingDigraph` computes the corresponding bipolar-valued *outranking digraph* `g` and a codual transform gives us the requested strict outranking digraph `gcd` with the pairwise $r(x \succnsim y)$ characteristic values (see Lines 11–19 below).

Listing 8.1 Random bipolar-valued strict outranking relation characteristics

```
1  >>> from randomPerfTabs import RandomCBPerformanceTableau
2  >>> t = RandomCBPerformanceTableau(numberOfActions=9,\
3  ...           numberOfCriteria=13,seed=200)
4  >>> from outrankingDigraphs import BipolarOutrankingDigraph
5  >>> g = BipolarOutrankingDigraph(t)
6  >>> gcd = ~(-g) # strict (codual) outranking digraph
7  >>> gcd.showRelationTable(ReflexiveTerms=False)
8   * ---- Relation Table -----
9    r(>)  |  'a1'  'a2'  'a3'  'a4'  'a5'  'a6'  'a7'  'a8'  'a9'
10   -----|------------------------------------------------------
11   'a1'  |    -   0.00 +0.10 -1.00 -0.13 -0.57 -0.23 +0.10 +0.00
12   'a2'  | -1.00    -   0.00 +0.00 -0.37 -0.42 -0.28 -0.32 -0.12
13   'a3'  | -0.10  0.00    -  -0.17 -0.35 -0.30 -0.17 -0.17 +0.00
14   'a4'  |  0.00  0.00 -0.42    -  -0.40 -0.20 -0.60 -0.27 -0.30
15   'a5'  | +0.13 +0.22 +0.10 +0.40    -  +0.03 +0.40 -0.03 -0.07
16   'a6'  | -0.07 -0.22 +0.20 +0.20 -0.37    -  +0.10 -0.03 -0.07
17   'a7'  | -0.20 +0.28 -0.03 -0.07 -0.40 -0.10    -  +0.27 +1.00
18   'a8'  | -0.10 -0.02 -0.23 -0.13 -0.37 +0.03 -0.27    -  +0.03
19   'a9'  |  0.00 +0.12 -1.00 -0.13 -0.03 -0.03 -1.00 -0.03    -
```

Some ranking rules will work on the associated CONDORCET digraph, i.e. the corresponding *median cut* polarised strict outranking digraph.

Listing 8.2 Median cut polarised strict outranking relation characteristics

```
>>> ccd = PolarisedOutrankingDigraph(gcd,\
...                  level=g.valuationdomain['med'],\
...                  KeepValues=False,StrictCut=True)
>>> ccd.showRelationTable(ReflexiveTerms=False,\
...                  IntegerValues=True)
 *---- Relation Table -----
  r(>)_med | 'a1' 'a2' 'a3' 'a4' 'a5' 'a6' 'a7' 'a8' 'a9'
  ---------|-------------------------------------------
     'a1'  |   -    0   +1   -1   -1   -1   -1   +1    0
     'a2'  |  -1    -   +0    0   -1   -1   -1   -1   -1
     'a3'  |  -1    0    -   -1   -1   -1   -1   -1    0
     'a4'  |   0    0   -1    -   -1   -1   -1   -1   -1
     'a5'  |  +1   +1   +1   +1    -   +1   +1   -1   -1
     'a6'  |  -1   -1   +1   +1   -1    -   +1   -1   -1
     'a7'  |  -1   +1   -1   -1   -1   -1    -   +1   +1
     'a8'  |  -1   -1   -1   -1   -1   +1   -1    -   +1
     'a9'  |   0   +1   -1   -1   -1   -1   -1   -1    -
```

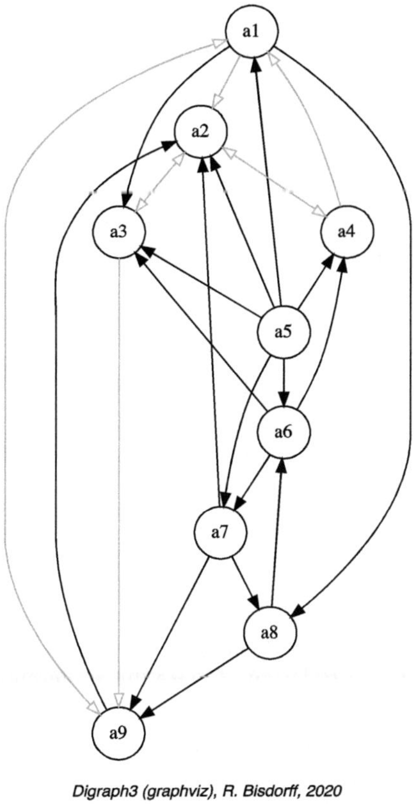

Fig. 8.1 *The strict outranking relation* $\succnsim$. The relation shown here is, for instance, *not transitive*: alternative a8 outranks alternative a6 and alternative a6 outranks a4, however, a8 does not outrank a4. Furthermore, alternatives a6, a7, and a8 show a cyclic outranking relation

Unfortunately, such crisp median cut CONDORCET digraphs, associated with a given strict outranking digraph, only exceptionally model a linear ordering. Usually, pairwise majority comparisons do not render a *complete* or, at least, a *transitive* partial order. There may even frequently appear *cyclic* outranking situations (see Sect. 7.4).

To discover how *difficult* this ranking problem can get, let us have a look in Fig. 8.1 at the corresponding strict outranking digraph *graphviz* drawing.[1]

```
>>> gcd.exportGraphViz(fileName='rankingTutorial')
 *---- exporting a dot file for GraphViz tools ---------*
  Exporting to rankingTutorial.dot
  dot -Grankdir=BT -Tpng rankingTutorial.dot\
                    -o rankingTutorial.png
```

The computeTransitivityDegree() method computes the *transitivity degree* of the outranking digraph gcd shown in Fig. 8.1, i.e. the ratio of the number

[1] The exportGraphViz() method is depending on drawing tools from graphviz software (https://graphviz.org/).

of closed transitive triples—$x \succnsim y$, $y \succnsim z$ and $x \succnsim z$—over the number of triples $x \succnsim y$, $y \succnsim z$ (see below Lines 3–4).

```
1 >>> gcd.computeTransitivityDegree(Comments=True)
2  Transitivity degree of graph <codual_rel_randomCBperftab>
3   triples x>y>z: 78, closed: 38, open: 40
4   closed/triples = 0.487
```

With only 49% of the required transitive arcs, the strict outranking digraph `gcd` is hence very far from being transitive; a serious problem when a linear ordering of the decision alternatives is looked for.

The `computeChordlessCircuits()` method, followed by the `showChordlessCircuits()` method, can check furthermore whether there appear any cyclic outranking situations.[2]

```
>>> gcd.computeChordlessCircuits()
>>> gcd.showChordlessCircuits()
  1 circuit(s).
  *---- Chordless circuits ----*
  1: ['a6', 'a7', 'a8'] , credibility : 0.033
```

There is one chordless circuit detected in the given strict outranking digraph `gcd`, namely alternative `a6` outranks alternative `a7`, the latter outranks `a8`, and `a8` outranks again alternative `a6` (see Fig. 8.1 on the previous page). Any potential linear ordering of these three alternatives will, in fact, always contradict somehow the given outranking relation.

Now, several heuristic ranking rules have been proposed for constructing a linear ordering which is closest in some specific sense to a given outranking relation. The DIGRAPH3 resources provide some of the most common of these ranking rules, like the COPELAND, KEMENY, SLATER, KOHLER, and the RANKEDPAIRS ranking rules.

8.2 The COPELAND Ranking

Algorithm 8.1 COPELAND ranking rule (Copeland 1951)

in : median cut strict outranking digraph $G(X, \succnsim_{>0})$,
out : linear ranking of the set X of decision actions.

1. For each alternative$x \in X$, the COPELAND ranking rule computes a score resulting from the sum over all non-reflexive pairs (x, y) of the differences between the crisp *strict outranking* characteristics $r(x \succnsim_{>0} y)$ and the crisp *strict outranked* characteristics $r(y \succnsim_{>0} x)$.
2. The alternatives are ranked in decreasing order of these COPELAND scores; ties, the case given, being resolved with a lexicographical rule applied to the identifiers of the decision actions.

[2] The `computeChordlessCircuits()` and `showChordlessCircuits()` methods are separate because there are various methods available for enumerating the chordless circuits in a digraph (Bisdorff 2010).

In Listing 8.3 we show the ranking of the given strict outranking digraph `gcd` obtained with the `CopelandRanking` class from the `linearOrders` module.

Listing 8.3 Computing a COPELAND ranking

```
1  >>> from linearOrders import CopelandRanking
2  >>> cop = CopelandRanking(gcd,Comments=True)
3   Copeland decreasing scores
4       a5 : +12
5       a1 :  +2
6       a6 :  +2
7       a7 :  +2
8       a8 :   0
9       a4 :  -3
10      a9 :  -3
11      a3 :  -5
12      a2 :  -7
13   Copeland Ranking:
14   ['a5','a1','a6','a7','a8','a4','a9','a3','a2']
```

Alternative `a5` obtains here the best COPELAND score (+12), followed by alternatives `a1`, `a6`, and `a7` with same score (+2); following the lexicographic rule, `a1` is hence ranked before `a6` and `a6` before `a7`. Same situation is observed for `a4` and `a9` with a score of −3 (see Listing 8.3 Lines 4–12). The COPELAND ranking rule is in fact *invariant* under the *codual* transform (see Sect. 2.6) and renders a same linear order indifferently from digraphs `g` or `gcd`.

The COPELAND rule works well, furthermore on any strict outranking digraph that models a linear (partial) order on the *median cut* strict outranking digraph `ccd` (Dias and Lamboray 2010). In Listing 8.4, the `computeRankingCorrelation()` method, coupled with the `showCorrelation()` method, indicates the ordinal correlation of the COPELAND ranking result, shown in Listing 8.3 Line 14, with the given outranking digraph `g` (see Chap. 16 and Bisdorff 2012).

Listing 8.4 Checking the ordinal quality of the COPELAND ranking

```
1  >>> corr = g.computeRankingCorrelation(cop.copelandRanking)
2  >>> g.showCorrelation(corr)
3   Correlation indexes:
4     Crisp ordinal correlation : +0.463
5     Valued equivalalence      : +0.107
6     Epistemic determination   :  0.230
```

With an epistemic determination level of 0.230 (Line 6), the crisp ordinal correlation—KENDALL τ—index of +0.463 is in fact supported by 61.5%$(100.0x(1.0 + 0.23)/2)$ of the strict outranking situations. Furthermore, the bipolar-valued *relational equivalence* between the strict outranking relation and the COPELAND ranking equals +0.107, i.e. a *majority* of 55.35% of the criteria significance supports the relational equivalence between the given strict outranking relation and the COPELAND ranking (see Chap. 16).

The COPELAND scores, shown in Listing 8.3 deliver actually only a *weak ranking*, i.e. a ranking with ties. This weak ranking may be constructed with the `WeakCopelandOrder` class from the `transitiveDigraphs` module.

Listing 8.5 Computing a weak COPELAND ranking

```
>>> from transitiveDigraphs import WeakCopelandOrder
>>> wcop = WeakCopelandOrder(g)
>>> wcop.showRankingByChoosing()
 Ranking by Choosing and Rejecting
   1st ranked ['a5']
     2nd ranked ['a1', 'a6', 'a7']
       3rd ranked ['a8']
       3rd last ranked ['a4', 'a9']
     2nd last ranked ['a3']
   1st last ranked ['a2']
```

In Listing 8.5, the `WeakCopelandOrder` class models the weak ranking actually delivered by the COPELAND scores (see Listing 8.3 on the preceding page). We may draw its corresponding skeleton.[3]

```
>>> wcop.exportGraphViz(fileName='weakCopelandRanking')
 *---- exporting a dot file for GraphViz tools ---------*
  Exporting to weakCopelandRanking.dot
  dot -Grankdir=TB -Tpng weakCopelandRanking.dot\
                    -o weakCopelandRanking.png
```

Let us now consider a similar ranking rule, but working directly on the criteria *significance majority margins*, i.e. the *bipolar-valued* outranking characteristic valuation.

8.3 The NETFLOWS Ranking

The bipolar-valued version of the COPELAND ranking rule, we call NETFLOWS,[4] computes for each alternative $x \in X$ a *net flow* score.

Algorithm 8.2 NETFLOWS ranking rule

in : bipolar-valued strict outranking digraph $G(X, \succnsim)$,
out : linear ranking of the set X of decision actions.

1. The NETFLOWS rule computes for each alternative $x \in X$ the sum over all non-reflexive $y \in X$ of the differences between the *strict outranking* characteristics $r(x \succnsim y)$ and the *strict outranked* characteristics $r(y \succnsim x)$.
2. The alternatives are ranked in decreasing order of these NETFLOWS scores; ties, the case given, being resolved with a lexicographical rule applied to the identifiers of the decision actions.

---

[3] The skeleton of a transitive relation drops the transitivity induced arcs.

[4] This ranking rule is also known under the name PROMETHEE ranking rule (Brans and Vincke 1985).

The `NetFlowsRanking` class from the `linearOrders` module computes this linear ranking from the given strict outranking digraph `gcd`.

Listing 8.6 Computing a NETFLOWS ranking

```
1 >>> from linearOrders import NetFlowsRanking
2 >>> nf = NetFlowsRanking(gcd,Comments=True)
3   Net flow scores :
4     a5 : +3.600
5     a7 : +2.800
6     a6 : +1.300
7     a3 : +0.033
8     a1 : -0.400
9     a8 : -0.567
10    a4 : -1.283
11    a9 : -2.600
12    a2 : -2.883
13  NetFlows Ranking:
14   ['a5','a7','a6','a3','a1','a8','a4','a9','a2']
15 >>> cop.copelandRanking # comparing both
16   ['a5','a1','a6','a7','a8','a4','a9','a3','a2']
```

In Listing 8.6 here, the NETFLOWS scores actually deliver directly a linear ranking *without ties* which is rather different from the one delivered by the COPELAND rule (compare Lines 14 and 16). It may happen, however, that we obtain, as with the COPELAND scores above, only a ranking with ties, which may then be resolved similarly by following a lexicographic rule applied to the identifiers of the decision alternatives. In such cases, it is possible to construct again a *weak ranking* with the corresponding `WeakNetFlowsOrder` class from the `transitiveDigraphs` module.

It is worthwhile noticing again, that similar to the COPELAND ranking rule seen before, the NETFLOWS ranking rule is also *invariant* under the codual transform (see Sect. 2.6) and delivers the same ranking result indifferently from digraph `g` or `gcd`.

The NETFLOWS ranking result appears to be better correlated (+0.638 vs. +0.463) with the given strict outranking relation than its crisp cousin, the COPELAND ranking (see Line 4 in Listing 8.7).

Listing 8.7 Checking the quality of the NETFLOWS ranking

```
1 >>> corr = gcd.computeOrdinalCorrelation(nf)
2 >>> gcd.showCorrelation(corr)
3  Correlation indexes:
4    Extended Kendall tau        : +0.638
5    Epistemic determination     :  0.230
6    Bipolar-valued equivalence  : +0.147
```

Indeed, the ordinal correlation index of +0.638 leads to a bipolar-valued *relational equivalence* characteristics of +0.147, i.e. a majority of 57.35% of the criteria significance supports the relational equivalence between the given outranking digraphs *g* or *gcd* and the corresponding NETFLOWS ranking (see

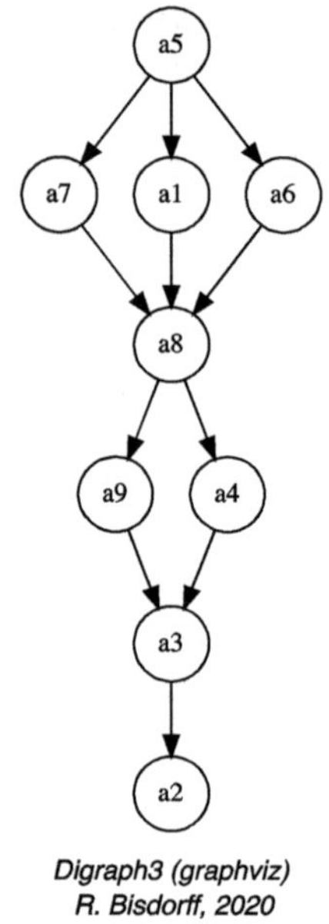

Fig. 8.2 *Drawing of the weak* COPELAND *ranking*. The drawing shows the skeleton of the weak ranking produced by the corresponding ties of the COPELAND scores

Chap. 16). The weaker ordinal ranking quality of the COPELAND rule (+0.463) stems in this example here essentially from the *weakness* of the actual COPELAND ranking result and our subsequent *arbitrary* lexicographic resolution of the many ties given by the COPELAND scores (see Fig. 8.2).

To appreciate now the ordinal correlations of both the COPELAND and the NETFLOWS rankings with the underlying strict outranking relation, it is useful to consider the '*optimal*' KEMENY and SLATER ranking rules.

8.4 KEMENY Rankings

A KEMENY ranking k is a linear ranking without ties of the set of n decision alternatives X which is *closest*, in the sense of the ordinal KENDALL distance,[5] to the given valued outranking digraph $\mathcal{G}$ (Kemeny 1959).

Formally:

$$k = \operatorname{argmax}_{p \in \mathscr{P}(X)} \sum_{i \neq j} \big(r(p[i] \succsim p[j]) - r(p[j] \succsim p[i]) \big), \qquad (8.1)$$

where $\mathscr{P}(X)$ denotes the set of all permutations of X and $i, j = 0, ...n$.

[5] See Chap. 16 and Bisdorff (2012).

The `KemenyRanking` class from the `linearOrders` module computes such a ranking which is highest possible correlated with the underlying strict outranking relation. The KEMENY rule is also *invariant* under the codual transform.

Mind that the `KemenyRanking` class constructor, in order to find a KEMENY ranking, has to compute NETFLOWS scores for every permutation of the list of decision alternatives (see Eq. 8.1 on the facing page). Therefore the class is limited, by default, to digraphs of order up to 7 (Bisdorff 2021). In Listing 8.8 Line 2, the `orderLimit` parameter allows to raise this limit.

Listing 8.8 Computing a KEMENY ranking

```
1 >>> from linearOrders import KemenyRanking
2 >>> ke = KemenyRanking(gcd,orderLimit=9)
3 >>> # default orderLimit is 7
4 >>> ke.showRanking()
5  ['a5','a6','a7','a3','a9','a4','a1','a8','a2']
6 >>> corr = gcd.computeOrdinalCorrelation(ke)
7 >>> gcd.showCorrelation(corr)
8  Correlation indexes:
9    Extended Kendall tau        : +0.779
10   Epistemic determination     :  0.230
11   Bipolar-valued equivalence  : +0.179
```

So, +0.779 represents the *highest possible* ordinal correlation index—*fitness*—any potential linear ranking can achieve with the given pairwise outranking digraph (see Listing 8.8 Line 9).

A KEMENY ranking may not be unique. In our example here, we obtain in fact two such KEMENY rankings with a same *maximal* KEMENY index of 12.92.

Listing 8.9 Optimal KEMENY rankings

```
>>> ke.maximalRankings
  [['a5','a6','a7','a3','a8','a9','a4','a1','a2'],
   ['a5','a6','a7','a3','a9','a4','a1','a8','a2']]
>>> ke.maxKemenyIndex
 Decimal('12.9166667')
```

Figure 8.3 on the next page shows the partial order defined by the epistemic disjunction of both optimal KEMENY rankings obtained with the `RankingsFusionDigraph` class (see Sect. 2.5).

Listing 8.10 Computing the epistemic disjunction of all optimal KEMENY rankings

```
>>> from transitiveDigraphs import RankingsFusionDigraph
>>> wke = RankingsFusionDigraph(ke,ke.maximalRankings)
>>> wke.exportGraphViz(fileName='tutorialKemeny')
 *---- exporting a dot file for GraphViz tools ---------*
  Exporting to tutorialKemeny.dot
  dot -Grankdir=TB -Tpng tutorialKemeny.dot\
                    -o tutorialKemeny.png
```

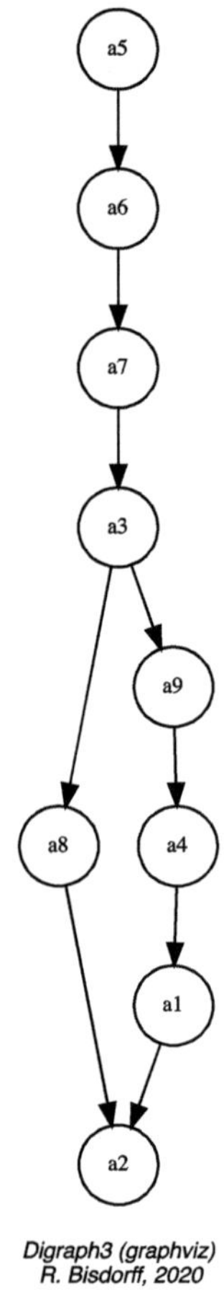

Fig. 8.3 *Epistemic disjunction of optimal* KEMENY *rankings*. It is interesting to notice that both KEMENY rankings only differ in their respective positioning of alternative `a8`; either before or after alternatives `a9`, `a4`, and `a1`

To retain now a specific representative among all the potential rankings with maximal KEMENY index, we will choose, with the help of the `showRankingConsensusQuality()` method, the one proposing the best criteria consensus.

Listing 8.11 Computing the consensus quality of the first KEMENY ranking

```
>>> g.showRankingConsensusQuality(ke.maximalRankings[0])
 Consensus quality of ranking:
  ['a5','a6','a7','a3','a8','a9','a4','a1','a2']
   crit. (weight):  tau    | crit. (weight):  tau
  -----------------------|-----------------------
   b09 (0.050)  : +0.361 | b04 (0.050)  : +0.333
   b08 (0.050)  : +0.292 | b01 (0.050)  : +0.264
   c01 (0.167)  : +0.250 | b03 (0.050)  : +0.222
   b07 (0.050)  : +0.194 | b05 (0.050)  : +0.167
   c02 (0.167)  : +0.000 | b10 (0.050)  : +0.000
   b02 (0.050)  : -0.042 | b06 (0.050)  : -0.097
   c03 (0.167)  : -0.167
  Summary:
    Weighted mean marginal correlation (a): +0.099
    Standard deviation (b)                : +0.177
    Ranking fairness (a)-(b)              : -0.079
```

In Listing 8.11 on the facing page are shown the ordinal correlation of the first Kemeny ranking with all the marginal strict outranking relations per performance criterion (see Lines 6–12). The Benefit criterion `b09` is highest correlated (+0.361, Line 6), whereas the Costs criterion `c03` is lowest correlated (−167, Line 12) with this KEMENY ranking.

The ranking consensus quality of the second KEMENY ranking is shown in Listing 8.12 below.

Listing 8.12 Computing the consensus quality of the second KEMENY ranking

```
1  >>> g.showRankingConsensusQuality(ke.maximalRankings[1])
2   Consensus quality of ranking:
3     ['a5','a6','a7','a3','a9','a4','a1','a8','a2']
4     crit. (weight):   tau    | crit. (weight):   tau
5    -----------------------|----------------------
6     b09  (0.050)    : +0.306 | b08  (0.050)    : +0.236
7     c01  (0.167)    : +0.194 | b07  (0.050)    : +0.194
8     c02  (0.167)    : +0.167 | b04  (0.050)    : +0.167
9     b03  (0.050)    : +0.167 | b01  (0.050)    : +0.153
10    b05  (0.050)    : +0.056 | b02  (0.050)    : +0.014
11    b06  (0.050)    : -0.042 | c03  (0.167)    : -0.111
12    b10  (0.050)    : -0.111
13   Summary:
14     Weighted mean marginal correlation (a): +0.099
15     Standard deviation (b)                : +0.132
16     Ranking fairness (a)-(b)              : -0.033
```

The Benefit criterion `b09` is again highest correlated (+0.306, Line 6) with this KEMENY ranking, but the weakest correlated criterion is now Benefit criterion `b10` (−0.111, Line 12).

Both KEMENY rankings show the same *weighted mean marginal correlation* (+0.099) with all 13 performance criteria. However, the second ranking shows a slightly lower *standard deviation*: +0.132 versus +0.177, resulting in a slightly *fairer* ranking result: −0.033 versus −0.079 (see Listing 8.11 on the preceding page Lines 14–16 and Listing 8.12 Lines 14–16).

When several rankings with maximal KEMENY index are given, the `Kemeny-Ranking` class constructor instantiates the ranking with *highest* mean marginal correlation and, in case of ties, with *lowest* weighted standard deviation. Here we obtain ranking: `[a5, a6, a7, a3, a9, a4, a1, a8, a2]` (see Line 5 in Listing 8.8 on page 105).

8.5 SLATER Rankings

The SLATER ranking rule is identical to the KEMENY rule, but it is working, instead, on the CONDORCET—*median cut polarised*—digraph `ccd` (Slater 1961). The SLATER rule is again *invariant* under the codual transform and the `SlaterRanking` class from the `linearOrders` module delivers hence indifferently on `g` or `gcd` the following results:

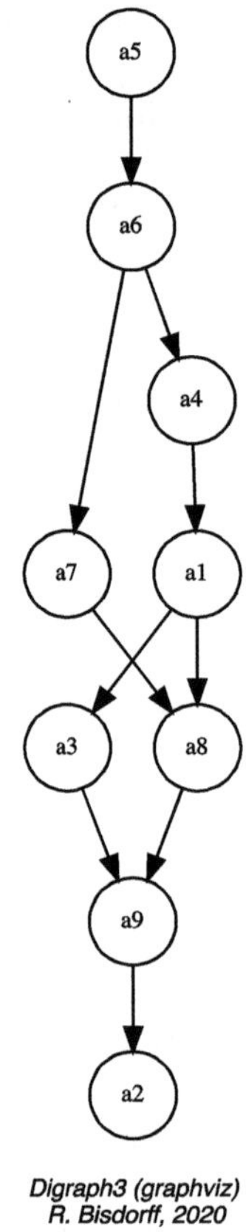

Fig. 8.4 *Epistemic disjunction of optimal* SLATER *rankings*. What precise SLATER ranking result should we hence adopt?

Listing 8.13 Computing a SLATER ranking

```
1  >>> from linearOrders import SlaterRanking
2  >>> sl = SlaterRanking(gcd,orderLimit=9)
3  >>> sl.slaterRanking
4    ['a5','a6','a4','a1','a3','a7','a8','a9','a2']
5  >>> corr = gcd.computeRankingCorrelation(sl.slaterRanking)
6  >>> sl.showCorrelation(corr)
7    Correlation indexes:
8     Extended Kendall tau         : +0.676
9     Epistemic determination      :  0.230
10    Bipolar-valued equivalence   : +0.156
11 >>> len(sl.maximalRankings)
12   7
```

We notice in Listing 8.13 above that the SLATER ranking shown in Line 4 represents a rather good fit (+0.676), slightly better apparently than the NETFLOWS ranking result (+0.638). However, there are in fact 7 such optimal SLATER rankings (see Line 12). The corresponding epistemic disjunction, again with the `RankingsFusionDigraph` class from the `transitiveDigraphs` module, gives the partial ordering shown in Fig. 8.4:

Listing 8.14 Computing the epistemic disjunction of optimal SLATER rankings

```
>>> from transitiveDigraphs import RankingsFusionDigraph
>>> slw = RankingsFusionDigraph(sl,sl.maximalRankings)
>>> slw.exportGraphViz(fileName='tutorialSlater')
 *---- exporting a dot file for GraphViz tools ----*
  Exporting to tutorialSlater.dot
  dot -Grankdir=TB -Tpng tutorialSlater.dot\
                     -o tutorialSlater.png
```

The KEMENY and SLATER ranking rules are furthermore computationally *difficult* problems and effective ranking results are only computable for tiny outranking digraphs (<20 objects) (see Eq. 8.1 on page 104).

More computationally efficient ranking heuristics, like the COPELAND and NETFLOWS rules, are therefore needed in practice. Let us finally, after these *ranking-by-scoring* strategies, also present two popular *ranking-by-choosing* rules.

8.6 The KOHLER Ranking-by-Choosing Rule

Algorithm 8.3 The KOHLER *ranking-by-choosing* rule (Kohler 1978)

in : bipolar-valued digraph $G(X, \succsim)$,
out : linear ranking of the set X of decision actions.

At step i (i goes from 1 to n) do the following:

1. Compute for each row of the bipolar-valued *strict* outranking relation table (see Listing 8.1 on page 98) the smallest value;
2. Select the row where this minimum is maximal. Ties are resolved in lexicographic order;
3. Put the selected decision alternative at rank i;
4. Delete the corresponding row and column from the relation table and restart until the table is empty.

Listing 8.15 Computing a KOHLER ranking

```
>>> from linearOrders import KohlerRanking
>>> kocd = KohlerRanking(gcd)
>>> kocd.showRanking()
  ['a5','a7','a6','a3','a9','a8','a4','a1','a2']
>>> corr = gcd.computeOrdinalCorrelation(kocd)
>>> gcd.showCorrelation(corr)
  Correlation indexes:
    Extended Kendall tau       : +0.747
    Epistemic determination    :  0.230
    Bipolar-valued equivalence : +0.172
```

With this *min-max* lexicographic ranking-by-choosing strategy, we obtain a correlation result (+0.747) that is until now clearly the nearest to an optimal

KEMENY ranking (see Listing 8.8 on page 105). Only two adjacent pairs: (a6, a7) and (a8, a9) are actually inverted here. Notice that the KOHLER ranking rule, contrary to the previously mentioned rules, is **not** invariant under the codual transform and requires to work on the strict outranking digraph gcd for a better correlation result.

```
>>> ko = KohlerRanking(g)
>>> corr = g.computeOrdinalCorrelation(ko)
>>> g.showCorrelation(corr)
 Correlation indexes:
  Crisp ordinal correlation   : +0.483
  Epistemic determination     :  0.230
  Bipolar-valued equivalence  : +0.111
```

But the KOHLER rule has a *dual* version, the prudent *Arrow-Raynaud* ordering-by-choosing rule, where a corresponding *max-min* strategy, when used on the *non-strict* dual outranking digraph $-g$, for ordering from *last* to *first* produces a similar ranking result (Arrow and Raynaud 1986).

Noticing that the NETFLOWS score of an alternative x represents in fact a bipolar-valued characteristic of the assertion "*alternative x is ranked first*", we may enhance the KOHLER rule by replacing the simple *min-max* strategy with an *iterated* maximal NETFLOWS score selection.

Algorithm 8.4 The iterated NETFLOWS ranking-by-choosing rule

in : bipolar-valued digraph $G(X, \succsim)$,
out : linear ranking of the set X of n decision actions.

At step i (i goes from 1 to n) do the following:

1. Compute for each row of the bipolar-valued outranking relation table (see Listing 8.1 on page 98) the corresponding NETFLOWS score;
2. Select the row where this score is *maximal*, ties being resolved by lexicographic order;
3. Put the corresponding decision alternative at rank i;
4. Delete the corresponding row and column from the relation table and restart until the table is empty.

The IteratedNetFlowsRanking class from the linearOrders module computes this ranking result.

Listing 8.16 Ranking-by-choosing with iterated maximal NETFLOWS scores

```
>>> from linearOrders import IteratedNetFlowsRanking
>>> inf = IteratedNetFlowsRanking(g)
>>> inf
 *------- Digraph instance description ------*
   Instance class      : IteratedNetFlowsRanking
   Instance name       : rel_randomCBperftab_ranked
   Digraph Order       : 9
   Digraph Size        : 36
```

```
9      Valuation domain    : [-1.00;1.00]
10     Determinateness (%)  : 100.00
11     Attributes      : ['valuedRanks', 'valuedOrdering',
12                        'iteratedNetFlowsRanking',
13                        'iteratedNetFlowsOrdering',
14                        'name', 'actions', 'order',
15                        'valuationdomain', 'relation',
16                        'gamma', 'notGamma']
17 >>> inf.iteratedNetFlowsRanking
18    ['a5','a7','a6','a3','a4','a1','a8','a9','a2']
19 >>> corr = g.computeRankingCorrelation(\
20 ...             inf.iteratedNetFlowsRanking)
21 >>> g.showCorrelation(corr)
22   Correlation indexes:
23     Crisp ordinal correlation  : +0.743
24     Epistemic determination    :  0.230
25     Bipolar-valued equivalence : +0.171
```

Like the KOHLER rule, the iterated NETFLOWS rule has also a dual *ordering-by-choosing* version, where instead of selecting at each step i the row with maximal NETFLOWS score, we choose the row with the *minimal* NETFLOWS score. Both the ranking and ordering result are computed by the `IteratedNetFlowsRanking` class (see Lines 12– 13 in Listing 8.16 on the facing page).

```
>>> inf.iteratedNetFlowsOrdering
   ['a2','a9','a1','a4','a3','a8','a7','a6','a5']
>>> corr = g.computeOrderCorrelation(\
...                inf.iteratedNetFlowsOrdering)
>>> g.showCorrelation(corr)
  Correlation indexes:
    Crisp ordinal correlation  : +0.751
    Epistemic determination    : 0.230
    Bipolar-valued equivalence : +0.173
```

The iterated NETFLOWS ranking and its dual, the iterated NETFLOWS ordering, do not usually deliver both the same result. In our example outranking digraph g, for instance, it is the *ordering-by-choosing* result that obtains a slightly better correlation with the given outranking digraph (+0.751), a result that is also slightly better than the original KOHLER ranking result (+0.747, see Listing 8.15 on page 109 Line 8 on page 109).

With different *ranking-by-choosing* and *ordering-by-choosing* results, it may be useful to *fuse* now, similar to what we have done before with KEMENY's and SLATER's optimal rankings, both, the iterated NETFLOWS ranking and ordering into a partial ranking. But we are hence back to the practical problem of what linear ranking should we eventually retain?

Let us finally mention another interesting *ranking-by-choosing* approach.

8.7 The RankedPairs Ranking Rule

N. Tideman's ranking-by-choosing heuristic, the RankedPairs rule, working best this time on the non-strict outranking digraph g, is based on a *prudent incremental* construction of linear orders that avoids on the fly any cycling outranking situations (Tideman 1987).

Algorithm 8.5 The RankedPairs ranking rule

in : bipolar-valued digraph $G(X, \succsim)$,
out : linear ranking of the set X of decision actions.

1. Rank the ordered pairs (x, y) of alternatives in decreasing order of $r(x \succsim y) + r(y \not\succsim x)$;
2. Consider the pairs in that order (ties are resolved by a lexicographic rule):
   - if the next pair does not create a *circuit* with the pairs already blocked, block this pair;
   - if the next pair creates a *circuit* with the already blocked pairs, skip it.

With our didactic outranking digraph g, we get the following result.

Listing 8.17 Computing a RankedPairs ranking

```
>>> from linearOrders import RankedPairsRanking
>>> rp = RankedPairsRanking(g)
>>> rp.showRanking()
  ['a5','a6','a7','a3','a8','a9','a4','a1','a2']
```

The RankedPairs rule renders in our example here luckily one of the two optimal Kemeny ranking, as we may verify below.

```
>>> ke.maximalRankings
  [['a5','a6','a7','a3','a8','a9','a4','a1','a2'],
   ['a5','a6','a7','a3','a9','a4','a1','a8','a2']]
>>> corr = g.computeOrdinalCorrelation(rp)
>>> g.showCorrelation(corr)
 Correlation indexes:
   Extended Kendall tau       : +0.779
   Epistemic determination    :  0.230
   Bipolar-valued equivalence : +0.179
```

Similar to Kohler's rule, the RankedPairs rule has also a prudent dual version, the *Dias-Lamboray ordering-by-choosing* rule, which produces, when working this time on the dual outranking digraph −g a similar ranking result (see Dias and Lamboray 2010; Lamboray 2009).

Besides of not providing a unique linear ranking, the *ranking-by-choosing* rules, as well as their duals, the *ordering-by-choosing* rules, are unfortunately not scalable to larger outranking digraphs (of orders >100). For such bigger outranking digraphs, with several hundred or thousands of alternatives, only the Copeland and the NetFlows *ranking-by-scoring* rules, with a polynomial complexity of $O(n^2)$, where n is the order of the outranking digraph, remain in fact tractable. Furthermore, as computing the COPELAND and NETFLOWS scores may be done separately per alternative, the latter ranking rules can right away be processed in parallel when multiprocessing resources are available.

The physical necessity to write down a list of items in a linear sequence renders the ranking decision problem very important in practice. However, a relative rating of such items into performance quantiles classes would be, from the very preference modelling perspective, more expressive and faithful. This is the subject of the following Chap. 9.

References

Arrow KJ, Raynaud H (1986) Social choice and multicriterion decision-making. The MIT Press, Cambridge

Bisdorff R (2010) Enumerating chordless circuits in directed graphs. In: ORBEL24-2010, 24th annual conference of the Belgian Operational Research Society (ORBEL aka Sogesci-B.V.W.B.), January 28–29, Liège (BE), Université de Liège (BE), pp 1–12. http://hdl.handle.net/10993/23926

Bisdorff R (2012) On measuring and testing the ordinal correlation between bipolar outranking relations. In: Mousseau V, Pirlot M (eds) DAP'2012 from multiple criteria decision aid to preference learning, University of Mons (Belgium), pp 91–100. http://hdl.handle.net/10993/23909

Bisdorff R (2013) On polarizing outranking relations with large performance differences. J Multi-Criteria Decis Anal Wiley 20:3–12. http://hdl.handle.net/10993/245

Bisdorff R (2021) Technical documentation of the Digraph3 collection of Python modules. https://digraph3.readthedocs.io/en/latest/techDoc.html

Brans JP, Vincke P (1985) A preference ranking organisation method: the PROMETHEE method for MCDM. Manag Sci 31(6):647–656

Copeland A (1951) A reasonable social welfare function. Seminar on mathematics in social sciences, University of Michigan

Dias L, Lamboray C (2010) Extensions of the prudence principle to exploit a valued outranking relation. Eur J Oper Res 201(3):828–837

Kemeny J (1959) Mathematics without numbers. Daedalus 88:577–591

Kohler G (1978) Choix multicritère et analyse algébrique de données ordinales. PhD thesis, Université scientifique et médicale de Grenoble

Lamboray C (2009) A prudent characterization of the Ranked-Pairs rule. Soc Choice Welf 32:129–155

Sen A (2009) Idea of justice. Allen Lane, London

Slater P (1961) Inconsistencies in a schedule of paired comparisons. Biometrica 48:303–312

Tideman N (1987) Independence of clones as a criterion for voting rules. Soc Choice Welf 4(3):185–206

Chapter 9
Rating by Sorting into Relative Performance Quantiles

Contents

Abstract In this chapter, we apply order statistics for sorting a set X of n potential decision alternatives, evaluated on m incommensurable performance criteria, into q quantile equivalence classes. The sorting algorithm is based on pairwise outranking characteristics involving the quantile class limits observed on each criterion. Thus we may implement a weak ordering algorithm of complexity $O(nmq)$.

9.1 Quantile Sorting on a Single Performance Criterion

A single criterion sorting category K is a (usually) lower-closed interval $[m_k; M_k[$ on a real-valued performance measurement scale, with $m_k \leq M_k$. If x is a measured performance on this scale, we may distinguish three sorting situations:

1. $x < m_k$ and $x < M_k$: The performance x is lower than category K.
2. $x \geqslant m_k$ and $x < M_k$: The performance x belongs to category K.
3. $x > m_k$ and $x \geqslant M_k$: The performance x is higher than category K.

As the relation $<$ is the dual of $\geqslant$ ($\not\geqslant$), it will be sufficient to check that $x \geqslant m_k$ as well as $x \not\geqslant M_k$ are true for x to be considered a member of category K.

Upper-closed categories (in a more mathematical integration style) may as well be considered. In this case it is sufficient to check that $m_k \not\geqslant x$ as well as $M_k \geqslant x$ are true for x to be considered a member of category K. It is worthwhile noticing that a category K such that $m_k = M_k$ is hence always empty by definition. In order to be able to properly sort over the complete range of values to be sorted, we will need to use a special, two-sided closed last, respectively first, category.

R. Bisdorff, *Algorithmic Decision Making with Python Resources*, International Series in Operations Research & Management Science 324, https://doi.org/10.1007/978-3-030-90928-4_9

Let $K = K_1, ..., K_q$ be a non-trivial partition of the criterion's performance measurement scale into $q \geq 2$ ordered categories K_k—i.e. lower-closed intervals $[m_k; M_k[$—such that $m_k < M_k$, $M_k = m_{k+1}$ for $k = 0, ..., q-1$ and $M_q = \infty$. And, let $A = \{a_1, a_2, a_3, ...\}$ be a finite set of not all equal performance measures observed on the scale in question.

Property For all performance measure $x \in A$ there exists now a unique k such that $x \in K_k$. If we assimilate, like in descriptive statistics, all the measures gathered in a category K_k to the central value of the category—i.e. $(m_k + M_k)/2$—the sorting result will hence define a weak order (complete preorder) on A.

Let $Q = \{Q_0, Q_1, ..., Q_q\}$ denote the set of $q+1$ increasing order-statistical quantiles—like quartiles or deciles—we may compute from the ordered set A of performance measures observed on a performance scale. If $Q_0 = \min(X)$, the following intervals: $[Q_0; Q_1[$, $[Q_1; Q_2[$, ..., $[Q_{q-1}; \infty[$ define a set of q lower-closed sorting categories. And, in the case of upper-closed categories, if $Q_q = \max(X)$, we obtain the intervals $]-\infty; Q_1]$, $]Q_1; Q_2]$, ..., $]Q_{q-1}; Q_q]$. The corresponding sorting of A will result, in both cases, in a repartition of all measures x into the q quantile categories K_k for $k = 1, ..., q$.

Example Let $A = \{ a_7 = 7.03, a_{15} = 9.45, a_{11} = 20.35, a_{16} = 25.94, a_{10} = 31.44, a_9 = 34.48, a_{12} = 34.50, a_{13} = 35.61, a_{14} = 36.54, a_{19} = 42.83, a_5 = 50.04, a_2 = 59.85, a_{17} = 61.35, a_{18} = 61.61, a_3 = 76.91, a_6 = 91.39, a_1 = 91.79, a_4 = 96.52, a_8 = 96.56, a_{20} = 98.42 \}$ be a set of 20 increasing performance measures observed on a given criterion. The lower-closed category limits we obtain with quartiles ($q = 4$) are: $Q_0 = 7.03 = a_7$, $Q_1 = 34.485$, $Q_2 = 54.945$ (median performance), and $Q_3 = 91.69$. And the sorting into these four categories defines on A a complete preorder with the following four equivalence classes: $K_1 = \{a_7, a_{10}, a_{11}, a_{10}, a_{15}, a_{16}\}$, $K_2 = \{a_5, a_9, a_{13}, a_{14}, a_{19}\}$, $K_3 = \{a_2, a_3, a_6, a_{17}, a_{18}\}$, and $K_4 = \{a_1, a_4, a_8, a_{20}\}$.

9.2 Sorting into Quantiles with Multiple Performance Criteria

Let us now suppose that we are given a performance tableau with a set X of n decision alternatives evaluated on a coherent family of m performance criteria associated with the corresponding outranking relation $\succsim$ defined on X. We denote x_j the performance of alternative x observed on criterion j.

Suppose furthermore that we want to sort the decision alternatives into q upper-closed quantile equivalence classes. We therefore consider a series : $k = k/q$ for $k = 0, ..., q$ of $q+1$ equally spaced quantiles, like quartiles: 0, 0.25, 0.5, 0.75, 1; quintiles: 0, 0.2, 0.4, 0.6, 0.8, 1: or deciles: 0, 0.1, 0.2, ..., 0.9, 1, for instance.

The upper-closed $\mathbf{q}^k$ class corresponds to the m quantile intervals $]q_j(p_{k-1});\ q_j(p_k)]$ observed on each criterion j, where $k = 2, ..., q$, $q_j(p_q) = \max_X(x_j)$, and the first class gathers all performances below or equal to $Q_j(p_1)$.

The lower-closed $\mathbf{q}_k$ class corresponds to the m quantile intervals $[q_j(p_{k-1});\ q_j(p_k)[$ observed on each criterion j, where $k = 1, ..., q-1$, $q_j(p_0) = \min_X(x_j)$, and the last class gathers all performances above or equal to $Q_j(p_{q-1})$.

We call **q-tiles** a complete series of $k = 1, ..., q$ upper-closed $\mathbf{q}^k$, respectively, lower-closed $\mathbf{q}_k$, multiple-criteria quantile classes.

Property With the help of the bipolar-valued characteristic of the outranking relation $r(x \succsim y)$ we may compute as follows the bipolar-valued characteristic of the assertion: x belongs to upper-closed q-tiles class $\mathbf{q}^k$ class, respectively, lower-closed class $\mathbf{q}_k$ (Bisdorff 2020):

$$r(x \in \mathbf{q}^k) = \min\big[-r\big(\mathbf{q}(p_{q-1}) \succsim x\big),\ r\big(\mathbf{q}(p_q) \succsim x\big)\big] \tag{9.1}$$

$$r(x \in \mathbf{q}_k) = \min\big[r\big(x \succsim \mathbf{q}(p_{q-1})\big),\ -r\big(x \succsim \mathbf{q}(p_q)\big)\big] \tag{9.2}$$

The `min` operator implements the logical conjunction and, the outranking relation $\succsim$ verifying the coduality principle, $-r\big(\mathbf{q}(p_{q-1}) \succsim x\big) = r\big(\mathbf{q}(p_{q-1}) \prec x\big)$, respectively, $-r\big(x \succsim \mathbf{q}(p_q)\big) = r\big(x \prec \mathbf{q}(p_q)\big)$.

The `QuantilesSortingDigraph` class from the `sortingDigraphs` module can compute, for instance, such a quintiling of a given random performance tableau (Bisdorff 2021).

Listing 9.1 Computing a quintiles sorting result

```
>>> from randomPerfTabs import RandomPerformanceTableau
>>> t = RandomPerformanceTableau(numberOfActions=50,seed=5)
>>> from sortingDigraphs import QuantilesSortingDigraph
>>> qs = QuantilesSortingDigraph(t,limitingQuantiles=5)
>>> qs
  *----- Object instance description -----------*
   Instance class   : QuantilesSortingDigraph
   Instance name    : sorting_with_5-tile_limits
   Actions          : 50
   Criteria         : 7
   Categories       : 5
   Lowerclosed      : False
   Size             : 841
   Valuation domain  :  [-1.00;1.00]
   Determinateness (%)  : 81.39
   Attributes           :  ['actions','actionsOrig',
        'criteria','evaluation','runTimes','name',
        'limitingQuantiles','LowerClosed',
        'categories','criteriaCategoryLimits',
        'profiles','profileLimits','hasNoVeto',
        'valuationdomain','nbrThreads','relation',
        'categoryContent','order','gamma','notGamma']
  *------  Constructor run times (in sec.) ------*
   # Threads          : 1
   Total time         : 0.03120
   Data input         : 0.00300
   Compute profiles : 0.00075
   Compute relation : 0.02581
   Weak Ordering     : 0.00052
```

5-Tiling the 50 decision alternatives results in 5 ordered upper-closed categories of decision alternatives—the quintiles—supported by a 81.39% majority of criteria significance (see Line 15 above).

The `showCriteriaQuantileLimits()` method shows in Listing 9.2 the quintile limits separating on each criterion the respective evaluations of the 50 decision alternatives into the 5 quintiles. Highest evaluations are observed on criterion `g1` (see Line 6 below).

Listing 9.2 Inspecting the quantile limits

```
1  >>> qs.showCriteriaQuantileLimits()
2       Quantile Class Limits (q = 5)
3       Upper-closed classes
4       crit.       0.20    0.40    0.60    0.80    1.00
5       *-----------------------------------------------
6        g1         31.35   41.09   58.53   71.91   98.08
7        g2         27.81   39.19   49.87   61.66   96.18
8        g3         25.10   34.78   49.45   63.97   92.59
9        g4         24.61   37.91   53.91   71.02   89.84
10       g5         26.94   36.43   52.16   72.52   96.25
11       g6         23.94   44.06   54.92   67.34   95.97
12       g7         30.94   47.40   55.46   69.04   97.10
```

And with the `showSorting()` method we can inspect the actual quintiles sorting result in Listing 9.3 below.

Listing 9.3 Computing a quintiles sorting result

```
>>> qs.showSorting()
    *--- Sorting results in descending order ---*
     ]0.80 - 1.00]: ['a22']
     ]0.60 - 0.80]: ['a03','a07','a08','a11','a14','a17',
                     'a19','a20','a29','a32','a33','a37',
                     'a39','a41','a42','a49']
     ]0.40 - 0.60]: ['a01','a02','a04','a05','a06','a08',
                     'a09','a16','a17','a18','a19','a21',
                     'a24','a27','a28','a30','a31','a35',
                     'a36','a40','a43','a46','a47','a48',
                     'a49','a50']
     ]0.20 - 0.40]: ['a04','a10','a12','a13','a15','a23',
                     'a25','a26','a34','a38','a43','a44',
                     'a45','a49']
     ]  < - 0.20]: ['a44']
```

Most of the decision alternatives (26) are gathered in the median quintile]0.40 – 0.60] class, whereas the highest quintile]0.80 – 1.00] and the lowest quintile] < –0.20] classes gather each one a unique decision alternative (`a22`, respectively, `a44`).

Inspecting the details of the corresponding sorting characteristics may be done with the `showSortingCharacteristics()` method.

Listing 9.4 Bipolar-valued sorting characteristics (extract)

```
1  >>> qs.showSortingCharacteristics()
2       x  in  q^k              r(q^k-1 < x)  r(q^k >= x)  r(x in q^k)
3       a22 in ]< - 0.20]          1.00         -0.86        -0.86
4       a22 in ]0.20 - 0.40]       0.86         -0.71        -0.71
5       a22 in ]0.40 - 0.60]       0.71         -0.71        -0.71
6       a22 in ]0.60 - 0.80]       0.71         -0.14        -0.14
7       a22 in ]0.80 - 1.00]       0.14          1.00         0.14
8       ...
9       ...
10      a44 in ]< - 0.20]          1.00          0.00         0.00
11      a44 in ]0.20 - 0.40]       0.00          0.57         0.00
12      a44 in ]0.40 - 0.60]      -0.57          0.86        -0.57
13      a44 in ]0.60 - 0.80]      -0.86          0.86        -0.86
14      a44 in ]0.80 - 1.00]      -0.86          0.86        -0.86
15      ...
16      ...
17      a49 in ]< - 0.20]          1.00         -0.43        -0.43
18      a49 in ]0.20 - 0.40]       0.43          0.00         0.00
19      a49 in ]0.40 - 0.60]       0.00          0.00         0.00
20      a49 in ]0.60 - 0.80]       0.00          0.57         0.00
21      a49 in ]0.80 - 1.00]      -0.57          0.86        -0.57
```

In Listing 9.4 Line 7, alternative `a22` verifies indeed positively both sorting conditions only for the highest quintile [0.80−1.00] class. Whereas alternatives `a44` and `a49`, for instance, weakly verify both sorting conditions for two, respectively, three, adjacent quintile classes (Lines 10–11 and 18–20).

Quantiles sorting results, indeed, verify always the following properties (Bisdorff 2020):

Property 9.1 (Formal Properties of a Quantiles Sorting Result)

1. *Coherence*: Each object is sorted into a non-empty subset of *adjacent* q-tiles classes. An alternative that would *miss* evaluations on all the criteria will be sorted conjointly in all q-tiled classes.
2. *Uniqueness*: When $r(x \in \mathbf{q}^k) \neq 0.0$ for $k = 1, ..., q$, the performance x is sorted into *exactly one single* q-tiled class.
3. *Separability*: Computing the sorting result for performance x is *independent* from the computing of the other performances' sorting results. This property gives access to efficient parallel processing of class membership characteristics.

The q-tiles sorting result leaves us hence with more or less *overlapping* ordered quantile equivalence classes. For constructing now a linearly ranked q-tiles partition of X , we may apply three strategies:

1. *Average* (default): In decreasing lexicographic order of the average of the lower and upper quantile limits and the upper quantile class limit;
2. *Optimistic*: In decreasing lexicographic order of the upper and lower quantile class limits;
3. *Pessimistic*: In decreasing lexicographic order of the lower and upper quantile class limits;

Listing 9.5 Weakly ranking the quintiles sorting result

```
1  >>> qs.showQuantileOrdering(strategy='average')
2       ]0.80-1.00]  :  ['a22']
3       ]0.60-0.80]  :  ['a03','a07','a11','a14','a20','a29',
4                        'a32','a33','a37','a39','a41','a42']
5       ]0.40-0.80]  :  ['a08','a17','a19']
6       ]0.20-0.80]  :  ['a49']
7       ]0.40-0.60]  :  ['a01','a02','a05','a06','a09','a16',
8                         'a18','a21','a24','a27','a28','a30',
9                         'a31','a35','a36','a40','a46','a47',
10                        'a48','a50']
11      ]0.20-0.60]  :  ['a04','a43']
12      ]0.20-0.40]  :  ['a10','a12','a13','a15','a23','a25',
13                        'a26','a34','a38','a45']
14      ]  < -0.40]  :  ['a44']
```

Following, for instance, the *average* ranking strategy, we find confirmed in the weak ranking shown in Listing 9.5 that alternative `a49` is indeed sorted into three adjacent quintiles classes, namely]0.20 − 0.80] and precedes the]0.40 − 0.60] class, of same average of lower and upper limits (see Line 6).

The `QuantilesSortingDigraph` class constructor models hence a linearly ordered decomposition of the corresponding bipolar-valued outranking digraph. This decomposition leads us to a new *sparse pre-ranked* outranking digraph model.

9.3 The Sparse Pre-ranked Outranking Digraph Model

Following the methodological requirements of the outranking approach, a given outranking digraph is, necessarily associated with a corresponding performance tableau (see Chap. 3). And, we can use this associated performance tableau for linearly decomposing the set of potential decision alternatives into ordered quantiles equivalence classes by using the quantiles sorting algorithm seen in the previous Section.

In the coding example shown in Listing 9.6 below, we generate, first a simple performance tableau of 75 decision alternatives and, secondly, construct with the `PreRankedOutrankingDigraph` class from `sparseOutrankingDigraphs` module the corresponding sparse outranking digraph called `prg`. Notice by the way in Line 3 the `BigData` flag used for generating a parsimoniously commented performance tableau (Bisdorff 2021).

Listing 9.6 Computing a *pre-ranked* sparse outranking digraph

```
1 >>> from randomPerfTabs import RandomPerformanceTableau
2 >>> tp = RandomPerformanceTableau(numberOfActions=75,\
3 ...                          BigData=True,seed=100)
4 >>> from sparseOutrankingDigraphs import\
5 ...               PreRankedOutrankingDigraph
6 >>> prg = PreRankedOutrankingDigraph(Tp,quantiles=5)
7 >>> prg
```

```
8     *----- Object instance description ------*
9      Instance class      : PreRankedOutrankingDigraph
10     Instance name       : randomperftab_pr
11     Actions             : 75
12     Criteria            : 7
13     Sorting by          : 5-Tiling
14     Ordering strategy   : average
15     Components          : 9
16     Minimal order       : 1
17     Maximal order       : 25
18     Average order       : 8.3
19     fill rate           : 20.432%
20     Attributes          :
           ['actions','criteria','evaluation','NA','name',
21         'order','runTimes','dimension','sortingParameters',
22         'valuationdomain','profiles','categories','sorting',
23         'decomposition','nbrComponents','components',
24         'fillRate','minimalComponentSize','maximalComponentSize',
25              ... ]
```

The ordering of the 5-tiling result is following the *average* lower and upper quintile limits strategy (see Listing 9.6 on the preceding page Line 14). We obtain here 9 ordered components of minimal order 1 and maximal order 25. With the `showDecomposition()` method, the corresponding *pre-ranked decomposition* may be inspected as follows:

Listing 9.7 The quantiles decomposition of a pre-ranked outranking digraph

```
1  >>> prg.showDecomposition()
2      *--- quantiles decomposition in decreasing order---*
3       c1. ]0.80-1.00] : [5, 42, 43, 47]
4       c2. ]0.60-1.00] : [73]
5       c3. ]0.60-0.80] : [1, 4, 13, 14, 22, 32, 34, 35, 40,
6                          41, 45, 61, 62, 65, 68, 70, 75]
7       c4. ]0.40-0.80] : [2, 54]
8       c5. ]0.40-0.60] : [3, 6, 7, 10, 15, 18, 19, 21, 23, 24,
9                          27, 30, 36, 37, 48, 51, 52, 56, 58,
10                         63, 67, 69, 71, 72, 74]
11      c6. ]0.20-0.60] : [8, 11, 25, 28, 64, 66]
12      c7. ]0.20-0.40] : [12, 16, 17, 20, 26, 31, 33, 38, 39,
13                         44, 46, 49, 50, 53, 55]
14      c8. ]   <-0.40] : [9, 29, 60]
15      c9. ]   <-0.20] : [57, 59]
```

The highest quintile class (]80% − 100%]) contains decision alternatives `5`, `42`, `43` and `47`. Lowest quintile class (] − 20%]) gathers alternatives `57` and `59` (see Listing 9.7 Lines 3 and 15).

The resulting sparse outranking relation is shown with the `showHTMLRelationMap()` method in a heatmap browser view.

```
   >>> prg.showHTMLRelationMap()
```

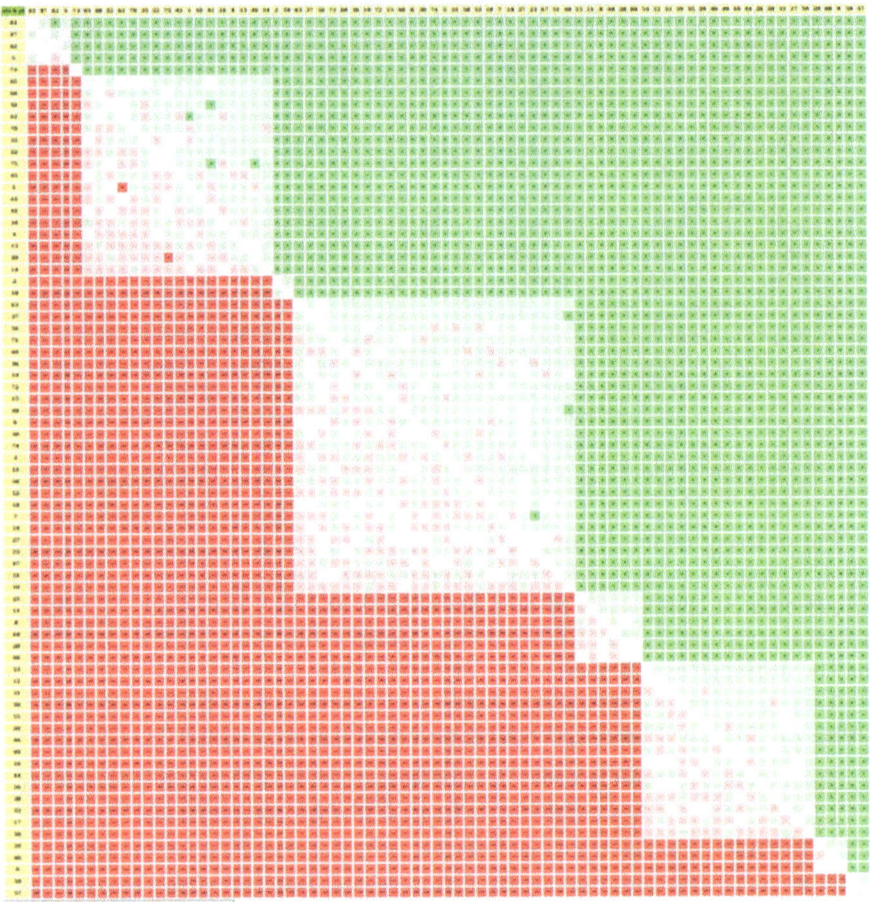

Fig. 9.1 The relation map of a sparse outranking digraph

In Fig. 9.1 we easily recognise the 9 linearly ordered quantile equivalence classes. *Green* and *light-green* pixels show positive *outranking* situations, whereas positive *outranked*—negative outranking—situations are shown in *red* and *light-red*. Indeterminate situations appear in white. In each one of the 9 quantile equivalence classes we recover the corresponding bipolar-valued outranking *sub-relation*, which leads to an actual *fill-rate* of 20.4% (see Listing 9.6 on page 120 Line 19).

Computing the ordinal correlation between the sparse and the standard outranking digraph allows to check how faithful the sparse model represents the complete outranking relation.

```
>>> g = BipolarOutrankingDigraph(tp)
>>> corr = prg.computeOrdinalCorrelation(g)
>>> g.showCorrelation(corr)
  Correlation indexes:
   Crisp ordinal correlation    : +0.863
   Epistemic determination      :  0.315
   Bipolar-valued equivalence   : +0.272
```

The ordinal correlation index between the standard and the sparse outranking digraphs is quite high (+0.863) and their bipolar-valued equivalence is supported by a mean criteria significance majority of $(1.0 + 0.272)/2 = 64\%$.

It is worthwhile noticing in Listing 9.6 on page 120 (Lines 20 and following) that sparse pre-ranked outranking digraphs do not contain a `relation` attribute. The access to pairwise outranking characteristic values is here provided via a corresponding `relation()` function.

Listing 9.8 Functional binary relation characteristics

```
1  def relation(self,x,y):
2      """
3      Dynamic construction of the global
4      outranking characteristic function r(x,y).
5      """
6      Min = self.valuationdomain['min']
7      Med = self.valuationdomain['med']
8      Max = self.valuationdomain['max']
9      if x == y:
10         return Med
11     cx = self.actions[x]['component']
12     cy = self.actions[y]['component']
13     if cx == cy:
14         return self.components[cx]['subGraph'].relation[x][y]
15     elif self.components[cx]['rank'] >
           self.components[cy]['rank']:
16         return Min
17     else:
18         return Max
```

In Listing 9.8 Lines 9–10, all reflexive situations are set to the *indeterminate* value. When two decision alternatives belong to a same component—quantile equivalence class—we access the relation attribute of the corresponding outranking sub-digraph (Lines 13–14). Otherwise we just check the respective ranks of the components.

9.4 Ranking Pre-ranked Sparse Outranking Digraphs

Each one of these nine ordered components is locally ranked by using a suitable ranking rule. Best operational results, both in run times and quality, are more or less equally obtained with the COPELAND and the NETFLOWS rules (see Sects. 8.2 and 8.3). The eventually obtained linear ordering (from the worst to best) is stored in a `prg.boostedOrder` attribute. A reversed linear order (from the best to the worst) is stored in a `prg.boostedRanking` attribute.

```
>>> prg.boostedRanking
  [43, 47, 42, 5, 73, 65, 68, 32, 62, 70, 35, 22, 75, 45, 1,
   61, 41, 34, 4, 13, 40, 14, 2, 54, 63, 37, 56, 71, 69, 36,
   19, 72, 15, 48, 6, 30, 74, 3, 21, 58, 52, 18, 7, 24, 27,
   23, 67, 51, 10, 25, 11, 8, 64, 28, 66, 53, 12, 31, 39, 55,
   20, 46, 49, 16, 44, 26, 38, 33, 17, 50, 29, 60, 9, 59, 57]
```

Alternative `43` appears *first-ranked*, whereas alternative `57` is *last ranked* (see above).

The quality of this ranking result may be assessed by computing with the `computeRankingCorrelation()` method its ordinal correlation index with the standard outranking digraph `g`.

```
>>> corr = g.computeRankingCorrelation(prg.boostedRanking)
>>> g.showCorrelation(corr)
 Correlation indexes:
   Crisp ordinal correlation   : +0.807
   Epistemic determination     :  0.315
   Bipolar-valued equivalence  : +0.254
```

One may also verify below that the COPELAND ranking obtained, for instance, from the standard outranking digraph `g` is as well highly correlated (+0.822) with the one obtained from the sparse outranking digraph `prg`.

```
>>> from linearOrders import CopelandOrder
>>> cop = CopelandOrder(g)
>>> print(cop.computeRankingCorrelation(prg.boostedRanking))
 {'correlation': 0.822, 'determination': 1.0}
```

Noticing the computational efficiency of the quantiles sorting construction, coupled with the separability property of the quantile class membership characteristics computation, we will make usage of the `PreRankedOutrankingDigraph` constructor in Chap. 11 for HPC ranking big and even huge performance tableaux (Bisdorff 2016). The next Chap. 10 addresses now the problem of absolute rating into learned quantile performance norms.

References

Bisdorff R (2016) Computing linear rankings from trillions of pairwise outranking situations. In: Busa-Fekete R, Hüllermeier E, Mousseau V, Pfannschmidt K (eds) From multiple criteria decision aid to preference learning, DAP'2016, University of Paderborn (Germany), pp 1–16. http://hdl.handle.net/10993/28613

Bisdorff R (2020) Lecture 10: On quantiles rating with multiple incommensurable criteria. In: Lectures of the algorithmic decision theory course, University of Luxembourg. http://hdl.handle.net/10993/37933

Bisdorff R (2021) Technical documentation of the Digraph3 collection of Python modules. https://digraph3.readthedocs.io/en/latest/techDoc.html

Chapter 10
Rating-by-Ranking with Learned Performance Quantile Norms

Contents

Abstract We address in this chapter the problem of rating multiple-criteria performances of a set of potential decision alternatives with respect to performance quantiles learned from historical performance data gathered from similar decision alternatives observed in the past. We show how to learn performance quantiles from such historical performance tableaux. New performance records may now be rated with respect to these quantile norms.

10.1 The Absolute Rating Problem

To illustrate the *absolute rating* problem we face, consider for a moment that, in a given decision making problem we observe, for instance, in Table 10.1 below, the multi-criteria performance evaluations of two potential decision alternatives, named `a1001` and `a1010`, evaluated on 7 *incommensurable* performance criteria: 2 *Costs* criteria `c1` and `c2` (to *minimise*) and 5 *Benefits* criteria `b1` to `b5` (to *maximise*).

The performance on *Benefits* criteria `b1`, `b4`, and `b5` is measured on a cardinal scale from 0.0 (worst) to 100.0 (best) whereas, the performance on the *Benefits* criteria `b2` and `b3` and on the *Costs* criterion `c1` is measured on an ordinal scale from 0 (worst) to 10 (best), respectively, −10 (worst) to 0 (best). The performance

Table 10.1 Multi-criteria performances of two potential decision alternatives

Criterion (weight)	`b1` (2)	`b2` (2)	`b3` (2)	`b4` (2)	`b5` (2)	`c1` (5)	`c2` (5)
`a1001`	37.0	2	2	61.0	31.0	−4	−40.0
`a1010`	32.0	9	6	55.0	51.0	−4	−35.0

R. Bisdorff, *Algorithmic Decision Making with Python Resources*,
International Series in Operations Research & Management Science 324,
https://doi.org/10.1007/978-3-030-90928-4_10

on the *Costs* criterion `c2` is eventually measured on a cardinal *negative* scale from −100.00 (worst) to 0.0 (best). The two *Costs* criteria are equi-significant (weight 5). Similarly, the five Benefits criteria are also equi-significant (weight 2). The *importance* (sum of significance weights: $2 \times 5 = 10$) of the *Costs* criteria is hence *equivalent* to the *importance* (sum of significance weights: $5 \times 2 = 10$) of the *Benefits* criteria.

The non-trivial decision problem we now face, is to decide, how the previous two multiple-criteria performance records of alternatives `a1001`, respectively, `a1010`, may be rated: *excellent*? *good*?, or *fair*?; perhaps even, *weak*? Or *very weak*? When compared with similar multi-criteria performance records one has already rated into quantiles in the past.

To solve this *absolute* rating problem, first, we need to estimate multi-criteria *performance quantiles* from historical records (Bisdorff 2020).

10.2 Incremental Learning of Historical Performance Quantiles

Suppose that we see flying in random multiple-criteria performances from a given model of random performance tableau (see Chap. 5). The question we address here is to estimate empirical performance quantiles on the basis of so far observed performance vectors. For this task, we are inspired by Chambers et al. (2006), who present an efficient algorithm for incrementally updating a quantile-binned cumulative distribution function (CDF) with newly observed CDFs.[1]

The `PerformanceQuantiles` class, using the `IncrementalQuantilesEstimator` class from the `randomNumbers` module, implements such a performance quantiles estimation based on a given performance tableau (Bisdorff 2021).

Its main attributes are:

- Ordered `objectives` and a `criteria` dictionaries from a valid performance tableau instance;
- A list `quantileFrequencies` of quantile frequencies like:
  - *quartiles* [0.0, 0.25, 05, 0.75, 1.0],
  - *quintiles* [0.0, 0.2, 0.4, 0.6, 0.8, 1.0], or
  - *deciles* [0.0, 0.1, 0.2, ...1.0] for instance;
- An ordered dictionary `limitingQuantiles` of so far estimated *lower* (default) or *upper* quantile class limits for each frequency per criterion;

[1] We have adapted in Python a C++ implementation published by Press et al. (2007, Chap. 5).

- An ordered dictionary `historySizes` for keeping track of the number of evaluations seen so far per criterion. Missing data, the case given, make these sizes vary from criterion to criterion.

Below, we show an example Python session concerning 900 decision alternatives randomly generated with a *Cost-Benefit* performance tableau model (see Sect. 5.3) from which are also drawn the performances of alternatives `a1001` and `a1010` shown in Table 10.1 on page 125 above.

Listing 10.1 Computing performance quantiles from a given performance tableau

```
1 >>> from performanceQuantiles import PerformanceQuantiles
2 >>> from randomPerfTabs import RandomCBPerformanceTableau
3 >>> nbrActions=900
4 >>> nbrCrit = 7
5 >>> seed = 100
6 >>> pt = RandomCBPerformanceTableau(numberOfActions=nbrActions,\
7 ...                      numberOfCriteria=nbrCrit,seed=seed)
8 >>> pq = PerformanceQuantiles(pt,\
9 ...                       numberOfBins = 'quartiles',\
10 ...                       LowerClosed=True)
11 >>> pq
12   *------- PerformanceQuantiles instance description ------*
13    Instance class    : PerformanceQuantiles
14    Instance name     : 4-tiled_performances
15    Objectives        : 2
16    Criteria          : 7
17    Quantiles         : 4
18    History sizes     : {'c1': 887,'b1': 888,'b2': 891,'b3': 895,
19                          'b4': 892,'c2': 893,'b5': 887}
20    Attributes        : ['perfTabType','valueDigits',
21                         'actionsTypeStatistics',
22                         'objectives', 'BigData',
23                         'missingDataProbability',
24                         'criteria', 'LowerClosed', 'name',
25                         'quantilesFrequencies', 'historySizes',
26                         'limitingQuantiles', 'cdf']
```

In Line 10 above, the `PerformanceQuantiles` class parameter `numberOfBins` we use for setting the wished number of quantile frequencies, may be either *quartiles* (4 bins), *quintiles* (5 bins), *deciles* (10 bins), *dodeciles* (20 bins) or any other integer number of quantile bins. The quantile bins may be either *lower-closed* (default) or *upper-closed.*

Inspecting the estimated quartile limits may be operated with the `showLimitingQuantiles()` method.

Listing 10.2 Printing out the estimated quartile limits

```
>>> pq.showLimitingQuantiles(ByObjectives=True)
    ----  Historical performance quantiles -----*
    Costs
    criteria | weights |   '0.00' '0.25' '0.50' '0.75' '1.00'
```

```
 ---------|-------------------------------------------------
    'c1'  |    5    |  -10     -7       -5      -3       0
    'c2'  |    5    | -96.37 -70.65 -50.10 -30.00   -1.43
 Benefits
 criteria | weights | '0.00'  '0.25' '0.50' '0.75' '1.00'
 ---------|---------------------------------------------------
    'b1'  |    2    |  1.99     29.82 49.44   70.73  99.83
    'b2'  |    2    |   0         3      5       7     10
    'b3'  |    2    |   0         3      5       7     10
    'b4'  |    2    |  3.27    30.10  50.82   70.89  98.05
    'b5'  |    2    |  0.85    29.08  48.55   69.98  97.56
```

Both objectives are *equally important*; the sum of weights (10) of the *Costs* criteria balance the sum of weights (10) of the *Benefits* criteria (see Listing 10.2 on the previous page column 2). The preference direction of the *Costs* criteria `c1` and `c2` is *negative*; the lesser the costs, the better it is, whereas all the *Benefits* criteria `b1` to `b5` show *positive* preference directions, i.e. the higher the benefits, the better it is. The columns entitled `0.00`, respectively, `1.00` show the *quartile* `Q0`, respectively, `Q4`, i.e. the *worst*, respectively, *best* performance observed so far on each criterion. Column `0.50` shows the *median* (`Q2`) performance observed on the criteria.

New decision alternatives with random multiple-criteria performance vectors from the same random performance tableau model as `pt` (see Listing 10.1 on the preceding page) may now be generated with a generic `RandomPerformanceGenerator` class from the `randomPerfTabs` module (Bisdorff 2021).[2]

Listing 10.3 Generating 100 new random decision alternatives of the same model

```
>>> from randomPerfTabs import RandomPerformanceGenerator
>>> rpg = RandomPerformanceGenerator(pt,seed=seed)
>>> newTab = rpg.randomPerformanceTableau(100)
```

The so far estimated historical quantile limits must, first, be updated with this newly arriving 100 data records:

```
1 >>> # Updating the quartile norms shown above
2 >>> pq.updateQuantiles(newTab,historySize=None)
```

Parameter `historySize` of the `updateQuantiles()` method (Line 2 above) allows to *balance* the *new* evaluations against the *historical* ones.

With `historySize = None` (the default setting), the balance in the example above is 900/1000 (90%, the weight of historical data) against 100/1000 (10%, the weight of the new incoming observations). Setting `historySize = 0`, for instance, will ignore all historical data (0/100 against 100/100) and restart building the quantile estimation with solely the new incoming data. The

[2] The `RandomPerformanceGenerator` class works for the *standard* performance tableau model (see Sect. 5.2), the *Cost-Benefit* model (see Sect. 5.3), and the 3-objectives model (see Sect. 5.4).

Performance quantiles

Sampling sizes between 986 and 995.

criterion	0.00	0.25	0.50	0.75	1.00
b1	1.99	28.77	49.63	75.27	99.83
b2	0.00	2.94	4.92	6.72	10.00
b3	0.00	2.90	4.86	8.01	10.00
b4	3.27	35.91	58.58	72.00	98.05
b5	0.85	32.84	48.09	69.75	99.00
c1	-10.00	-7.35	-5.39	-3.38	0.00
c2	-96.37	-72.22	-52.27	-33.99	-1.43

Fig. 10.1 *Showing updated quartiles limits per criterion.* The 0.25 column shows the first quartile (`Q1`) limits, the 0.50 column shows the second quartile (`Q2`) limits and the 0.75 column shows the third quartile (`Q3`) limits. Column 0.00 (respectively, 1.00) shows the minimum (respectively, maximum) performance on each criterion

`showHTMLLimitingQuantiles()` method shows the updated quantile limits in a browser view (see Fig. 10.1).

```
>>> # showing the updated quantile limits in a browser view
>>> pq.showHTMLLimitingQuantiles(Transposed=True)
```

10.3 Rating-by-Ranking New Performances with Quantile Norms

For *rating* a newly given set of decision alternatives with the help of empirical performance quantiles estimated from historical data, the `sortingDigraphs` module provides the `LearnedQuantilesRatingDigraph` class, a specialisation of the `QuantilesSortingDigraph` class (see Chap. 9). The absolute rating result is computed by *ranking* the new performance records together with the learned historical quantile limits.

The class constructor requires a valid `PerformanceQuantiles` instance and, by default, uses the COPELAND or the NETFLOWS ranking rule, whichever fits best in an ordinal correlation sense with the underlying outranking digraph.

It is important to notice that the `LearnedQuantilesRatingDigraph` class, contrary to the generic `OutrankingDigraph` class, does not inherit from the generic `PerformanceTableau` class, but instead from the `PerformanceQuantiles` class (Bisdorff 2021).

The `actions` attribute in such a `LearnedQuantilesRatingDigraph` class instance contains not only the newly given decision alternatives, but also the historical quantile limits (attribute `profiles`) obtained from a given `PerformanceQuantiles` class instance, i.e. estimated quantile bins' performance limits from historical performance data.

We reconsider now the `PerformanceQuantiles` object instance `pq` as computed in the previous section. Let `newActions` be a list of 10 new decision alternatives generated with the same random performance tableau generator `rpq` and including, for our didactic purpose, the two decision alternatives `a1001` and `a1010` mentioned at the beginning.

Listing 10.4 Computing the absolute rating of 10 new decision alternatives

```
1  >>> from sortingDigraphs import LearnedQuantilesRatingDigraph
2  >>> newActions = rpg.randomActions(10)
3  >>> lqr = LearnedQuantilesRatingDigraph(pq,newActions,\
4  ...                                  rankingRule='best')
5  >>> lqr
6   *---- Object instance description
7    Instance class        : LearnedQuantilesRatingDigraph
8    Instance name         : normedRatingDigraph
9    Criteria              : 7
10   Quantile profiles     : 4
11   Lower-closed bins     : True
12   New actions           : 10
13   Size                  : 93
14   Determinateness (%)   : 76.1
15   Ranking rule          : Copeland
16   Ordinal correlation : +0.95
17   Attributes: ['runTimes','objectives','criteria',
18     'LowerClosed','quantilesFrequencies',
19     'limitingQuantiles','historySizes','cdf','name',
20     'newActions','evaluation','categories',
21     'criteriaCategoryLimits','profiles','profileLimits',
22     'hasNoVeto','actions','completeRelation','relation',
23     'concordanceRelation','valuationdomain','order',
24     'gamma','notGamma','rankingRule','rankingCorrelation',
25     'rankingScores','actionsRanking','ratingCategories',
26     'ratingRelation','relationOrig']
```

Data input to the `LearnedQuantilesRatingDigraph` class constructor is a valid `PerformanceQuantiles` object `pq` and a `newActions` list of newly generated decision alternatives with the same random generator `rpg` (see Listing 10.4 Lines 3–4).

The `actionsSubset` parameter of the `showPerformanceTableau()` method allows a look at the digraph's nodes, here called `newActions`.

Listing 10.5 Performance tableau of the new incoming decision alternatives

```
>>> lqr.showPerformanceTableau(actionsSubset=lqr.newActions)
 *----  performance tableau -----*
 criteria | a1001 a1002 a1003 a1004 a1005 a1006 a1007 a1008 a1009 a1010
 ---------|-----------------------------------------------------------
    'b1'  |  37.0  27.0  24.0  16.0  42.0  33.0  39.0  64.0  42.0  32.0
    'b2'  |   2.0   5.0   8.0   3.0   3.0   3.0   6.0   5.0   4.0   9.0
    'b3'  |   2.0   4.0   2.0   1.0   6.0   3.0   2.0   6.0   6.0   6.0
    'b4'  |  61.0  54.0  74.0  25.0  28.0  20.0  20.0  49.0  44.0  55.0
    'b5'  |  31.0  63.0  61.0  48.0  30.0  39.0  16.0  96.0  57.0  51.0
    'c1'  |  -4.0  -6.0  -8.0  -5.0  -1.0  -5.0  -1.0  -6.0  -6.0  -4.0
    'c2'  | -40.0 -23.0 -37.0 -37.0 -24.0 -27.0 -73.0 -43.0 -94.0 -35.0
```

Among the 10 new incoming decision alternatives, we recognise alternatives `a1001` (see column 2) and `a1010` (see last column) we have shown in Table 10.1 on page 125.

The `actions` dictionary of a `LearnedQuantilesRatingDigraph` class instance includes, besides the 10 performance evaluations of the ten new alternatives, also the closed lower limits of the four quartile classes: `m1` = $[0.0 - [$, `m2` = $[0.25 - [$, `m3` = $[0.5 - [$, `m4` = $[0.75 - [$. We find these limits in the `profiles` attribute (see Listing 10.6 below).

Listing 10.6 Showing the limiting profiles of the rating quantiles

```
>>> lqr.showPerformanceTableau(actionsSubset=lqr.profiles)
    *----  Quartiles limit profiles -----*
    criteria |  'm1'    'm2'    'm3'    'm4'
    ---------|-----------------------------
       'b1'  |  2.0     28.8    49.6    75.3
       'b2'  |  0.0      2.9     4.9     6.7
       'b3'  |  0.0      2.9     4.9     8.0
       'b4'  |  3.3     35.9    58.6    72.0
       'b5'  |  0.8     32.8    48.1    69.7
       'c1'  | -10.0    -7.4    -5.4    -3.4
       'c2'  | -96.4   -72.2   -52.3   -34.0
```

The main run time is spent by the `LearnedQuantilesRatingDigraph` class constructor in computing a bipolar-valued outranking relation on the extended actions set including both the new alternatives as well as the quartile class limits. In case of large volumes, i.e. many new decision alternatives and centile classes, for instance, a multi-threading version may be used when multiple processing cores are available (Bisdorff 2021).

The actual rating procedure will rely on a complete ranking of the new decision alternatives as well as the quantile class limits obtained from the corresponding bipolar-valued outranking digraph. Two efficient and scalable ranking rules, the COPELAND and its valued version, the NETFLOWS rule may be used for this purpose. The `rankingRule` parameter allows to choose one of both. With `rankingRule='best'` (see Listing 10.4 on the preceding page Line 4) the `LearnedQuantilesRatingDigraph` constructor will choose the ranking rule that results in the highest ordinal correlation with the given outranking relation (see Chap. 16 and Bisdorff 2012).

In this rating example, the COPELAND rule appears to be the more appropriate ranking rule.

Listing 10.7 COPELAND ranking of new alternatives and historical quartile limits

```
>>> lqr.rankingRule
  'Copeland'
>>> lqr.actionsRanking
  ['m4', 'a1005', 'a1010', 'a1002', 'a1008', 'a1006', 'a1001',
   'a1003', 'm3', 'a1007', 'a1004', 'a1009', 'm2', 'm1']
>>> lqr.showCorrelation(lqr.rankingCorrelation)
  Correlation indexes:
    Crisp ordinal correlation  : +0.945
    Epistemic determination    :  0.522
    Bipolar-valued equivalence : +0.493
```

We achieve here in Listing 10.7 on the preceding page a linear ranking without ties (from best to worst) of the digraph's actions set, i.e. including the new decision alternatives as well as the quartile limits $m1$ to $m4$, which is very close in an ordinal sense ($\tau = 0.945$) to the underlying strict outranking relation.

The eventual rating procedure is based in this example on the *lower* quartile limits, such that the quartile contents are filtered out in increasing order of the *quartiles*.

```
>>> lqr.ratingCategories
 OrderedDict([
 ('m2', ['a1007','a1004','a1009']),
 ('m3', ['a1005','a1010','a1002','a1008',
         'a1006','a1001','a1003'])
 ])
```

We notice above that no new decision alternative is actually Rated into the lowest [0.0 − 0.25[, respectively, highest [0.75 − [quartile. Indeed, the absolute rating result is shown with the `showQuantilesRating()` method:

Listing 10.8 Absolute quartiles rating result

```
>>> lqr.showQuantilesRating()
  *-------- Quartiles rating result --------
   [0.50 - 0.75[ ['a1005', 'a1010', 'a1002', 'a1008',
                  'a1006', 'a1001', 'a1003']
   [0.25 - 0.50[ ['a1007', 'a1004', 'a1009']
```

The same result may also be shown in a browser view with the `showHTMLRatingHeatmap()` method using a specialised rating heatmap format (see Fig. 10.2 on the next page):

```
>>> lqr.showHTMLRatingHeatmap(\
...           pageTitle='Heatmap of Quartiles Rating',\
...           Correlations=True,colorLevels=5)
```

Using furthermore a specialised version of the `exportGraphViz()` method allows drawing in Fig. 10.3 on page 134 the same rating result in a HASSE diagram format.[3]

```
>>> lqr.exportRatingGraphViz('quartileRatingDigraph')
 *---- exporting a dot file for GraphViz tools ---------*
  Exporting to quartileRatingDigraph.dot
  dot -Grankdir=TB -Tpng quartileRatingDigraph.dot -o
      quartileRatingDigraph.png
```

We have actually solved now the *absolute rating* problem stated at the beginning. Decision alternatives `a1001` and `a1010` (see Table 10.1 on page 125) are both rated into the same third quartile `Q3` class (see Fig. 10.3 on page 134), even if the COPELAND ranking, obtained from the underlying strict outranking digraph (see

[3] Note that the graphviz dot file was post-edited in order to mark in blue alternatives `a1001` and `a1010`.

Heatmap of Quartiles Rating

Ranking rule: **Copeland**; *Ranking correlation:* **0.938**

criteria	c2	b3	c1	b4	b1	b2	b5
weights	5	2	5	2	2	2	2
tau(*)	+0.64	+0.54	+0.43	+0.37	+0.37	+0.35	+0.34
[0.75 -	-30.00	7.00	-3.00	70.89	70.73	7.00	69.98
a1005c	-24.00	6.00	-1.00	28.00	42.00	3.00	30.00
a1010n	-35.00	6.00	-4.00	55.00	32.00	9.00	51.00
a1002c	-23.00	4.00	-6.00	54.00	27.00	5.00	63.00
a1008n	-43.00	6.00	-6.00	49.00	64.00	5.00	96.00
a1006c	-27.00	3.00	-5.00	20.00	33.00	3.00	39.00
a1001c	-40.00	2.00	-4.00	61.00	37.00	2.00	31.00
a1003a	-37.00	2.00	-8.00	74.00	24.00	8.00	61.00
[0.50 -	-50.10	5.00	-5.00	50.82	49.44	5.00	48.55
a1007c	-73.00	2.00	-1.00	20.00	39.00	6.00	16.00
a1004c	-37.00	1.00	-5.00	25.00	16.00	3.00	48.00
a1009n	-94.00	6.00	-6.00	44.00	42.00	4.00	57.00
[0.25 -	-70.65	3.00	-7.00	30.10	29.82	3.00	29.08
[0.00 -	-96.37	0.00	-10.00	3.27	1.99	0.00	0.85

Color legend:

quantile	20.00%	40.00%	60.00%	80.00%	100.00%

(*) tau: Ordinal (Kendall) correlation between marginal criterion and global ranking relation.

Fig. 10.2 *Heatmap of absolute quartiles ranking.* The quartile equivalence classes appear lower-closed. No alternative is rated into the `Q1` class ([0.00 – 0.25[) and no alternative is rated into the Q4 class ([0.75 – 1.00])

Fig. 10.2), suggests that alternative `a1010` is effectively *better performing than* alternative `a1001`.

A *preciser* rating result may indeed be achieved when using *deciles* instead of *quartiles* for estimating the historical marginal cumulative distribution functions.

Listing 10.9 Absolute deciles rating result

```
>>> pq1 = PerformanceQuantiles(pt, numberOfBins='deciles',\
...                       LowerClosed=True)
>>> pq1.updateQuantiles(newTab,historySize=None)
>>> lqr1 = LearnedQuantilesRatingDigraph(pq1,newActions,\
...                                      rankingRule='best')
>>> lqr1.showQuantilesRating()
  *-------- Deciles rating result ---------
   [0.60 - 0.70[ ['a1005', 'a1010', 'a1008', 'a1002']
   [0.50 - 0.60[ ['a1006', 'a1001', 'a1003']
   [0.40 - 0.50[ ['a1007', 'a1004']
   [0.30 - 0.40[ ['a1009']
```

Compared with the quartiles rating result, we notice in Listing 10.9 that the seven alternatives (`a1001`, `a1002`, `a1003`, `a1005`, `a1006`, `a1008`, and `a1010`), rated before into the third quartile class [0.50 – 0.75[, are now divided up: alternatives `a1002`, `a1005`, `a1008`, and `a1010` attain now the 7th decile class [0.60 – 0.70[, whereas alternatives `a1001`, `a1003` and `a1006` attain only the 6th decile class [0.50 – 0.60[. Of the three `Q2` [0.25 – 0.50[rated alternatives (`a1004`, `a1007`

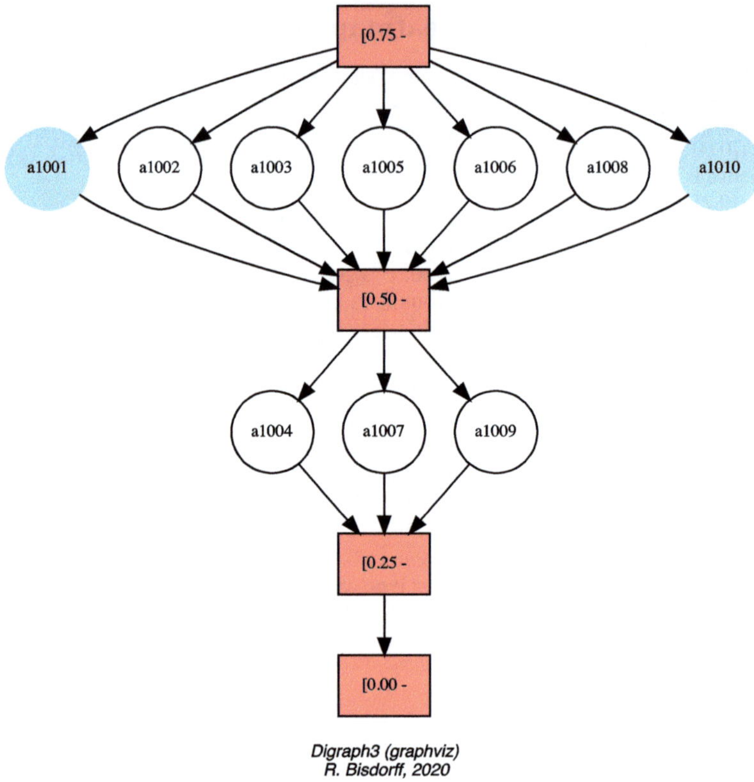

Fig. 10.3 Absolute quartiles rating digraph

and a1009), alternatives a1004 and a1007 are now rated into the 5th decile class $[0.40 - 0.50[$ and a1009 is lowest rated into the 4th decile class $[0.30 - 0.40[$.

A browser heatmap view in Fig. 10.4 on the next page more conveniently illustrates this refined rating result.

```
>>> lqr1.showHTMLRatingHeatmap(\
...         pageTitle='Heatmap of Deciles rating',\
...         colorLevels=5, Correlations=True)
```

To avoid having to recompute performance deciles from historical data when wishing to refine a rating result, it is useful, depending on the actual size of the historical data, to initially compute performance quantiles with a relatively high number of bins, for instance, *dodeciles* or even *centiles*. It is then possible to interpolate on the fly *quartiles* or *deciles*, for instance, when constructing the rating digraph.

Fig. 10.4 *Absolute deciles rating*. Decision alternatives `a1001` and `a1010` are now, as expected, rated in the 6th decile (`D6`), respectively, in the 7th decile (`D7`)

Heatmap of Deciles rating

Ranking rule: **NetFlows**; *Ranking correlation:* **0.960**

criteria	c2	b3	c1	b1	b5	b2	b4
weights	5	2	5	2	2	2	2
tau(*)	0.67	0.65	0.58	0.57	0.53	0.53	0.48
[0.90 -	-20.32	7.73	-2.53	86.83	82.16	7.66	82.04
[0.80 -	-29.70	7.26	-3.35	79.30	75.15	6.64	74.66
[0.70 -	-37.97	6.67	-4.14	70.95	60.20	5.88	69.76
a1005c	-24.00	6.00	-1.00	42.00	30.00	3.00	28.00
a1010n	-35.00	6.00	-4.00	32.00	51.00	9.00	55.00
a1008n	-43.00	6.00	-6.00	64.00	96.00	5.00	49.00
a1002c	-23.00	4.00	-6.00	27.00	63.00	5.00	54.00
[0.60 -	-44.23	5.92	-5.04	60.56	56.01	5.37	62.23
a1006c	-27.00	3.00	-5.00	33.00	39.00	3.00	20.00
a1001c	-40.00	2.00	-4.00	37.00	31.00	2.00	61.00
a1003a	-37.00	2.00	-8.00	24.00	61.00	8.00	74.00
[0.50 -	-52.22	4.64	-6.02	49.56	48.07	4.83	58.45
a1007c	-73.00	2.00	-1.00	39.00	16.00	6.00	20.00
a1004c	-37.00	1.00	-5.00	16.00	48.00	3.00	25.00
[0.40 -	-60.50	3.84	-6.69	39.61	40.16	4.25	49.82
a1009n	-94.00	6.00	-6.00	42.00	57.00	4.00	44.00
[0.30 -	-67.14	3.12	-7.32	30.85	34.33	3.30	40.89
[0.20 -	-77.07	2.55	-7.94	23.84	29.57	2.27	30.45
[0.10 -	-83.04	1.99	-8.48	16.64	16.91	1.58	24.78
[0.00 -	-96.37	0.00	-10.00	1.99	0.85	0.00	3.27

Color legend:

quantile	20.00%	40.00%	60.00%	80.00%	100.00%

(*) tau: Ordinal (Kendall) correlation between marginal criterion and global ranking relation.

Listing 10.10 From deciles interpolated quartiles rating result

```
1 >>> lqr2 = LearnedQuantilesRatingDigraph(pq1,newActions,
2 ...                         quantiles='quartiles')
3 >>> lqr2.showQuantilesRating()
4   *-------- Deciles rating result ---------
5    [0.50 - 0.75[ ['a1005', 'a1010', 'a1002', 'a1008',
6                  'a1006', 'a1001', 'a1003']
7    [0.25 - 0.50[ ['a1004', 'a1007', 'a1009']
```

With the `quantiles` parameter (see Listing 10.10 Line 2), we recover by interpolation the same quartiles rating as obtained directly with historical performance quartiles (see Listing 10.8 on page 132). Mind that a correct interpolation of quantiles from a given cumulative distribution function requires more or less uniform distributions of observations in each bin.

More generally, in the case of industrial production monitoring problems, for instance, where large volumes of historical performance data may be available, it can be of interest to estimate even more precisely the marginal cumulative distribution functions, especially when *tail* rating results, i.e. distinguishing *very best*, or *very weak* multiple-criteria performances, become a critical issue. Similarly, the `historySize` parameter may be used for monitoring on the fly *unstable* random multiple-criteria performance data (Chambers et al. 2006).

The next Chap. 11 is tackling the challenging problem of ranking big performance tableaux with thousands and millions of multiple-criteria records on HPC clusters.

References

Bisdorff R (2012) On measuring and testing the ordinal correlation between bipolar outranking relations. In: Mousseau V, Pirlot M (eds) DAP'2012 from multiple criteria decision aid to preference learning, University of Mons (Belgium), pp 91–100. http://hdl.handle.net/10993/23909

Bisdorff R (2020) Lecture 5: Simulating from arbitrary empirical random distributions. In: Lectures of the computational statistics course, University of Luxembourg. http://hdl.handle.net/10993/37870

Bisdorff R (2021) Technical documentation of the Digraph3 collection of Python modules. https://digraph3.readthedocs.io/en/latest/techDoc.html

Chambers J, James D, Lambert D, Vander Wiel S (2006) Monitoring networked applications with incremental quantile estimation. Stat Sci 21(4):463–475

Press W, Teukolsky S, Vetterling W, Flannery B (2007) Single-pass estimation of arbitrary quantiles. In: Numerical recipes: the art of scientific computing, Sect 5.8.2, 3rd edn. Cambridge University Press, Cambridge, pp 435–438

Chapter 11
HPC Ranking of Big Performance Tableaux

Contents

Abstract The sparse outranking digraph, introduced in Chap. 9, is suitable for tackling the ranking of big multiple-criteria performance tableaux with thousands or millions of records. To effectively compute rankings from performance tableaux of these sizes, we propose in this chapter a collection of C-compiled and optimised DIGRAPH3 modules that may be run on HPC equipment as available, for instance, at the University of Luxembourg.

11.1 C-compiled Python Modules

The DIGRAPH3 collection provides cythonized,[1] i.e. C-compiled and optimised versions of the main python modules for tackling multiple-criteria decision problems facing very large sets of decision alternatives (> 10,000). Such problems appear usually with a combinatorial organisation of the potential decision alternatives, as is frequently the case in bioinformatics, for instance. If HPC facilities with nodes supporting numerous cores (>20) and big RAM (>50 GB) are available, ranking up to several millions of alternatives becomes effectively tractable (Bisdorff 2016).

Four cythonized DIGRAPH3 modules, prefixed with the letter `c` and taking a `pyx` extension, are provided with their corresponding setup tools in the `cython` directory of the DIGRAPH3 resources, namely:

---

[1] CYTHON C-extension for Python, https://cython.org/.

R. Bisdorff, *Algorithmic Decision Making with Python Resources*,
International Series in Operations Research & Management Science 324,
https://doi.org/10.1007/978-3-030-90928-4_11

`cRandPerfTabs.pyx`, `cIntegerOutrankingDigraphs.pyx`, `cIntegerSortingDigraphs.pyx`, and `cSparseIntegerOutrankingDigraphs.pyx`.

Their automatic compilation and installation, alongside the standard DIGRAPH3 python3 modules, requires the *cython* compiler:
`...$ python3 -m pip install cython`
and a C compiler:
`...$ sudo apt install gcc` on Ubuntu, for instance.

These cythonized modules, specifically designed for being run on HPC clusters, require the Unix *forking* start method of subprocesses and therefore, due to forking problems on Mac OS platforms, may only operate safely on Linux platforms (see start methods of the Python multiprocessing library).

11.2 Big Data Performance Tableaux

In order to efficiently type the C variables, the `cRandPerfTabs` module provides the root `cPerformanceTableau` class, but, with *integer* action keys, *float* performance evaluations, *integer* criteria weights and *float* discrimination thresholds. And, to limit as much as possible memory occupation of class instances, all the usual verbose comments are dropped from the description of the `actions` and `criteria` dictionaries. A random instance may be generated with the `cRandomPerformanceTableau` class from the `cRandPerfTabs` module:

Listing 11.1 Big data performance tableau format

```
>>> from cRandPerfTabs import cRandomPerformanceTableau
>>> t = cRandomPerformanceTableau(numberOfActions=4,\
...                                 numberOfCriteria=2)
>>> t
  *------- PerformanceTableau instance description ------*
    Instance class     : cRandomPerformanceTableau
    Seed               : None
    Instance name      : cRandomperftab
    Actions            : 4
    Criteria           : 2
    Attributes         : ['randomSeed', 'name', 'actions',
                          'criteria', 'evaluation', 'weightPreorder']
>>> t.actions
   OrderedDict([(1, {'name': '#1'}), (2, {'name': '#2'}),
                (3, {'name': '#3'}), (4, {'name': '#4'})])
>>> t.criteria
   OrderedDict([
   ('g1', {'name': 'RandomPerformanceTableau() instance',
               'thresholds': {'ind': (10.0, 0.0),
                              'pref': (20.0, 0.0),
                              'veto': (80.0, 0.0)},
               'scale': (0.0, 100.0),
               'weight': 1,
               'preferenceDirection': 'max'}),
   ('g2', {'name': 'RandomPerformanceTableau() instance',
               'thresholds': {'ind': (10.0, 0.0),
```

```
                                'pref': (20.0, 0.0),
                                'veto': (80.0, 0.0)},
                'scale': (0.0, 100.0),
                'weight': 1,
                'preferenceDirection': 'max'})])
>>> t.evaluation
  {'g1': {1: 35.17, 2: 56.4, 3: 1.94, 4: 5.51},
   'g2': {1: 95.12, 2: 90.54, 3: 51.84, 4: 15.42}}
>>> t.showPerformanceTableau()
        Criteria |  'g1'    'g2'
        Actions  |    1       1
        ---------|---------------
           '#1'  |  91.18   90.42
           '#2'  |  66.82   41.31
           '#3'  |  35.76   28.86
           '#4'  |   7.78   37.64
```

Several methods for converting from the Big Data model to the standard model and vice versa are provided by the `cPerformanceTableau` class.

```
>>> t1 = t.convert2Standard()
>>> t1.convertWeight2Decimal()
>>> t1.convertEvaluation2Decimal()
>>> t1
  *------- PerformanceTableau instance description ------*
   Instance class    : PerformanceTableau
   Seed              : None
   Instance name     : std_cRandomperftab
   Actions           : 4
   Criteria          : 2
   Attributes        : ['name', 'actions', 'criteria',
                        'weightPreorder', 'evaluation',
                        'randomSeed']
```

11.3 C-implemented Integer-Valued Outranking Digraphs

The C-compiled version of the `BipolarOutrankingDigraph` class takes integer r-characteristic values.

Listing 11.2 Constructing big bipolar-valued outranking digraphs

```
>>> from cRandPerfTabs import cRandomPerformanceTableau
>>> t = cRandomPerformanceTableau(numberOfActions=1000,\
...                                 numberOfCriteria=2,seed=100)
>>> from cIntegerOutrankingDigraphs import\
...                    IntegerBipolarOutrankingDigraph
>>> g = IntegerBipolarOutrankingDigraph(t,\
...                    Threading=True, nbrCores=4)
>>> g
  *------- Object instance description ------*
   Instance class    : IntegerBipolarOutrankingDigraph
   Instance name     : rel_cRandomperftab
   Actions           : 1000
```

```
  Criteria            : 2
  Size                : 460795
  Determinateness     : 55.975
  Valuation domain : {'min': -2, 'med': 0, 'max': 2,
                      'hasIntegerValuation': True}
  Attributes          : ['name', 'actions', 'criteria',
                      'totalWeight', 'valuationdomain',
                      'methodData', 'evaluation',
                      'order', 'runTimes', 'nbrThreads',
                      'relation', 'gamma', 'notGamma']
  ----  Constructor run times (in sec.) ----
  Total time          : 2.29166
  Data input          : 0.01170
  Compute relation : 1.73179
  Gamma sets          : 0.54816
  Threads             : 4
```

On a common PC with four single threaded cores, the `IntegerBipolarOutrankingDigraph` class constructor takes about 2.3 seconds for computing a *million* pairwise outranking characteristic values (see Listing 11.2 on the preceding page). In a similar setting, the standard `BipolarOutrankingDigraph` constructor operates more than three times slower.

```
>>> from outrankingDigraphs import BipolarOutrankingDigraph
>>> t1 = t.convert2Standard()
>>> g1 = BipolarOutrankingDigraph(t1,\
...                    Threading=True,nbrCores=4)
>>> g1
 *------- Object instance description ------*
  Instance class          : BipolarOutrankingDigraph
  Instance name           : rel_std_cRandomperftab
  Actions                 : 1000
  Criteria                : 2
  Size                    : 460795
  Determinateness (%)     : 77.99
  Valuation domain        : [-1.00;1.00]
  Attributes              : ['name','actions','valuationdomain',
                             'criteria','methodData','evaluation',
                             'NA','order','runTimes','nbrThreads',
                             'relation','gamma','notGamma']
  ----  Constructor run times (in sec.) ----
  Total time          : 8.35748
  Data input          : 0.01407
  Compute relation : 7.30510
  Gamma sets          : 1.03831
  Threads             : 4
```

By far, most of the run time is in each case needed for computing the individual pairwise outranking characteristic values: 1.7 second versus 7.3 second. Notice also below the memory occupations—about 108 MB for `g` and 212 MB for `g1`—of both outranking digraph instances (see Line 3 and 5 below).

```
1 >>> from digraphsTools import total_size
2 >>> total_size(g)
```

```
    107503580
>>> total_size(g1)
    211519995
>>> total_size(g.relation)/total_size(g)
    0.34
>>> total_size(g.gamma)/total_size(g)
    0.46
```

The standard `Decimal` valued `BipolarOutrankingDigraph` instance `g1` thus nearly doubles the memory occupation of the corresponding `IntegerBipolarOutrankingDigraph` `g` instance. About 80% of this memory occupation is due to the `g.relation` (34%) and the `g.gamma` (46%) dictionaries. And this ratio grows quadratically with the outranking digraph order. To limit the object sizes for really big outranking digraphs, we need to furthermore abandon the implementation of complete adjacency tables and gamma functions.

11.4 The Sparse Implementation of Big Outranking Digraphs

The idea is to first decompose the complete outranking relation into an ordered collection of equivalent quantile performance classes. Let us consider for this illustration a random performance tableau with 100 decision alternatives evaluated on 7 criteria.

```
>>> from cRandPerfTabs import cRandomPerformanceTableau
>>> t = cRandomPerformanceTableau(numberOfActions=100,\
...                               numberOfCriteria=7,seed=100)
```

The `cSparseIntegerOutrankingDigraphs` module provides the `SparseIntegerOutrankingDigraph` class for sorting the 100 decision alternatives into overlapping quartile classes and rank all 100 with respect to the average quantile limits.

Listing 11.3 Constructing the sparse integer outranking digraph

```
>>> from cSparseIntegerOutrankingDigraphs import\
...                        SparseIntegerOutrankingDigraph
>>> sg = SparseIntegerOutrankingDigraph(t,quantiles=4)
>>> sg
  *----- Object instance description --------------*
   Instance class     : SparseIntegerOutrankingDigraph
   Instance name      : cRandomperftab_mp
   Actions            : 100
   Criteria           : 7
   Sorting by         : 4-Tiling
   Ordering strategy  : average
   Ranking rule       : Copeland
   Components         : 6
   Minimal order      : 1
```

```
  Maximal order       : 35
  Average order       : 16.7
  fill rate           : 24.970%
 *----  Constructor run times (in sec.) ----
  Nbr of threads      : 1
  Total time          : 0.08212
  QuantilesSorting    : 0.01481
  Preordering         : 0.00022
  Decomposing         : 0.06707
  Ordering            : 0.00000
  Attributes         : ['runTimes', 'name', 'actions',
      'criteria', 'evaluation', 'order', 'dimension',
      'sortingParameters', 'nbrOfCPUs',
      'valuationdomain', 'profiles', 'categories',
      'sorting', 'minimalComponentSize',
      'decomposition', 'nbrComponents', 'nd',
      'components', 'fillRate',
      'maximalComponentSize', 'componentRankingRule',
      'boostedRanking']
```

We obtain in this example here a decomposition into 6 linearly ordered components with a maximal component size of 35 for component `c3`.

Listing 11.4 The 6 components of a sparse outranking digraph

```
>>> sg.showDecomposition()
 *--- quantiles decomposition in decreasing order---*
  c1. ]0.75-1.00]  : [3,22,24,34,41,44,50,53,56,62,93]
  c2. ]0.50-1.00]  : [7,29,43,58,63,81,96]
  c3. ]0.50-0.75]  : [1,2,5,8,10,11,20,21,25,28,30,33,
                      35,36,45,48,57,59,61,65,66,68,70,
                      71,73,76,82,85,89,90,91,92,94,95,97]
  c4. ]0.25-0.75]  : [17,19,26,27,40,46,55,64,69,87,98,100]
  c5. ]0.25-0.50]  : [4,6,9,12,13,14,15,16,18,23,31,32,
                      37,38,39,42,47,49,51,52,54,60,67,72,
                      74,75,77,78,80,86,88,99]
  c6. ]<-0.25]     : [79,83,84]
```

A restricted outranking relation is stored for each component with more than one alternative. The `showRelationMap()` prints out below the relation map of the sparse digraph `sg` for the 75 first-ranked alternatives (see Fig. 11.1 on the next page):

```
>>> sg.showRelationMap(toIndex=75)
```

With a fill rate of 25%, the memory occupation of this sparse outranking digraph `sg` instance takes now only 769 kB, compared to the 1.7 MB required by a corresponding standard `BipolarOutrankingDigraph` instance (see Listing 11.3 on the preceding page).

```
>>> print('%.0fkB' % (total_size(sg)/1024) )
    769kB
```

For sparse outranking digraphs, the adjacency table is implemented as a dynamic `relation()` function instead of a double dictionary.

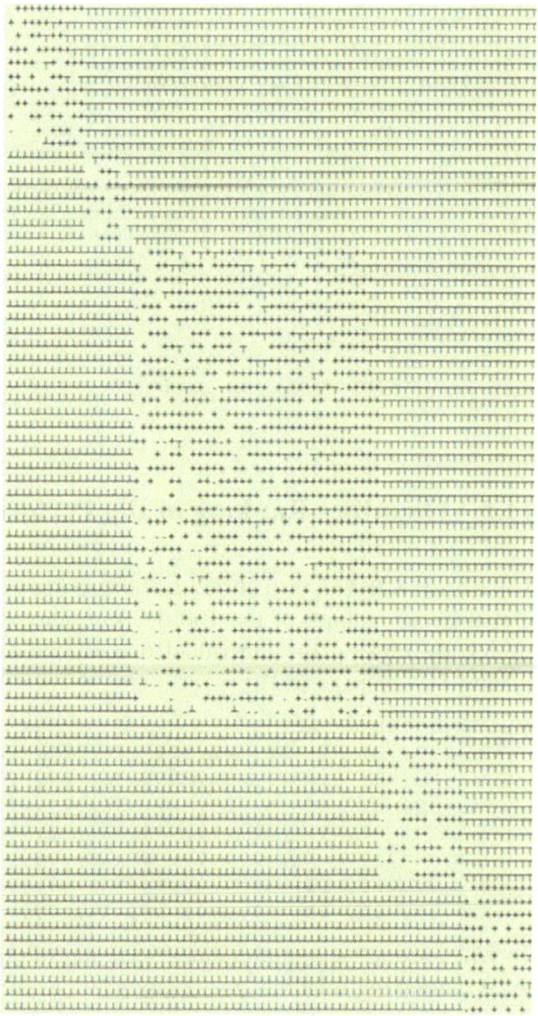

Fig. 11.1 *Sparse quartiles-sorting decomposed outranking relation* (extract). Legend: *outranking* for certain ($\top$); *outranked* for certain ($\bot$); more or less *outranking* (+); more or less *outranked* (−); *indeterminate* ( )

Listing 11.5 The `relation()` function of a sparse outranking digraph

```
def relation(self, int x, int y):
    """
    *Parameters*:
        * x (int action key),
        * y (int action key).
    Dynamic construction of the global outranking
    characteristic function *r(x S y)*.
    """
    cdef int Min, Med, Max, rx, ry
    Min = self.valuationdomain['min']
    Med = self.valuationdomain['med']
    Max = self.valuationdomain['max']
    if x == y:
        return Med
    cx = self.actions[x]['component']
```

```
        cy = self.actions[y]['component']
        #print(self.components)
        rx = self.components[cx]['rank']
        ry = self.components[cy]['rank']
        if rx == ry:
            try:
                rxpg = self.components[cx]['subGraph'].relation
                return rxpg[x][y]
            except AttributeError:
                componentRanking = self.components[cx]['componentRanking']
                if componentRanking.index(x) < componentRanking.index(x):
                    return Max
                else:
                    return Min
        elif rx > ry:
            return Min
        else:
            return Max
```

11.5 Quantiles Ranking of Big Performance Tableaux

The `boostedRanking` attribute of the sparse outranking digraph `sg` contains the global linear ranking of the 100 decision alternatives. This ranking is computed by locally ranking the individual 6 components with the COPELAND rule (see Listing 11.4 on page 142).

```
>>> sg.boostedRanking
  [22, 53, 3, 34, 56, 62, 24, 44, 50, 93, 41, 63, 29, 58,
   96, 7, 43, 81, 91, 35, 25, 76, 66, 65, 8, 10, 1, 11, 61,
   30, 48, 45, 68, 5, 89, 57, 59, 85, 82, 73, 33, 94, 70,
   97, 20, 92, 71, 90, 95, 21, 28, 2, 36, 87, 40, 98, 46, 55,
   100, 64, 17, 26, 27, 19, 69, 6, 38, 4, 37, 60, 31, 77, 78,
   47, 99, 18, 12, 80, 54, 88, 39, 9, 72, 86, 42, 13, 23, 67,
   52, 15, 32, 49, 51, 74, 16, 14, 75, 79, 83, 84]
```

When actually computing linear rankings of the complete set of alternatives, the local outranking relations per component are of no practical usage, and we may furthermore reduce the memory occupation of the resulting digraph by:

1. refining the ordering of the quantile classes by taking into account how well an alternative is outranking the lower limit of its quantile class, respectively, the upper limit of its quantile class is *not* outranking the alternative;
2. dropping the local outranking digraphs and keeping for each component only a locally ranked list of alternatives.

We provide therefore the `cQuantilesRankingDigraph` class from the `cSparseIntegerOutrankingDigraphs` module.

Listing 11.6 Ranking the sparse integer outranking digraph

```
>>> from cSparseIntegerOutrankingDigraphs import\
...             cQuantilesRankingDigraph
>>> qr = cQuantilesRankingDigraph(t,4)
```

```
4 >>> qr
5  *----- Object instance description --------------*
6   Instance class      : cQuantilesRankingDigraph
7   Instance name       : cRandomperftab_mp
8   Actions             : 100
9   Criteria            : 7
10  Sorting by          : 4-Tiling
11  Ordering strategey  : optimal
12  Ranking rule        : Copeland
13  Components          : 47
14  Minimal order       : 1
15  Maximal order       : 10
16  Average order       : 2.1
17  fill rate           : 2.566%
18 *----  Constructor run times (in sec.) ----*
19  Nbr of threads      : 1
20  Total time          : 0.03702
21  QuantilesSorting    : 0.01785
22  Preordering         : 0.00022
23  Decomposing         : 0.01892
24  Attributes        : ['runTimes', 'name',
25  'actions', 'order', 'dimension', 'sortingParameters',
26  'nbrOfCPUs','valuationdomain', 'profiles', 'categories',
27  'sorting', 'minimalComponentSize','decomposition',
28  'nbrComponents', 'nd', 'components', 'fillRate',
29  'maximalComponentSize', 'componentRankingRule',
        'boostedRanking']
```

With this *optimised* quantile ordering strategy, we obtain now 47 performance equivalence classes (see Listing 11.6 on the facing page Line 13).

Listing 11.7 The ordered components of the sparse outranking digraph

```
>>> qr.components
    OrderedDict([
    ('c01', {'rank': 1,
             'lowQtileLimit': ']0.75',
             'highQtileLimit': '1.00]',
             'componentRanking': [53]}),
    ('c02', {'rank': 2,
             'lowQtileLimit': ']0.75',
             'highQtileLimit': '1.00]',
             'componentRanking': [3, 23, 63, 50]}),
    ('c03', {'rank': 3,
             'lowQtileLimit': ']0.75',
             'highQtileLimit': '1.00]',
             'componentRanking': [34, 44, 56, 24, 93, 41]}),
    ...
    ...
    ...
    ('c45', {'rank': 45,
             'lowQtileLimit': ']0.25',
             'highQtileLimit': '0.50]',
             'componentRanking': [49]}),
```

```
    ('c46', {'rank': 46,
             'lowQtileLimit': ']0.25',
             'highQtileLimit': '0.50]',
             'componentRanking': [52, 16, 86]}),
    ('c47', {'rank': 47,
             'lowQtileLimit': ']<',
             'highQtileLimit': '0.25]',
             'componentRanking': [79, 83, 84]})])
>>> print('%.0fkB' % (total_size(qr)/1024) )
    208kB
```

We observe in the last Line of Listing 11.7 on the previous page an even more considerably less voluminous memory occupation 208 kB, compared to the 769 kB of the `SparseIntegerOutrankingDigraph` instance.

It is opportune, however, to measure the loss of quality of the resulting COPELAND ranking when working with sparse outranking digraphs.

Listing 11.8 Measuring the loss of quality with respect to the standard outranking digraph

```
1  >>> from cIntegerOutrankingDigraphs import\
2  ...                         IntegerBipolarOutrankingDigraph
3  >>> ig = IntegerBipolarOutrankingDigraph(t)
4  >>> print('Complete outranking : %+.3f'\
5  ...    % (ig.computeOrderCorrelation(\
6  ...              ig.computeCopelandOrder())['correlation']))
7      Complete outranking : +0.747
8
9  >>> print('Sparse 4-tiling : %+.3f'\
10 ...     % (ig.computeOrderCorrelation(\
11 ...         list(reversed(sg.boostedRanking)))['correlation']))
12     Sparse 4-tiling           : +0.717
13
14 >>> print('Optimzed sparse 4-tiling: %+.3f'\
15 ...   % (ig.computeOrderCorrelation(\
16 ...     list(reversed(qr.boostedRanking)))\
17 ...     ['correlation']))
18     Optimzed sparse 4-tiling: +0.705
```

The best ranking correlation with the pairwise outranking situations (+0.747) is naturally given when we apply the COPELAND rule to the standard outranking digraph (see Listing 10.8 on page 132 Line 7). When applying the same rule to the sparse 4-tiled outranking digraph, one gets a correlation of +0.717 (Line 12), and when applying the COPELAND rule to the optimised sparse 4-tiled digraph, we still obtain a correlation of +0.705 (Line 18). These optimistic results actually depend on the number of quantiles we use as well as on the given model of random performance tableau.

In case of `Random3ObjectivesPerformanceTableau` instances we would get in a similar setting an even better standard outranking correlation of +0.86, a sparse 4-tiling correlation of +0.82, and an optimised sparse 4-tiling correlation of +0.81.

Fig. 11.2 *HPC-UL Ranking Performance Records (Spring 2018).* On the big memory equipped Gaia-183 node with 64 cores we were able to rank one million, respectively, 6 million decision alternatives in about 2 minutes, respectively, 41 minutes

$\gtrapprox^q$ outranking relation order	size	q	fill rate	nbr. cores	run time
5 000	25×10^6	4	0.005%	28	0.5"
10 000	1×10^8	4	0.001%	28	1"
100 000	1×10^{10}	5	0.002%	28	10"
1 000 000	1×10^{12}	6	0.001%	64	2'
3 000 000	9×10^{12}	15	0.004%	64	13'
6 000 000	36×10^{12}	15	0.002%	64	41'

11.6 HPC Quantiles Ranking Records

Following from the separability property of the q-tiles sorting of each action into a q-tiles class, the q-sorting algorithm may be safely split into as much threads as are multiple processing cores available for working in parallel (see Proposition 9.1). Furthermore, the ranking procedure being local to each diagonal component, these procedures may as well be safely processed in parallel threads on each component restricted outranking digraph.

Using the HPC platform of the University of Luxembourg—https://hpc-docs.uni.lu/ (Varrette et al. 2014)—the run times shown in Fig. 11.2 for very big ranking problems could be achieved both on:

- *Iris -skylake* nodes with 28 cores:
  (see https://hpc-docs.uni.lu/systems/iris), and
- the 3TB -bigmem *Gaia-183* node with 64 cores
  (see https://hpc.uni.lu/systems/gaia),

by running the cythonized Python modules in an Intel compiled virtual Python 3.6.5 environment [GCC Intel(R) 17.0.1—enable-optimisations c++ gcc 6.3 mode] on Debian 8 Linux.

The next Part III presents three real case studies of building a best choice recommendation—Chap. 12, ranking incommensurable multiple-criteria performance records—Chap. 13, and rating student enrolment qualities—Chap. 14. Chapter 15 eventually list some exercises for home work and exam questions.

References

Bisdorff R (2016) Computing linear rankings from trillions of pairwise outranking situations. In: Busa-Fekete R, Hüllermeier E, Mousseau V, Pfannschmidt K (eds) From multiple criteria decision aid to preference learning , DAP'2016, University of Paderborn (Germany), pp 1–16. http://hdl.handle.net/10993/28613

Varrette S, Bouvry P, Cartiaux H, Georgatos F (2014) Management of an academic HPC cluster: the UL experience. In: Intl. conf. on high performance computing & simulation (HPCS 2014), Bologna (Italy). IEEE, New York, pp 959–967

Part III
Evaluation and Decision Case Studies

The third part with four chapters proposes decision making case studies illustrating different performance evaluation models and decision problems. Chapter 12 concerns building a best choice recommendation for helping a young student to select a foreign language study program. Chapter 13, inspired by the THE World University Rankings, presents and discusses the mutiple incommensurable criteria ranking of Academic Computer Science Departments. Inspired by the 2004 data published by DER SPIEGEL magazine, Chap. 14 showcases a multiple-criteria rating problem about the student enrolment quality of German universities. Chapter 15 lists finally some exercises for students taking a course on Algorithmic Decision Theory.

Chapter 12
Alice's Best Choice: A Selection Case Study

Contents

Abstract The chapter presents a case study concerning the building of a best choice recommendation for Alice, a German student who wants some advice concerning the choice of her future University studies. We present Alice's performance tableau—potential foreign language study programs, her decision objectives, performance criteria and performance evaluations—and build a best choice recommendation for her. A thorough robustness analysis confirms a very best choice.

12.1 The Decision Problem

The chapter content is inspired by a multiple criteria decision analysis case study (Düsch 2001).

Alice will probably receive her "Abitur" with satisfactory and/or good marks and wants to start her further studies thereafter (Fig. 12.1 on the following page). She would not mind staying in Köln, yet is ready to move elsewhere if opportune. The length of the higher studies do concern her, as she wants to earn her life as soon as possible. Her parents, however, agree to financially support her study fees as well as her living costs during her studies.

Alice has already identified 10 potential study programs.

In Table 12.1 on the next page we notice that Alice considers three *Graduate Interpreter* studies (8 or 9 Semesters), respectively, in Köln, in Saarbrücken or in Heidelberg; and five *Qualified translator* studies (8 or 9 Semesters), respectively, in

R. Bisdorff, *Algorithmic Decision Making with Python Resources*,
International Series in Operations Research & Management Science 324,
https://doi.org/10.1007/978-3-030-90928-4_12

Fig. 12.1 Alice D., 19 years old German student finishing her secondary studies in Köln (Germany), desires to undertake foreign languages studies

Table 12.1 The potential study programs

ID	Diploma	Institution	City
T-UD	Qualified translator (T)	University (UD)	Düsseldorf
T-FHK	Qualified translator (T)	Higher Technical School (FHK)	Köln
T-FHM	Qualified translator (T)	Higher Technical School (FHM)	München
I-FHK	Graduate interpreter (I)	Higher Technical School (FHK)	Köln
T-USB	Qualified translator (T)	University (USB)	Saarbrücken
I-USB	Graduate interpreter (I)	University (USB)	Saarbrücken
T-UHB	Qualified translator (T)	University (UHB)	Heidelberg
I-UHB	Graduate interpreter (I)	University (UHB)	Heidelberg
S-HKK	Specialised secretary (S)	Chamber of Commerce (HKK)	Köln
C-HKK	Foreign correspondent (C)	Chamber of Commerce (HKK)	Köln

Köln, in Düsseldorf, in Saarbrücken, in Heidelberg or in Munich. She also considers two short (4 Semesters) study programs at the Chamber of Commerce in Köln.

Four **decision objectives** of more or less equal importance are guiding Alice's choice:

1. *maximise* the attractiveness of the study place (GEO),
2. *maximise* the attractiveness of her further studies (LEA),
3. *minimise* her financial dependency on her parents (FIN),
4. *maximise* her professional perspectives (PRA).

The decision consequences Alice wishes to take into account for evaluating the potential study programs with respect to each of the four objectives are modelled by the following *coherent family of criteria* (Roy 1991; Roy and Bouyssou 1993). Such a family of performance criteria verifies:

1. *Exhaustiveness*: No argument acceptable to Alice can be put forward to justify a preference in favour of study program *x* versus program *y* when *x* and *y* have the same performance level on each of the performance criteria;

Table 12.2 Alice's family of performance criteria

ID	Name	Comment	Objective	Weight
DH	Proximity	Distance in km to her home (min)	GEO	3
BC	Big City	Number of inhabitants (max)	GEO	3
AS	Studies	Attractiveness of the studies (max)	LEA	6
SF	Fees	Annual study fees (min)	FIN	2
LC	Living	Monthly living costs (min)	FIN	2
SL	Length	Length of the studies (min)	FIN	2
AP	Profession	Attractiveness of the profession (max)	PRA	2
AI	Income	Annual income after studying (max)	PRA	2
PR	Prestige	Occupational prestige (max)	PRA	2

2. *Cohesiveness*: Alice recognises that program x must be preferred to program y whenever the performance level of x is significantly better than that of y on one of the criteria of positive weight, performance levels of x and y being the same on each of the other criteria;
3. *Non-redundancy*: One of the above requirements is violated if one of the performance criteria is left out from the family.

Within each decision objective, the performance criteria are considered to be *equi-significant*. Hence, the four decision objectives get a same *importance weight* of 6 (see Table 12.2 Column 5).

12.2 The Performance Tableau

The actual evaluations of Alice's potential study programs are stored in a file named `AliceChoice.py` of `PerformanceTableau` format.[1]

Listing 12.1 Alice's performance tableau

```
>>> from perfTabs import PerformanceTableau
>>> t = PerformanceTableau('AliceChoice')
>>> t.showObjectives()
  *------ decision objectives -------"
    GEO: Geographical aspect
     DH Distance to parent's home 3
     BC Number of inhabitants      3
     Total weight: 6 (2 criteria)
    LEA: Learning aspect
     AS Attractiveness of the study program 6
     Total weight: 6.00 (1 criteria)
```

[1] The performance tableau `AliceChoice.py` may be found in the `examples` directory of the DIGRAPH3 resources (Bisdorff 2021).

AliceChoice: Family of Criteria

#	Identifyer	Name	Comment	Weight	Scale			Thresholds (ax + b)		
					direction	min	max	indifference	preference	veto
1	AI	Annual professional income after studying	Professional aspect measured in x / 1000 Euros	2.00	max	0.00	50.00	0.00x + 0.00	0.00x + 1.00	
2	AP	Attractiveness of the profession	Professional aspect subjectively measured on a three-level scale: 0 (weak), 1 (fair), 2 (good)	2.00	max	0.00	2.00	0.00x + 0.00	0.00x + 1.00	
3	AS	Attractiveness of the study program	Learning aspect subjectively measured from 0 (weak) to 10 (excellent)	6.00	max	0.00	10.00	0.00x + 0.00	0.00x + 1.00	0.00x + 7.00
4	BC	Number of inhabitants	Geographical aspect: measured in x / 1000	3.00	max	0.00	2000.00	0.01x + 0.00	0.05x + 0.00	
5	DH	Distance to parent's home	Geographical aspect measured in km	3.00	min	0.00	1000.00	0.00x + 0.00	0.00x + 10.00	
6	LC	Monthly living costs	Financial aspect measured in Euros	2.00	min	0.00	1000.00	0.00x + 0.00	0.00x + 100.00	
7	OP	Occupational Prestige	Professional aspect measured in SIOPS points	2.00	max	0.00	100.00	0.00x + 0.00	0.00x + 10.00	
8	SF	Annual registration fees	Financial aspect measured in Euros	2.00	min	400.00	4000.00	0.00x + 0.00	0.00x + 100.00	
9	SL	study time	Financial aspect measured in number of semesters	2.00	min	0.00	10.00	0.00x + 0.00	0.00x + 0.50	

Fig. 12.2 Alice's performance criteria

```
  FIN: Financial aspect
   SF Annual registration fees 2
   LC Monthly living costs     2
   SL Study time               2
   Total weight: 6.00 (3 criteria)
  PRA: Professional aspect
   AP Attractiveness of the profession        2
   AI Annual professional income after studying 2
   OP Occupational Prestige                   2
   Total weight: 6.00 (3 criteria)
```

Details of the performance criteria may be consulted in the browser view below.

```
>>> t.showHTMLCriteria()
```

It is worthwhile noticing in Fig. 12.2 that, on her subjective attractiveness scale of the study programs (criterion AS), Alice considers a performance differences of 7 points to be *considerable* and triggering, the case given, an outranking polarisation (Bisdorff 2013). Notice also the proportional *indifference* (1%) and *preference* (5%) discrimination thresholds shown on criterion BC (number of inhabitants).

Alice is subjectively evaluating the *Attractiveness* of the studies (criterion AS) on an ordinal scale from *weak* (0) to *excellent* (10). Similarly, she is subjectively evaluating the *Attractiveness* of the respective professions (criterion AP) on a three level ordinal scale from *weak* (0), *fair* (1) to *good* (2). Considering the *Occupational Prestige* (criterion OP), she looked up the SIOPS.[2] All the other evaluation data she found on the internet.

[2] Standard International Occupational Prestige Scale (Ganzeboom and Treiman 1996).

Heatmap of Performance Tableau 'AliceChoice'

criteria	AS	AP	SF	OP	AI	DH	LC	BC	SL
weights	+6.00	+2.00	+2.00	+2.00	+2.00	+3.00	+2.00	+3.00	+2.00
tau(*)	+0.71	+0.64	+0.36	+0.36	+0.24	+0.03	-0.04	-0.07	-0.24
I-FHK	8	2	-400	62	35	0	0	1015	-8
I-USB	8	2	-400	62	45	-269	-1000	196	-9
T-FHK	5	1	-400	62	35	0	0	1015	-8
I-UHB	8	2	-400	62	45	-275	-1000	140	-9
T-UD	5	1	-400	62	45	-41	-1000	567	-9
T-USB	5	1	-400	62	45	-260	-1000	196	-9
T-FHM	4	1	-400	62	35	-631	-1000	1241	-8
T-UHB	5	1	-400	62	45	-275	-1000	140	-9
C-HKK	2	0	-4000	44	30	0	0	1015	-4
S-HKK	1	0	-4000	44	30	0	0	1015	-4

Color legend:

quantile	20.00%	40.00%	60.00%	80.00%	100.00%

(*) tau: *Ordinal (Kendall) correlation between marginal criterion and global ranking relation*
Ranking rule: **NetFlows**
Ordinal (Kendall) correlation between global ranking and global outranking relation: **+0.692**

Fig. 12.3 Heatmap of Alice's performance tableau

In the *heatmap view* shown in Fig. 12.3, we may now consult Alice's performance evaluations.

```
>>> t.showHTMLPerformanceHeatmap(\
...         colorLevels=5,Correlations=True,ndigits=0)
```

Her ten potential study programs are ordered with the NETFLOWS ranking rule applied to the corresponding bipolar-valued outranking digraph (see Sect. 8.3). *Graduate interpreter* studies in Köln (`I-FHK`) or Saarbrücken (`I-USB`), followed by *Qualified Translator* studies in Köln (`T-FHK`) appear to be Alice's most preferred alternatives. The least attractive for her appear to be studies at the Chamber of Commerce of Köln (`C-HKK`, `S-HKK`). Notice by the way that evaluations on performance criteria to be *minimised*, like *Distance to Home* (criterion `DH`) or *Study time* (criterion `SL`), are registered as *negative* values, so that smaller measures are, in this case, preferred to larger ones.

It is finally interesting to observe in Fig. 12.3 (third row) that the *most significant* performance criteria, appear to be for Alice, on the one side, the *Attractiveness* of the study program (criterion `AS`, tau = +0.72) followed by the *Attractiveness* of the future profession (criterion `AP`, tau = +0.62). On the other side, *Study times* (criterion `SL`, tau = −0.24), *Big city* (criterion `BC`, tau = −0.07) as well as *Monthly living costs* (criterion `LC`, tau = −0.04) appear to be *not so significant* (see Chap. 16).

12.3 Building a Best Choice Recommendation

Let us now have a look at the resulting pairwise outranking situations.

Listing 12.2 Computing Alice's outranking digraph

```
1 >>> from outrankingDigraphs import BipolarOutrankingDigraph
2 >>> dg = BipolarOutrankingDigraph(t)
3 >>> dg
4   *------- Object instance description ------*
5    Instance class       : BipolarOutrankingDigraph
6    Instance name        : rel_AliceChoice
7    Actions              : 10
8    Criteria             : 9
9    Size                 : 67
10   Determinateness (%)  : 73.91
11   Valuation domain     : [-1.00;1.00]
12 >>> dg.computeSymmetryDegree(Comments=True)
13  Symmetry degree of graph <rel_AliceChoice> : 0.49
```

From Alice's performance tableau we obtain in the digraph `dg` 67 positively validated pairwise outranking situations, supported by a 74% majority of criteria significance (see Lines 9–10 in Listing 12.2).

Due to the poorly discriminating performance evaluations, nearly half of these outranking situations (see Line 12) are *symmetric* and reveal actually *more or less indifference* situations between the potential study programs. This is well illustrated in the *relation map* of the outranking digraph shown in Fig. 12.4 on the next page.

```
>>> dg.showHTMLRelationMap(\
...               tableTitle='Outranking relation map',\
...               rankingRule='Copeland')
```

We have mentioned that Alice considers a performance difference of 7 points on the *Attractiveness of studies* criterion `AS` to be considerable which triggers, the case given, a potential polarisation of the outranking characteristics. In Fig. 12.4 on the facing page, these polarisations appear in the last column and last row. The `showPolarisations()` method is useful for inspecting the occurrence of outranking polarisations.

Listing 12.3 Inspecting polarised outranking situations

```
>>> dg.showPolarisations()
  *----  Negative polarisations ----*
   number of negative polarisations : 3
     1: r(S-HKK >= I-FHK) = -0.17
       criterion: AS
       Considerable performance difference : -7.00
       Veto discrimination threshold        : -7.00
       Polarisation: r(S-HKK >= I-FHK) = -0.17 ==> -1.00
     2: r(S-HKK >= I-USB) = -0.17
       criterion: AS
       Considerable performance difference : -7.00
```

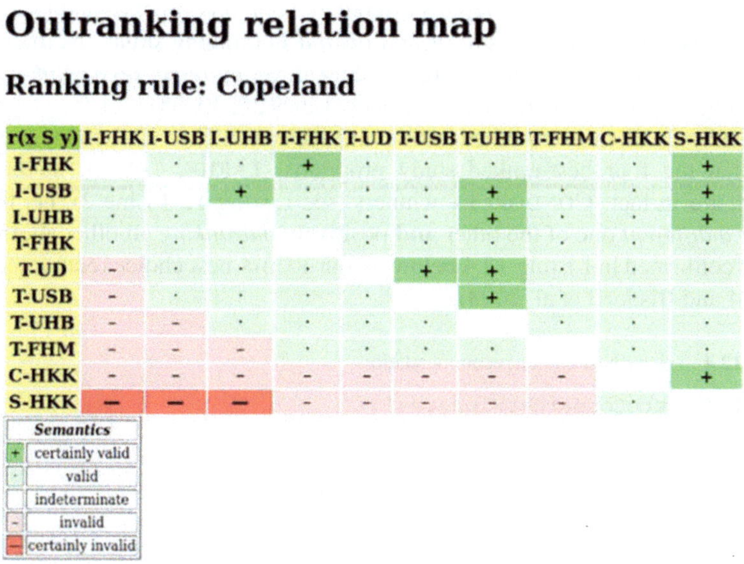

Fig. 12.4 COPELAND ranked outranking relation map

```
      Veto discrimination threshold          : -7.00
      Polarisation: r(S-HKK >= I-USB) = -0.17 ==> -1.00
     3: r(S-HKK >= I-UHB) = -0.17
      criterion: AS
      Considerable performance difference : -7.00
      Veto discrimination threshold          : -7.00
      Polarisation: r(S-HKK >= I-UHB) = -0.17 ==> -1.00
   *----  Positive polarisations ----*
    number of positive polarisations: 3
     1: r(I-FHK >= S-HKK) = 0.83
      criterion: AS
      Considerable performance difference : 7.00
      Counter-veto threshold                 : 7.00
      Polarisation: r(I-FHK >= S-HKK) = 0.83 ==> +1.00
     2: r(I-USB >= S-HKK) = 0.17
      criterion: AS
      Considerable performance difference : 7.00
      Counter-veto threshold                 : 7.00
      Polarisation: r(I-USB >= S-HKK) = 0.17 ==> +1.00
     3: r(I-UHB >= S-HKK) = 0.17
      criterion: AS
      Considerable performance difference : 7.00
      Counter-veto threshold                 : 7.00
      Polarisation: r(I-UHB >= S-HKK) = 0.17 ==> +1.00
```

In Listing 12.3 on the preceding page, we see that *considerable* performance differences concerning the *Attractiveness of the studies* (`AS` criterion) are indeed observed between the *Specialised Secretary* study program offered in Köln and the

Graduate Interpreter study programs offered in Köln, Saarbrücken and Heidelberg. They polarise, hence, three *more or less invalid* outranking situations to *certainly invalid* (Lines 8, 13, 18) and corresponding three *more or less valid* converse outranking situations to *certainly valid* ones (Lines 25, 30, 35).

One may finally notice in the relation map, shown in Fig. 12.4 on the preceding page, that the four best-ranked study programs, `I-FHK`, `I-USB`, `I-UHB`, and `T-FHK`, are in fact CONDORCET winners (see Listing 12.4 Line 2), i.e. they are all four *indifferent* one of the other **and** positively *outranking* all other alternatives, a result confirmed in Listing 12.4 below by our RUBIS best choice recommendation (Chap. 4 and Bisdorff et al. 2008).

Listing 12.4 Alice's best choice recommendation

```
1  >>> dg.computeCondorcetWinners()
2    ['I-FHK', 'I-UHB', 'I-USB', 'T-FHK']
3  >>> dg.showBestChoiceRecommendation()
4    Best choice recommendation(s) (BCR)
5      (in decreasing order of determinateness)
6     Credibility domain: [-1.00,1.00]
7     === >> potential first choice(s)
8     choice                  : ['I-FHK','I-UHB','I-USB','T-FHK']
9       independence          : 0.17
10      dominance             : 0.08
11      absorbency            : -0.83
12      covering (%)          : 62.50
13      determinateness (%)   : 68.75
14      most credible action(s) = {'I-FHK': 0.75,'T-FHK': 0.17,
15                                 'I-USB': 0.17,'I-UHB': 0.17}
16    === >> potential last choice(s)
17    choice                  : ['C-HKK', 'S-HKK']
18      independence          : 0.50
19      dominance             : -0.83
20      absorbency            : 0.17
21      covered (%)           : 100.00
22      determinateness (%)   : 58.33
23      most credible action(s) = {'S-HKK': 0.17,'C-HKK': 0.17}
```

Most credible best choice among the four best-ranked study programs eventually becomes the *Graduate Interpreter* study program at the Technical High School in Köln (see Line 14) supported by a $(0.75 + 1)/2.0 = 87.5\%(18/24)$ majority of global criteria significance.[3]

In the relation map, shown in Fig. 12.4 on the previous page, we see in the left lower corner that the *asymmetric part* of the outranking relation, i.e. the corresponding *strict* outranking relation, is actually *transitive* (see Line 3 in Listing 12.5 on the facing page below). Hence, a *graphviz* drawing of its *skeleton*, oriented by the previous *first*, respectively, *last* choice, may well illustrate our

[3] See Sect. 4.5 on computing best choice recommendations and Sect. 17.6 on solving kernel equation systems.

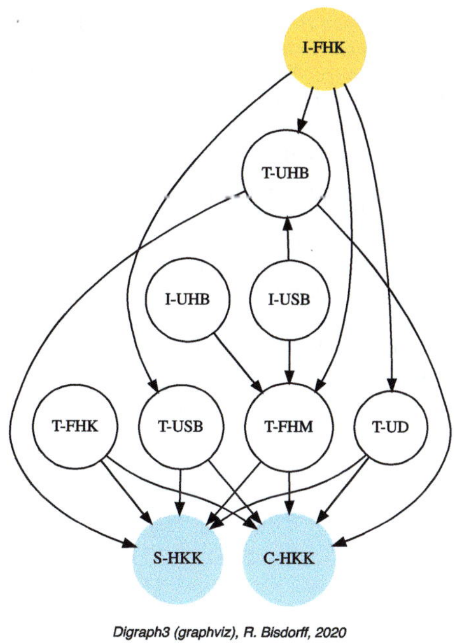

Fig. 12.5 *Alice's best choice recommendation.* We notice that the *Graduate Interpreter* studies come first, followed by the *Qualified Translator* studies. Last come the *Chamber of Commerce*'s specialised studies. This confirms again the high significance that Alice attaches to the *attractiveness* of her further studies and of her future profession (see criteria `AS` and `AP` in Fig. 12.3 on page 155)

best choice recommendation. The `Reverse=True` flag in Line 4 eliminates the transitivity induced arcs from digraph `dgcd` (Fig. 12.5).

Listing 12.5 Alice's strict best choice recommendation

```
1 >>> dgcd = ~(-dg)
2 >>> dgcd.isTransitive()
3     True
4 >>> dgcd.closeTransitive(Reverse=True,InSite=True)
5 >>> dgcd.exportGraphViz('aliceBestChoice',\
6 ...                     firstChoice=['I-FHK'],\
7 ...                     lastChoice=['S-HKK','C-HKK'])
8   *---- exporting a dot file for GraphViz tools ---------*
9    Exporting to aliceBestChoice.dot
10   dot -Grankdir=BT -Tpng aliceBestChoice.dot -o
         aliceBestChoice.png
```

Let us now, for instance, check the pairwise outranking situations observed between the first and second-ranked alternative, i.e. *Graduate Interpreter* studies in Köln versus *Graduate Interpreter* studies in Saarbrücken (see `I-FHK` and `I-USB` in Fig. 12.6 on the following page).

```
>>> dg.showHTMLPairwiseOutrankings('I-FHK','I-USB')
```

In a similar way, one may finally compute a *weak ranking* of all the potential study programs with the help of the `RankingByChoosingDigraph` class from the `transitiveDigraphs` module that computes a bipolar ranking by conjointly *best-choosing* and *last-rejecting* (Bisdorff 1999).

Pairwise Comparison

Comparing actions : (I-FHK,I-USB)

crit.	wght.	g(x)	g(y)	diff	ind	pref	concord	v	polarisation
AI	2.00	35.00	+45.00	-10	0.00	1.00	-2.00		
AP	2.00	2.00	+2.00	0	0.00	1.00	+2.00		
AS	6.00	8.00	+8.00	0	0.00	1.00	+6.00		
BC	3.00	1015.00	+196.00	819	10.15	50.75	+3.00		
DH	3.00	0.00	-269.00	269	0.00	10.00	+3.00		
LC	2.00	0.00	-1000.00	1000	0.00	100.00	+2.00		
OP	2.00	62.00	+62.00	0	0.00	10.00	+2.00		
SF	2.00	-400.00	-400.00	0	0.00	100.00	+2.00		
SL	2.00	-8.00	-9.00	1	0.00	0.50	+2.00		

Valuation in range: -24.00 to +24.00; global concordance: +20.00

Pairwise Comparison

Comparing actions : (I-USB,I-FHK)

crit.	wght.	g(x)	g(y)	diff	ind	pref	concord	v	polarisation
AI	2.00	45.00	+35.00	10	0.00	1.00	+2.00		
AP	2.00	2.00	+2.00	0	0.00	1.00	+2.00		
AS	6.00	8.00	+8.00	0	0.00	1.00	+6.00		
BC	3.00	196.00	+1015.00	-819	10.15	50.75	-3.00		
DH	3.00	-269.00	+0.00	-269	0.00	10.00	-3.00		
LC	2.00	-1000.00	+0.00	-1000	0.00	100.00	-2.00		
OP	2.00	62.00	+62.00	0	0.00	10.00	+2.00		
SF	2.00	-400.00	-400.00	0	0.00	100.00	+2.00		
SL	2.00	-9.00	-8.00	-1	0.00	0.50	-2.00		

Valuation in range: -24.00 to +24.00; global concordance: +4.00

Fig. 12.6 *Comparing the first and second best-ranked study programs*. The Köln alternative is *at least as well evaluated as* the Saarbrücken alternative on all the performance criteria, except the *Annual income* (of significance 2/24). Conversely, the Saarbrücken alternative is clearly *outperformed* from the geographical (0/6) as well as from the financial perspective (2/6)

Listing 12.6 Weakly ranking by bipolar best-choosing and last-rejecting

```
>>> from transitiveDigraphs import\
...                 RankingByChoosingDigraph
>>> rbc = RankingByChoosingDigraph(dg)
>>> rbc.showRankingByChoosing()
    Ranking by Choosing and Rejecting
     1st ranked ['I-FHK']
       2nd ranked ['I-USB']
         3rd ranked ['I-UHB']
           4th ranked ['T-FHK']
             5th ranked ['T-UD']
             5th last ranked ['T-UD']
           4th last ranked ['T-UHB', 'T-USB']
         3rd last ranked ['T-FHM']
       2nd last ranked ['C-HKK']
     1st last ranked ['S-HKK']
```

In Listing 12.6, we find confirmed that the *Interpreter* studies appear all preferred to the *Translator* studies. Furthermore, the *Interpreter* studies in Saarbrücken appear preferred to the same studies in Heidelberg. The Köln alternative is apparently the preferred one of all the *Translater* studies. And, the *Foreign Correspondent* and the *Specialised Secretary* studies appear second-last and last ranked.

Yet, how *robust* are our findings with respect to potential settings of the decision objectives' importance and the performance criteria significance weights?

12.4 Robustness Analysis

Alice considers her four decision objectives as being *more or less* equally important. Here we have, however, allocated *strictly equal* importance weights with *strictly equi-significant* criteria per objective. How robust is the previous best choice recommendation when, now, we would consider the importance of the objectives and, hence, the significance of the respective performance criteria to be *more or less uncertain*?

To answer this question, we consider the respective criteria significance weights w_j for $j = 1, ..., 9$ to be *triangular random variables* in the range 0 to $2w_j$ with mode = w_j. Computing a corresponding 90%-*confident* outranking digraph may be done with the help of the `ConfidentBipolarOutrankingDigraph` class (see Chap. 18).

Listing 12.7 Computing the 90% confident outranking digraph

```
1  >>> from outrankingDigraphs import\
2  ...         ConfidentBipolarOutrankingDigraph
3  >>> cdg = ConfidentBipolarOutrankingDigraph(t,\
4  ...         distribution='triangular',confidence=90.0)
5  >>> cdg
6    *------- Object instance description ------*
7     Instance class         : ConfidentBipolarOutrankingDigraph
8     Instance name          : rel_AliceChoice_CLT
9     Actions                : 10
10    Criteria               : 9
11    Size                   : 44
12    Valuation domain       : [-1.00;1.00]
13    Uncertainty model      : triangular(a=0,b=2w)
14    Likelihood domain      : [-1.0;+1.0]
15    Confidence level       : 90.0%
16    Confident majority     : 14/24 (58.3%)
17    Determinateness (%)    : 68.19
```

Of the original 67 valid outranking situations, 44 outranking situations are retained as being 90%-confident (see Listing 12.7 Line 11). The corresponding 90%-confident *qualified majority* of criteria significance amounts to $14/24 = 58.3\%$ (Line 16).

Concerning now a 90%-confident best choice recommendation, we are lucky.

Listing 12.8 Computing the 90%-confident best choice recommendation

```
>>> cdg.computeCondorcetWinners()
  ['I-FHK']
>>> cdg.showBestChoiceRecommendation()
  ***********************
  Best choice recommendation(s) (BCR)
   (in decreasing order of determinateness)
   Credibility domain: [-1.00,1.00]
  === >> potential best choice(s)
```

```
9   choice                  : ['I-FHK','I-UHB','I-USB',
10                             'T-FHK','T-FHM']
11      independence          : 0.00
12      dominance             : 0.42
13      absorbency            : 0.00
14      covering (%)          : 20.00
15      determinateness (%)   : 61.25
16   - most credible action(s) = { 'I-FHK': 0.75, }
```

The *Graduate Interpreter* studies in Köln remain indeed a 90%-confident CONDORCET winner (see Fig. 12.7 Line 2). Hence, the same study program also remains our 90%-confident most credible best choice supported by a comfortable 18/24(87.5%) majority of the global criteria significance (see Lines 9–10 and 16 in Listing 12.8 on the preceding page on the previous page).

When previously comparing the two best-ranked study programs (see Fig. 12.6 on page 160), we have observed that `I-FHK` actually positively outranks `I-USB` on all four decision objectives. When admitting equi-significant criteria significances

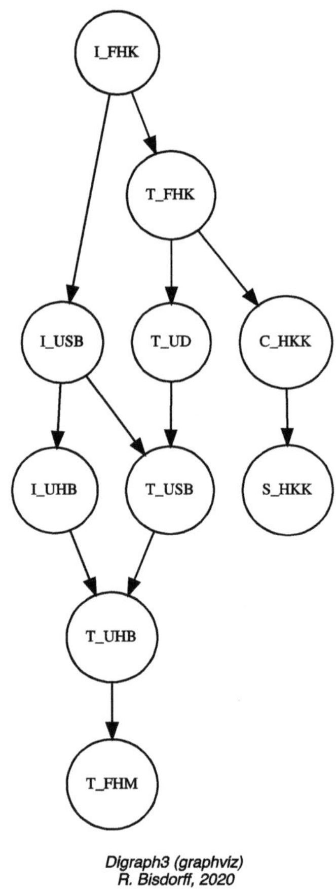

Fig. 12.7 *Unopposed partial ranking of the potential study programs*. Again, when *equally significant* performance criteria are assumed per decision objective, we observe in the figure here that $I - FHK$ remains the stable best choice, *independently* of the actual importance weights that Alice may wish to allocate to her four decision objectives

per objective, this outranking situation is hence valid independently of the importance weights Alice may allocate to each of her decision objectives.

The `UnOpposedBipolarOutrankingDigraph` class constructor from the `outrankingDigraphs` module computes these *unopposed* outranking situations (see Sect. 19.5).

Listing 12.9 Computing the unopposed outranking situations

```
1  >>> from outrankingDigraphs import\
2  ...                        UnOpposedBipolarOutrankingDigraph
3  >>> uop = UnOpposedBipolarOutrankingDigraph(t)
4  >>> uop
5    *------- Object instance description ------*
6     Instance class        : UnOpposedBipolarOutrankingDigraph
7     Instance name         : AliceChoice_unopposed_outrankings
8     Actions               : 10
9     Criteria              : 9
10    Size                  : 28
11    Oppositeness (%)      : 58.21
12    Determinateness (%)   : 62.94
13    Valuation domain      : [-1.00;1.00]
14 >>> uop.isTransitive()
15   True
```

28 out the 67 standard outranking situations remain valid, which leads to an *oppositeness degree* of $(1.0 - 28/67) = 58.21\%$ (see Lines 10–11 in Listing 12.9). Remarkable furthermore is that this unopposed outranking digraph uop is actually *transitive*, i.e. modelling a *partial ranking* of the study programs (Lines 14–15).

The `exportGraphViz()` method of the `TransitiveDigraph` class draws the corresponding partial ranking.

```
>>> from transitiveDigraphs import TransitiveDigraph
>>> TransitiveDigraph.exportGraphViz(uop,\
...           fileName='choiceUnopposed')
  *---- exporting a dot file for GraphViz tools ---------*
   Exporting to choiceUnopposed.dot
   dot -Grankdir=TB -Tpng choiceUnopposed.dot -o
       choiceUnopposed.png
```

In view of her performance tableau shown in Fig. 12.3 on page 155, *Graduate Interpreter* studies at the Technical High School Köln, thus, represent definitely **Alice's very best choice**.

For further reading about the RUBIS *Best Choice* methodology, one may consult in Bisdorff (2015) the study of a *real decision-aiding case* about choosing a best poster in a scientific conference.

References

Bisdorff R (1999) Bipolar ranking from pairwise fuzzy outrankings. Belg J Oper Res Stat Comp Sci 37(4):379–387. http://hdl.handle.net/10993/38738

Bisdorff R (2013) On polarizing outranking relations with large performance differences. J Multi-Crit Dec Anal Wiley 20:3–12. http://hdl.handle.net/10993/245
Bisdorff R (2015) The EURO 2004 best poster award: choosing the best poster in a scientific conference. In: Bisdorff R, Dias L, Meyer P, Mousseau V, Pirlot M (eds) Evaluation and decision models with multiple criteria: case studies. Springer, New York, pp 117–166
Bisdorff R (2021) Documentation of the Digraph3 collection of Python modules for Algorithmic Decision Theory. https://digraph3.readthedocs.io/en/latest/
Bisdorff R, Meyer P, Roubens M (2008) Rubis: a bipolar-valued outranking method for the best choice decision problem. Quart J Oper Res 6(2):143–165. http://hdl.handle.net/10993/23716
Düsch E (2001) Entscheidung für eine fremdsprachenausbildung. In: Eisenführ F, Langer T, Weber M (eds) Fallstudien zu rationalem Entscheiden. Fallstudie A. Springer, New York, pp 1–18
Ganzeboom H, Treiman D (1996) Internationally comparable measures of occupational status for the 1988 international standard classification of occupations. Soc Sci Res 25:201–239
Roy B (1991) The outranking approach and the foundations of electre methods. Theory Decis 31(1):49–73
Roy B, Bouyssou D (1993) Aide Multicritère à la Décision : Méthodes et Cas. Economica, Paris

Chapter 13
The Best Academic Computer Science Depts: A Ranking Case Study

Contents

Abstract In this case study, we are solving with our DIGRAPH3 resources a ranking decision problem based on published data from the *Times Higher Education* (THE) *World University Rankings* 2016 by *Computer Science* (CS) subject. Several hundred academic CS Departments, from all over the world, were ranked that year following an overall numerical score based on the weighted average of five performance criteria: *Teaching* (the learning environment, 30%), *Research* (volume, income and reputation, 30%), *Citations* (research influence, 27.5%), *International outlook* (staff, students, and research, 7.5%), and *Industry income* (innovation, 5%). To illustrate our DIGRAPH3 programming resources, we shall first have a look into the THE multiple-criteria ranking data with short Python scripts. In a second section, we shall relax the commensurability hypothesis of the ranking criteria and show how to similarly rank with multiple incommensurable performance criteria of solely ordinal significance. A third section is finally devoted to introduce quality measures for qualifying ranking results.

13.1 The THE Performance Tableau

For this decision making case study, we use an extract of the published TIMES HIGHER EDUCATION (THE) World University rankings 2016 by Computer Science (CS) subject, concerning the 75 first-ranked academic Institutions.[1] The multiple-

[1] THE World University Rankings 2016 by Computer Science subject.

R. Bisdorff, *Algorithmic Decision Making with Python Resources*,
International Series in Operations Research & Management Science 324,
https://doi.org/10.1007/978-3-030-90928-4_13

criteria performance tableau data, collected from the THE web pages, is stored in a file named `the-cs-2016.py` of `PerformanceTableau` format.[2]

Listing 13.1 Performance tableau of the 75 first-ranked academic Institutions

```
>>> from perfTabs import PerformanceTableau
>>> cspt = PerformanceTableau('the_cs_2016')
>>> cspt
  *------- PerformanceTableau instance description ------*
   Instance class      : PerformanceTableau
   Instance name       : the-cs-2016
   Actions             : 75
   Objectives          : 5
   Criteria            : 5
   NaN proportion (%)  : 0.0
   Attributes          : ['name','description','actions',
                            'objectives','criteria',
                            'weightPreorder','NA','evaluation']
```

Potential decision alternatives, in our case here, are the 75 THE best-ranked CS Departments in 2016, all of them located at world renowned Institutions, like California Institute of Technology, Swiss Federal Institute of Technology Zurich, Technical University München, University of Oxford or the National University of Singapore (see Listing 13.2 below).

Instead of using prefigured DIGRAPH3 `show...()` methods, readily available for inspecting `PerformanceTableau` instances, Listing 13.2 illustrates how to write small Python scripts for printing out its content.

Listing 13.2 Printing the CS Departments

```
>>> for x in cspt.actions:
...     print('%s:\t%s (%s)' %\
...        (x,cspt.actions[x]['name'],cspt.actions[x]['comment']) )
  albt: University of Alberta (CA)
  anu:  Australian National University (AU)
  ariz: Arizona State University (US)
  bju:  Beijing University (CN)
  bro:  Brown University (US)
  calt: California Institute of Technology (US)
  cbu:  Columbia University (US)
  chku: Chinese University of Hong Kong (HK)
  cihk: City University of Hong Kong (HK)
  cir:  University of California at Irvine (US)
  cmel: Carnegie Mellon University (US)
  cou:  Cornell University (US)
  csb:  University of California at Santa Barbara (US)
  csd:  University Of California at San Diego (US)
  dut:  Delft University of Technology (NL)
  eind: Eindhoven University of Technology (NL)
  ens:  Superior Normal School at Paris (FR)
  epfl: Swiss Federal Institute of Technology Lausanne (CH)
  epfr: Polytechnic school of Paris (FR)
  ethz: Swiss Federal Institute of Technology Zurich (CH)
```

[2] The performance tableau `the-cs-2016.py` is available in the `examples` directory of the DIGRAPH3 resources (Bisdorff 2021).

```
   frei: University of Freiburg (DE)
   git:  Georgia Institute of Technology (US)
   glas: University of Glasgow (UK)
   hels: University of Helsinki (FI)
   hkpu: Hong Kong Polytechnic University (CN)
   hkst: Hong Kong University of Science and Technology (HK)
   hku:  Hong Kong University (HK)
   humb: Berlin Humboldt University (DE)
   icl:  Imperial College London (UK)
   indis:Indian Institute of Science (IN)
   itmo: ITMO University (RU)
   kcl:  King's College London (UK)
   kist: Korea Adv. Institute of Science and Technology (KR)
   kit:  Karlsruhe Institute of Technology (DE)
   kth:  KTH Royal Institute of Technology (SE)
   kuj:  Kyoto University (JP)
   kul:  Catholic University Leuven (BE)
   lms:  Lomonosov Moscow State University (RU)
   man:  University of Manchester (UK)
   mcp:  University of Maryland College Park (US)
   mel:  University of Melbourne (AU)
   mil:  Polytechnic University of Milan (IT)
   mit:  Massachusetts Institute of Technology (US)
   naji: Nanjing University (CN)
   ntu:  Nanyang Technological University of Singapore (SG)
   ntw:  National Taiwan University (TW)
   nyu:  New York University (US)
   oxf:  University of Oxford (UK)
   pud:  Purdue University (US)
   qut:  Queensland University of Technology (AU)
   rcu:  Rice University (US)
   rwth: RWTH Aachen University (DE)
   shJi: Shanghai Jiao Tong University (CN)
   sing: National University of Singapore (SG)
   sou:  University of Southampton (UK)
   stut: University of Stuttgart (DE)
   tech: Technion - Israel Institute of Technology (IL)
   tlavu:Tel Aviv University (IR)
   tsu:  Tsinghua University (CN)
   tub:  Technical University of Berlin (DE)
   tud:  Technical University of Darmstadt (DE)
   tum:  Technical University of Muenchen (DE)
   ucl:  University College London (UK)
   ued:  University of Edinburgh (UK)
   uiu:  University of Illinois at Urbana-Champagne (US)
   unlu: University of Luxembourg (LU)
   unsw: University of New South Wales (AU)
   unt:  University of Toronto (CA)
   uta:  University of Texas at Austin (US)
   utj:  University of Tokyo (JP)
   utw:  University of Twente (NL)
   uwa:  University of Waterloo (CA)
   wash: University of Washington (US)
   wtu:  Vienna University of Technology (AUS)
   zhej: Zhejiang University (CN)
```

The THE authors base their 2016 ranking on five decision objectives (Times Higher Education 2016).

Listing 13.3 The THE ranking objectives

```
>>> for obj in cspt.objectives:
...     print('%s: %s (%.1f%%),\n\t%s' %\
...                (obj,cspt.objectives[obj]['name'],\
...                 cspt.objectives[obj]['weight'],
...                 cspt.objectives[obj]['comment'])\
```

```
...        )
 Teaching: Best learning environment (30.0%)
   Reputation survey; Staff-to-student ration;
   Doctorate-to-student ratio;
   Doctorate-to-academic-staff ratio;
   Institutional income.
 Research: Highest volume and reputation (30.0%)
   Reputation survey;
   Research income;
   Research productivity.
 Citations: Highest research influence (27.5%)
   Impact.
 International outlook: Most international staff,
                        students and research (7.5%)
   Proportions of international students;
   Proportions of international staff;
   International collaborations.
 Industry income: Best knowledge transfer (5.0%)
   Volume.
```

With a cumulated importance of 87% (see above), *Teaching*, *Research*, and *Citations* represent clearly the major ranking objectives. *International outlook* and *Industry income* are considered of minor importance (12.5%).

THE authors do, unfortunately, not publish the detail of their performance assessments for evaluating CS Depts with respect to each one of performance criteria per ranking objective.[3] The THE 2016 ranking publication reveals solely a compound assessment on a single performance criteria per ranking objective. The five retained performance criteria may be printed out as follows.

```
>>> for g in cspt.criteria:
...     print('%s:\t%s, %s (%.1f%%)' %\
...
    (g,cspt.criteria[g]['name'],cspt.criteria[g]['comment'],\
...        cspt.criteria[g]['weight']) )
  gtch: Teaching, The learning environment (30.0%)
  gres: Research, Volume, income and reputation (30.0%)
  gcit: Citations, Research influence (27.5%)
  gint: International outlook, In staff, students and research
     (7.5%)
  gind: Industry income, knowledge transfer (5.0%)
```

The largest part (87.5%) of criteria significance is, hence canonically, allocated to the major ranking criteria: *Teaching* (30%), *Research* (30%) and *Citations* (27.5%). The small remaining part (12.5%) goes to *International outlook* (7.5%) and *Industry income* (5%).

In order to render commensurable these performance criteria, the THE authors replace, per criterion, the actual performance evaluation obtained by each University with the corresponding *quantile* observed in the cumulative distribution of the performance evaluations obtained by all the surveyed institutions (Times Higher Education 2016). The THE ranking is eventually determined by an *overall score*

[3] THE gives some insight on the subject and significance of the actual ranking criteria used for evaluating along each ranking objective on her website (Times Higher Education 2016).

per University which corresponds to the weighted average of these five criteria quantiles, as illustrated in Listing 13.4.

Listing 13.4 Computing the THE overall scores

```
>>> theScores = []
>>> for x in cspt.actions:
...     xscore = Decimal('0')
...     for g in cspt.criteria:
...         xscore += cspt.evaluation[g][x] *\
...           (cspt.criteria[g]['weight']/Decimal('100'))
...       theScores.append((xscore,x))
```

In Listing 13.5 (Lines 15–16), we may thus notice that, in the 2016 edition of the THE World University rankings by CS subject, the Swiss Federal Institute of Technology Zürich is first-ranked with an overall score of 92.9; followed by the California Institute of Technology (overall score: 92.4).[4]

Listing 13.5 Printing the ranked performance table

```
1 >>> theScores.sort(reverse = True)
2 >>> print('##  Univ \tgtch  gres  gcit  gint  gind  overall')
3 >>> print('-----------------------------------------------')
4 >>> i = 1
5 >>> for it in theScores:
6 ...     x = it[1]
7 ...     xScore = it[0]
8 ...     print('%2d: %s' % (i,x), end=' \t')
9 ...     for g in cspt.criteria:
10 ...         print('%.1f ' % (cspt.evaluation[g][x]),end=' ')
11 ...         print(' %.1f' % xScore)
12 ...         i += 1
13    ##  Univ     gtch  gres  gcit  gint  gind  overall
14    -----------------------------------------------
15     1: ethz     89.2  97.3  97.1  93.6  64.1   92.9
16     2: calt     91.5  96.0  99.8  59.1  85.9   92.4
17     3: oxf      94.0  92.0  98.8  93.6  44.3   92.2
18     4: mit      87.3  95.4  99.4  73.9  87.5   92.1
19     5: git      87.2  99.7  91.3  63.0  79.5   89.9
20     6: cmel     88.1  92.3  99.4  58.9  71.1   89.4
21     7: icl      90.1  87.5  95.1  94.3  49.9   89.0
22     8: epfl     86.3  91.6  94.8  97.2  42.7   88.9
23     9: tum      87.6  95.1  87.9  52.9  95.1   87.7
24    10: sing     89.9  91.3  83.0  95.3  50.6   86.9
25    11: cou      81.6  94.1  99.7  55.7  45.7   86.6
26    12: ucl      85.5  90.3  87.6  94.7  42.4   86.1
27    13: wash     84.4  88.7  99.3  57.4  41.2   85.6
28    14: hkst     74.3  92.0  96.2  84.4  55.8   85.5
29    15: ntu      76.6  87.7  90.4  92.9  86.9   85.5
30    16: ued      85.7  85.3  89.7  95.0  38.8   85.0
31    17: unt      79.9  84.4  99.6  77.6  38.4   84.4
32    18: uiu      85.0  83.1  99.2  51.4  42.2   83.7
33    19: mcp      79.7  89.3  94.6  29.8  51.7   81.5
34    20: cbu      81.2  78.5  94.7  66.9  45.7   81.3
35    21: tsu      88.1  90.2  76.7  27.1  85.9   80.9
36    22: csd      75.2  81.6  99.8  39.7  59.8   80.5
```

4 The author's own Computer Science Dept at the University of Luxembourg was ranked on position 63 with an overall score of 58.0.

```
 23: uwa     75.3  82.6  91.3  72.9  41.5   80.0
 24: nyu     71.1  77.4  99.4  78.0  39.8   79.7
 25: uta     72.6  85.3  99.6  31.6  49.7   79.6
 26: kit     73.8  85.5  84.4  41.3  76.8   77.9
 27: bju     83.0  85.3  70.1  30.7  99.4   77.0
 28: csb     65.6  70.9  94.8  72.9  74.9   76.2
 29: rwth    77.8  85.0  70.8  43.7  89.4   76.1
 30: hku     77.0  73.0  77.0  96.8  39.5   75.4
 31: pud     76.9  84.8  70.8  58.1  56.7   75.2
 32: kist    79.4  88.2  64.2  31.6  92.8   74.9
 33: kcl     45.5  94.6  86.3  95.1  38.3   74.8
 34: chku    64.1  69.3  94.7  75.6  49.9   74.2
 35: epfr    81.7  60.6  78.1  85.3  62.9   73.7
 36: dut     64.1  78.3  76.3  69.8  90.1   73.4
 37: tub     66.2  82.4  71.0  55.4  99.9   73.3
 38: utj     92.0  91.7  48.7  25.8  49.6   72.9
 39: cir     68.8  64.6  93.0  65.1  40.4   72.5
 40: ntw     81.5  79.8  66.6  25.5  67.6   72.0
 41: anu     47.2  73.0  92.2  90.0  48.1   70.6
 42: rcu     64.1  53.8  99.4  63.7  46.1   69.8
 43: mel     56.1  70.2  83.7  83.3  50.4   69.7
 44: lms     81.5  68.1  61.0  31.1  87.8   68.4
 45: ens     71.8  40.9  98.7  69.6  43.5   68.3
 46: wtu     61.8  73.5  73.7  51.9  62.2   67.9
 47: tech    54.9  71.0  85.1  51.7  40.1   67.1
 48: bro     58.5  54.9  96.8  52.3  38.6   66.5
 49: man     63.5  71.9  62.9  84.1  42.1   66.3
 50: zhej    73.5  70.4  60.7  22.6  75.7   65.3
 51: frei    54.2  51.6  89.5  49.7  99.9   65.1
 52: unsw    60.2  58.2  70.5  87.0  44.3   63.6
 53: kuj     75.4  72.8  49.5  28.3  51.4   62.8
 54: sou     48.2  60.7  75.5  87.4  43.2   62.1
 55: shJi    66.9  68.3  62.4  22.8  38.5   61.4
 56: itmo    58.0  32.0  98.7  39.2  68.7   60.5
 57: kul     35.2  55.8  92.0  46.0  88.3   60.5
 58: glas    35.2  52.5  91.2  85.8  39.2   59.8
 59: utw     38.2  52.8  87.0  69.0  60.0   59.4
 60: stut    54.2  60.6  61.1  36.3  97.8   58.9
 61: naji    51.4  76.9  48.8  39.7  74.4   58.6
 62: tud     46.6  53.6  75.9  53.7  66.5   58.3
 63: unlu    35.2  44.2  87.4  99.7  54.1   58.0
 64: qut     45.5  42.6  82.8  75.2  63.0   58.0
 65: hkpu    46.8  36.5  91.4  73.2  41.5   57.7
 66: albt    39.2  53.3  69.9  91.9  75.4   57.6
 67: mil     46.4  64.3  69.2  44.1  38.5   57.5
 68: hels    48.8  49.6  80.4  50.6  39.5   57.4
 69: cihk    42.4  44.9  80.1  76.2  67.9   57.3
 70: tlavu   34.1  57.2  89.0  45.3  38.6   57.2
 71: indis   56.9  76.1  49.3  20.1  41.5   57.0
 72: ariz    28.4  61.8  84.3  59.3  42.0   56.8
 73: kth     44.8  42.0  83.6  71.6  39.2   56.4
 74: humb    48.4  31.3  94.7  41.5  45.5   55.3
 75: eind    32.4  48.4  81.5  72.2  45.8   54.4
```

It is important to notice that a ranking by weighted average scores requires *commensurable ranking criteria* of precise decimal significance and on which precise decimal performance evaluations are given. It is very unlikely that the THE 2016 performance assessments verify indeed these conditions. Here we show how to relax these methodological requirements—precise commensurable criteria and decimal evaluations—by following instead a bipolar-valued epistemic logic based ranking methodology (see Chap. 8).

13.2 Ranking with Multiple Criteria of Ordinal Significance

Let us, first, have a critical look in Fig. 13.1 at the THE performance criteria.

```
>>> cspt.showHTMLCriteria(Sorted=False)
```

Considering a very likely imprecision of the performance evaluation procedure, followed by some potential violation of uniform distributed quantile classes, we assume here that a performance quantile difference of up to 2.5% is *insignificant*, whereas a difference of 5% and more warrants a *clearly better*, respectively, *clearly less good* performance. With quantiles 94%, respectively, 87.3%, Oxford's CS teaching environment, for instance, is thus clearly better evaluated than that of the MIT (see Listing 13.5 on page 169 Lines 27–28). We shall furthermore assume that a *considerable* performance quantile difference of 60%, observed on the three major ranking criteria: *Teaching*, *Research*, and *Citations*, will trigger the polarisation of pairwise outranking, respectively, outranked situations (Bisdorff 2013).

The effect these performance discrimination thresholds induce on the outranking modelling can be inspected with the `showCriteria()` method.

Listing 13.6 Inspecting the performance discrimination thresholds

```
>>> cspt.showCriteria()
  *----  criteria -----*
    gtch 'Teaching'
      Scale = (Decimal('0.00'), Decimal('100.00'))
      Weight = 0.300
      Threshold ind : 2.50 + 0.00x ;   percentile:  8.07
      Threshold pref : 5.00 + 0.00x ;  percentile: 15.75
      Threshold veto : 60.00 + 0.00x ; percentile: 99.75
    gres 'Research'
      Scale = (Decimal('0.00'), Decimal('100.00'))
      Weight = 0.300
      Threshold ind : 2.50 + 0.00x ;   percentile:  7.86
      Threshold pref : 5.00 + 0.00x ;  percentile: 16.14
      Threshold veto : 60.00 + 0.00x ; percentile: 99.21
```

the_cs_2016: Family of Criteria

#	Identifyer	Name	Comment	Weight	Scale			Thresholds (ax + b)		
					direction	min	max	indifference	preference	veto
1	gtch	Teaching	The learning environment	30.00	max	0.00	100.00	0.00x + 2.50	0.00x + 5.00	0.00x + 60.00
2	gres	Research	Volume, income and reputation	30.00	max	0.00	100.00	0.00x + 2.50	0.00x + 5.00	0.00x + 60.00
3	gcit	Citations	Research influence	27.50	max	0.00	100.00	0.00x + 2.50	0.00x + 5.00	0.00x + 60.00
4	gint	International outlook	In staff, students and research	7.50	max	0.00	100.00	0.00x + 2.50	0.00x + 5.00	
5	gind	Industry income	Innovation	5.00	max	0.00	100.00	0.00x + 2.50	0.00x + 5.00	

Fig. 13.1 The THE ranking criteria

```
gcit 'Citations'
  Scale = (Decimal('0.00'), Decimal('100.00'))
  Weight = 0.275
  Threshold ind : 2.50 + 0.00x ;  percentile:  11.82
  Threshold pref : 5.00 + 0.00x ;  percentile:  22.99
  Threshold veto : 60.00 + 0.00x ; percentile: 100.00
gint 'International outlook'
  Scale = (Decimal('0.00'), Decimal('100.00'))
  Weight = 0.075
  Threshold ind : 2.50 + 0.00x ;  percentile:  6.45
  Threshold pref : 5.00 + 0.00x ; percentile: 11.75
gind 'Industry income'
  Scale = (Decimal('0.00'), Decimal('100.00'))
  Weight = 0.050
  Threshold ind : 2.50 + 0.00x ;  percentile: 11.82
  Threshold pref : 5.00 + 0.00x ; percentile: 21.51
```

Between 6% and 12% of the observed quantile differences are, hence, considered to be *insignificant*. Similarly, between 77% and 88% are considered to be *significant*. Less than 1% correspond to *considerable* quantile differences on both the *Teaching* and *Research* criteria; actually triggering an epistemic polarisation effect (Bisdorff 2013).

Besides the likely imprecise performance discrimination, the *precise decimal significance weights*, as allocated by the THE authors to the five ranking criteria are, as well, quite *questionable*. Criteria significance weights may carry sometimes hidden strategies for rendering the performance evaluations commensurable in view of a numerical computation of the overall ranking scores. The eventual ranking result is, the case given, as much depending on the precise values of the given criteria significance weights as, vice versa, the given precise significance weights are depending on the subjectively expected and accepted ranking results.[5] We will therefore drop these precise decimal weights and, instead, only require a corresponding criteria significance preorder: `gtch` = `gres` > `gcit` > `gint` > `gind`. *Teaching environment* and *Research volume and reputation* are equally considered most significant, followed by *Research influence*. Then comes *International outlook in staff, students and research* and, least significant finally, *Industry income and innovation*.

Both these working hypotheses:—performance discrimination thresholds and—solely ordinal criteria significance, give us way to a ranking methodology based on *robust pairwise outranking* situations (see Chap. 19 and (Bisdorff 2004)).

Definition 13.1 (Robust Outranking Situations with Ordinal Criteria Significance Weights)

- We say that CS Dept x *robustly outranks* CS Dept y when x positively outranks y with **all** significance weight vectors that are compatible with the significance preorder: `gtch` = `gres` > `gcit` > `gint` > `gind`;

[5] In a social choice context, this potential double bind between voting profiles and election result corresponds to voting manipulation strategies.

- We say that CS Dept x is *robustly outranked* by CS Dept y when x is positively outranked by y with **all** significance weight vectors that are compatible with the significance preorder: `gtch` = `gres` > `gcit` > `gint` > `gind`;
- Otherwise, CS Depts x and y are considered to be *incomparable*.

A corresponding digraph constructor is provided by the `RobustOutranking-Digraph` class.

Listing 13.7 Computing the robust outranking digraph

```
1 >>> from outrankingDigraphs import RobustOutrankingDigraph
2 >>> rdg = RobustOutrankingDigraph(cspt)
3 >>> rdg
4   *------- Object instance description ------*
5    Instance class        : RobustOutrankingDigraph
6    Instance name         : robust_the_cs_2016
7    Actions               : 75
8    Criteria              : 5
9    Size                  : 2993
10   Determinateness (%)   : 78.16
11   Valuation domain      : [-1.00;1.00]
12 >>> rdg.computeIncomparabilityDegree(Comments=True)
13  Incomparability degree (%) of digraph <robust_the_cs_2016>:
14   links x<->y y: 2775, incomparable: 102, comparable: 2673
15   (incomparable/links) =  0.037
16 >>> rdg.computeTransitivityDegree(Comments=True)
17  Transitivity degree of digraph <robust_the_cs_2016>:
18   triples x>y>z: 405150, closed: 218489, open: 186661
19   (closed/triples) =  0.539
20 >>> rdg.computeSymmetryDegree(Comments=True)
21  Symmetry degree (%) of digraph <robust_the_cs_2016>:
22   arcs x>y: 2673, symmetric: 320, asymmetric: 2353
23   (symmetric/arcs) =  0.12
```

In the resulting digraph instance `rdg` (see Line 9 in Listing 13.7), we observe 2993 such robust pairwise outranking situations validated with a mean significance of 78% (Line 10). Unfortunately, in the case here, they do not deliver any complete linear ranking. The robust outranking digraph `rdg` contains 102 incomparability situations (3.7%, Line 15); nearly half of its transitive closure is missing (46.1%, Line 19) and 12% of the positive outranking situations correspond in fact to symmetric indifference situations (Line 23).

Worse even, the digraph `rdg` admits a really high number of outranking circuits.[6]

Listing 13.8 Inspecting outranking circuits

```
>>> rdg.computeChordlessCircuits()
>>> rdg.showChordlessCircuits()
```

[6] The `computeChordlessCircuits()` and `showChordlessCircuits()` methods are separate because there are various methods available for enumerating the chordless circuits in a digraph (Bisdorff 2010).

```
*---- Chordless circuits ----*
 145 circuits.
 1:  ['albt','unlu','ariz','hels'], cred. : 0.300
 2:  ['albt','tlavu','hels'], cred. : 0.150
 3:  ['anu', 'man', 'itmo'], cred. : 0.250
 4:  ['anu', 'zhej', 'rcu'], cred. : 0.250
   ...
   ...
 82:  ['csb','epfr','rwth'], cred. : 0.250
 83:  ['csb','epfr','pud','nyu'], cred. : 0.250
 84:  ['csd','kcl','kist'], cred. : 0.250
   ...
   ...
 142:  ['kul','qut','mil'], cred. : 0.250
 143:  ['lms','rcu','tech'], cred. : 0.300
 144:  ['mil','stut','qut'], cred. : 0.300
 145:  ['mil','stut','tud'], cred. : 0.300
```

Among the 145 detected robust outranking circuits reported in Listing 13.8 on the previous page, we notice, for instance, two outranking circuits of length 4 (see circuits 1 and 83).

Let us inspect below the bipolar-valued robust outranking characteristics of the first circuit.

Listing 13.9 Showing the relation table with stability denotation

```
>>> rdg.showRelationTable(actionsSubset=\
...            ['albt','unlu','ariz','hels'],\
...            Sorted=False)
  * ---- Relation Table -----
  r/(stab) |  'albt' 'unlu' 'ariz' 'hels'
  -----|------------------------------
   'albt' |  +1.00   +0.30   +0.00   +0.00
          |   (+4)    (+2)    (-1)    (-1)
   'unlu' | +0.00   +1.00   +0.40   +0.00
          |   (+0)    (+4)    (+2)    (-1)
   'ariz' |  +0.00   -0.12   +1.00   +0.40
          |   (+1)    (-2)    (+4)    (+2)
   'hels' |  +0.45   +0.00   -0.03   +1.00
          |   (+2)    (+1)    (-2)    (+4)
  Valuation domain: [-1.0; 1.0]
  Stability denotation semantics:
  +4|-4 : unanimous outranking | outranked situation;
  +2|-2 : outranking | outranked situation validated
     with all potential significance weights that are
     compatible with the given significance preorder;
  +1|-1 : validated outranking | outranked situation
     with the given significance weights;
    0   : indeterminate relational situation.
```

In Listing 13.9, we notice that the robust outranking circuit `[albt, unlu, ariz, hels]` will reappear with all potential criteria significance weight vectors that are compatible with given preorder: `gtch` = `gres` > `gcit` > `gint` > `gind`.

Notice also the (± 1) marked outranking situations, like the one between `albt` and `ariz`. The statement that "*Arizona State University strictly outranks University of Alberta*" is in fact valid with the precise THE criteria significance weights, but not with all potential significance weights vectors that are compatible with the given significance preorder. All these (± 1) marked outranking situations become hence *doubtful* ($r(x \succsim y) = 0.00$) and the corresponding CS Depts, like University of Alberta and Arizona State University, become *incomparable* in a robust outranking sense.

Showing many incomparabilities and indifferences, not being transitive and containing many robust outranking circuits make that no ranking algorithm, applied to digraph `rdg`, does exist that would produce a unique optimal linear ranking result. Methodologically, we are only left with ranking heuristics. In Chap. 8 on ranking with multiple criteria we have now seen several potential heuristic ranking rules that can be used for ranking with an outranking digraph; yet, they may deliver potentially more or less diverging results. Considering the order of digraph `rdg` (75) and the largely unequal THE criteria significance weights, we rather opt, in this tutorial, for the NETFLOWS ranking rule.[7] Its complexity in $O(n^2)$ is indeed quite tractable and, by avoiding potential *tyranny of short majority* effects, the NETFLOWS rule may specifically take the ranking criteria significance weights into a more fairly balanced account.

The NETFLOWS ranking result of the CS Depts can be directly computed with the `computeNetFlowsRanking()` method.

Listing 13.10 Computing a robust NETFLOWS ranking

```
>>> nfRanking = rdg.computeNetFlowsRanking()
>>> nfRanking
 ['ethz','calt','mit', 'oxf',  'cmel','git', 'epfl',
  'icl', 'cou', 'tum', 'wash', 'sing','hkst','ucl',
  'uiu', 'unt', 'ued', 'ntu',  'mcp', 'csd', 'cbu',
  'uta', 'tsu', 'nyu', 'uwa',  'csb', 'kit', 'utj',
  'bju', 'kcl', 'chku','kist', 'rwth','pud', 'epfr',
  'hku', 'rcu', 'cir', 'dut',  'ens', 'ntw', 'anu',
  'tub', 'mel', 'lms', 'bro',  'frei','wtu', 'tech',
  'itmo','zhej','man', 'kuj',  'kul', 'unsw','glas',
  'utw', 'unlu','naji','sou',  'hkpu','qut', 'humb',
  'shJi','stut','tud', 'tlavu','cihk','albt','indis',
  'ariz','kth', 'hels','eind', 'mil']
```

We actually obtain in Listing 13.10 a very similar ranking result as the one obtained with the THE average scores. The same group of seven Depts: `ethz`, `calt`, `mit`, `oxf`, `cmel`, `git`, and `epfl` is top-ranked. And a same group of Depts: `tlavu`, `cihk`, `indis`, `ariz`, `kth`, `hels`, `eind`, and `mil` appears at the end of the list.

[7] The reader might try other ranking rules, like the COPELAND or KOHLER rules. Mind that the latter ranking-by-choosing rule is more complex (see Chap. 8).

We can print out the difference between the overall scores based THE ranking and the NETFLOWS ranking above with the following short Python script, where we make use of an ordered Python dictionary with net flow scores, stored in the `rdg.netFlowsRankingDict` attribute by the previous computation.

Listing 13.11 Comparing the robust NETFLOWS ranking with the THE ranking

```
>>> # rdg.netFlowsRankingDict: ordered dictionary with net flow
>>> # scores stored in rdg by the computeNetFlowsRanking() method
>>> # theScores = [(xScore_1,x_1), (xScore_2,x_2),... ]
>>> # is sorted in decreasing order of xscores_i
>>> print(\
...   ' NetFlows ranking   gtch  gres  gcit  gint  gind   THE ranking')
>>> for i in range(75):
...     x = nfRanking[i]
...     xScore = rdg.netFlowsRankingDict[x]['netFlow']
...     thexScore,thex = theScores[i]
...     print('%2d: %s (%.2f) ' % (i+1,x,xScore), end='\t')
...     for g in rdg.criteria:
...         print('%.1f ' % (t.evaluation[g][x]),end=' ')
...     print(' %s (%.2f)' % (thex,thexScore) )
 NetFlows ranking   gtch  gres  gcit  gint  gind   THE ranking
  1: ethz (116.95)  89.2  97.3  97.1  93.6  64.1   ethz (92.88)
  2: calt (116.15)  91.5  96.0  99.8  59.1  85.9   calt (92.42)
  3: mit (112.72)   87.3  95.4  99.4  73.9  87.5   oxf (92.20)
  4: oxf (112.00)   94.0  92.0  98.8  93.6  44.3   mit (92.06)
  5: cmel (101.60)  88.1  92.3  99.4  58.9  71.1   git (89.88)
  6: git (93.40)    87.2  99.7  91.3  63.0  79.5   cmel (89.43)
  7: epfl (90.88)   86.3  91.6  94.8  97.2  42.7   icl (89.00)
  8: icl (90.62)    90.1  87.5  95.1  94.3  49.9   epfl (88.86)
  9: cou (84.60)    81.6  94.1  99.7  55.7  45.7   tum (87.70)
 10: tum (80.42)    87.6  95.1  87.9  52.9  95.1   sing (86.86)
 11: wash (76.28)   84.4  88.7  99.3  57.4  41.2   cou (86.59)
 12: sing (73.05)   89.9  91.3  83.0  95.3  50.6   ucl (86.05)
 13: hkst (71.05)   74.3  92.0  96.2  84.4  55.8   wash (85.60)
 14: ucl (66.78)    85.5  90.3  87.6  94.7  42.4   hkst (85.47)
 15: uiu (64.80)    85.0  83.1  99.2  51.4  42.2   ntu (85.46)
 16: unt (62.65)    79.9  84.4  99.6  77.6  38.4   ued (85.03)
 17: ued (58.67)    85.7  85.3  89.7  95.0  38.8   unt (84.42)
 18: ntu (57.88)    76.6  87.7  90.4  92.9  86.9   uiu (83.67)
 19: mcp (54.08)    79.7  89.3  94.6  29.8  51.7   mcp (81.53)
 20: csd (46.62)    75.2  81.6  99.8  39.7  59.8   cbu (81.25)
 21: cbu (44.27)    81.2  78.5  94.7  66.9  45.7   tsu (80.91)
 22: uta (43.27)    72.6  85.3  99.6  31.6  49.7   csd (80.45)
 23: tsu (42.42)    88.1  90.2  76.7  27.1  85.9   uwa (80.02)
 24: nyu (35.30)    71.1  77.4  99.4  78.0  39.8   nyu (79.72)
 25: uwa (28.88)    75.3  82.6  91.3  72.9  41.5   uta (79.61)
 26: csb (18.18)    65.6  70.9  94.8  72.9  74.9   kit (77.94)
 27: kit (16.32)    73.8  85.5  84.4  41.3  76.8   bju (77.04)
 28: utj (15.95)    92.0  91.7  48.7  25.8  49.6   csb (76.23)
 29: bju (15.45)    83.0  85.3  70.1  30.7  99.4   rwth (76.06)
 30: kcl (11.95)    45.5  94.6  86.3  95.1  38.3   hku (75.41)
 31: chku (9.43)    64.1  69.3  94.7  75.6  49.9   pud (75.17)
 32: kist (7.30)    79.4  88.2  64.2  31.6  92.8   kist (74.94)
 33: rwth (5.00)    77.8  85.0  70.8  43.7  89.4   kcl (74.81)
 34: pud (2.40)     76.9  84.8  70.8  58.1  56.7   chku (74.23)
 35: epfr (-1.70)   81.7  60.6  78.1  85.3  62.9   epfr (73.71)
 36: hku (-3.83)    77.0  73.0  77.0  96.8  39.5   dut (73.44)
 37: rcu (-6.38)    64.1  53.8  99.4  63.7  46.1   tub (73.25)
 38: cir (-8.20)    68.8  64.6  93.0  65.1  40.4   utj (72.92)
 39: dut (-8.85)    64.1  78.3  76.3  69.8  90.1   cir (72.50)
 40: ens (-8.97)    71.8  40.9  98.7  69.6  43.5   ntw (72.00)
 41: ntw (-11.15)   81.5  79.8  66.6  25.5  67.6   anu (70.57)
 42: anu (-11.50)   47.2  73.0  92.2  90.0  48.1   rcu (69.79)
 43: tub (-12.20)   66.2  82.4  71.0  55.4  99.9   mel (69.67)
```

```
59 44: mel (-23.98)    56.1 70.2 83.7 83.3 50.4   lms (68.38)
60 45: lms (-25.43)    81.5 68.1 61.0 31.1 87.8   ens (68.35)
61 46: bro (-27.18)    58.5 54.9 96.8 52.3 38.6   wtu (67.86)
62 47: frei (-34.42)   54.2 51.6 89.5 49.7 99.9   tech (67.06)
63 48: wtu (-35.05)    61.8 73.5 73.7 51.9 62.2   bro (66.49)
64 49: tech (-37.95)   54.9 71.0 85.1 51.7 40.1   man (66.33)
65 50: itmo (-38.50)   58.0 32.0 98.7 39.2 68.7   zhej (65.34)
66 51: zhej (-43.70)   73.5 70.4 60.7 22.6 75.7   frei (65.08)
67 52: man (-44.03)    63.5 71.9 62.9 84.1 42.1   unsw (63.65)
68 53: kuj (-47.40)    75.4 72.8 49.5 28.3 51.4   kuj (62.77)
69 54: kul (-49.98)    35.2 55.8 92.0 46.0 88.3   sou (62.15)
70 55: unsw (-54.88)   60.2 58.2 70.5 87.0 44.3   shJi (61.35)
71 56: glas (-56.98)   35.2 52.5 91.2 85.8 39.2   itmo (60.52)
72 57: utw (-59.27)    38.2 52.8 87.0 69.0 60.0   kul (60.47)
73 58: unlu (-60.08)   35.2 44.2 87.4 99.7 54.1   glas (59.78)
74 59: naji (-60.52)   51.4 76.9 48.8 39.7 74.4   utw (59.40)
75 60: sou (-60.83)    48.2 60.7 75.5 87.4 43.2   stut (58.85)
76 61: hkpu (-62.05)   46.8 36.5 91.4 73.2 41.5   naji (58.61)
77 62: qut (-66.17)    45.5 42.6 82.8 75.2 63.0   tud (58.28)
78 63: humb (-68.10)   48.4 31.3 94.7 41.5 45.5   unlu (58.04)
79 64: shJi (-69.72)   66.9 68.3 62.4 22.8 38.5   qut (57.99)
80 65: stut (-69.90)   54.2 60.6 61.1 36.3 97.8   hkpu (57.69)
81 66: tud (-70.83)    46.6 53.6 75.9 53.7 66.5   albt (57.63)
82 67: tlavu (-71.50)  34.1 57.2 89.0 45.3 38.6   mil (57.47)
83 68: cihk (-72.20)   42.4 44.9 80.1 76.2 67.9   hels (57.40)
84 69: albt (-72.33)   39.2 53.3 69.9 91.9 75.4   cihk (57.33)
85 70: indis (-72.53)  56.9 76.1 49.3 20.1 41.5   tlavu (57.19)
86 71: ariz (-75.10)   28.4 61.8 84.3 59.3 42.0   indis (57.04)
87 72: kth (-77.10)    44.8 42.0 83.6 71.6 39.2   ariz (56.79)
88 73: hels (-79.55)   48.8 49.6 80.4 50.6 39.5   kth (56.36)
89 74: eind (-82.85)   32.4 48.4 81.5 72.2 45.8   humb (55.34)
90 75: mil (-83.67)    46.4 64.3 69.2 44.1 38.5   eind (54.36)
```

The first inversion we observe in Listing 13.11 on the facing page (Lines 18–19) concerns Oxford University and the MIT, switching positions 3 and 4. Most inversions are similarly short and concern only switching very close positions in either way. There are some slightly more important inversions concerning, for instance, the Hong Kong University CS Dept, ranked into position 30 in the THE ranking and here in the position 36 (Line 51). The opposite situation may also happen; the Berlin Humboldt University CS Dept, occupying the 74th position in the THE ranking, advances in the robust NETFLOWS ranking to position 63 (Line 78).

In our bipolar-valued epistemic framework, the NETFLOWS score of any CS Dept x corresponds to the criteria significance support for the logical statement "*x is first-ranked*". Formally

$$r(x \text{ is first-ranked}) = \sum_{y \neq x} r\big((x \succsim y) + (y \not\succsim x)\big) = \sum_{y \neq x} \big(r(x \succsim y) - r(y \succsim x)\big). \tag{13.1}$$

Using the robust outranking characteristics of digraph `rdg`, we can thus explicitly compute, for instance, ETH Zürich's NETFLOWS score, denoted `nfx` below.

```
>>> x = 'ethz'
>>> nfx = Decimal('0')
>>> for y in rdg.actions:
...     if x != y:
...         nfx += (rdg.relation[x][y]\
```

```
...                    - rdg.relation[y][x])
>>> print(x, nfx)
 ethz 116.950
```

In Listing 13.11 on page 176 (Line 16), one may now verify that ETH Zürich obtains indeed the highest NETFLOWS score 116.95 and gives hence the *most credible* first-ranked CS Dept of the 75 potential candidates.

Yet, how may we now convince the reader that the outranking based ranking result here appears more objective and trustworthy than the classic value theory based THE ranking by average quantile scores?

13.3 How to Judge the Quality of a Ranking Result?

In a multiple-criteria based ranking problem, inspecting pairwise marginal performance differences may give objectivity to global preferential statements. That a CS Dept x convincingly outranks Dept y can conveniently be checked with the `showPairwiseOutrankings()` method.. The ETH Zürich CS Dept is, for instance, first ranked before Caltech's Dept in both previous rankings. Lest us check the preferential reasons.

Listing 13.12 Comparing pairwise criteria performances

```
1  >>> rdg.showPairwiseOutrankings('ethz','calt')
2   *------------  pairwise comparisons ----*
3   Valuation in range: -100.00 to +100.00
4   Comparing actions : ('ethz', 'calt')
5   crit.    wght.  g(ethz)  g(calt) diff  | ind   pref     r()  |
6   -------------------------------------   ------------------
7   'gcit'  27.50   97.10    99.80   -2.70 | 2.50  5.00   +0.00 |
8   'gind'   5.00   64.10    85.90  -21.80 | 2.50  5.00   -5.00 |
9   'gint'   7.50   93.60    59.10  +34.50 | 2.50  5.00   +7.50 |
10  'gres'  30.00   97.30    96.00   +1.30 | 2.50  5.00  +30.00 |
11  'gtch'  30.00   89.20    91.50   -2.30 | 2.50  5.00  +30.00 |
12                                                       ------
13                                      r(x >= y):  +62.50
14  crit.    wght.  g(calt)  g(ethz) diff  | ind   pref     r()  |
15  -------------------------------------   ----------------
16  'gcit'  27.50   99.80    97.10   +2.70 | 2.50  5.00  +27.50 |
17  'gind'   5.00   85.90    64.10  +21.80 | 2.50  5.00   +5.00 |
18  'gint'   7.50   59.10    93.60  -34.50 | 2.50  5.00   -7.50 |
19  'gres'  30.00   96.00    97.30   -1.30 | 2.50  5.00  +30.00 |
20  'gtch'  30.00   91.50    89.20   +2.30 | 2.50  5.00  +30.00 |
21                                                      -------
22                                        r(y >= x): +85.00
```

A significant positive performance difference (+34.50), concerning the *International outlook* criterion (of 7, 5% significance), is observed in Listing 13.12 in favour of the ETH Zürich Dept (Line 9 above). Similarly, a significant positive performance difference (+21.80), concerning the *Industry income* criterion (of 5% significance), is observed, this time, in favour of the Caltech Dept (Line 17). The former, larger positive, performance difference, observed on a more significant criterion, gives so far a first convincing argument of 12.5% significance for putting

ETH Zürich first, before Caltech. Yet, the slightly positive performance difference (+2.70, Line 16) between Caltech and ETH Zürich on the *Citations* criterion (of 27.5% significance) confirms an "*at least as well evaluated as*" situation in favour of the Caltech Dept.

The inverse negative performance difference (−2.70, Line 7), however, is neither *significant* (< −5.00), nor *insignificant* (> −2.50), and does hence neither confirm nor infirm a "*not at least as well evaluated as*" situation in disfavour of ETH Zürich. We observe here a convincing argument of 27.5% significance for ranking Caltech first, before ETH Zürich.

Notice finally, that, on the *Teaching* and *Research* criteria of total significance 60%, both Depts do, with performance differences (<| 2.50 |), perform one as well as the other. As these two major performance criteria together necessarily admit always the highest significance with the imposed significance weight preorder: `gtch` = `gres` > `gcit` > `gint` > `gind`, both outranking situations get in fact globally confirmed at stability level +2 (see Chap. 19).

A browser view of the corresponding robust relation map, ordering the CS Depts again with the same NETFLOWS ranking rule, well illustrates all such *stable outranking* situations.

```
>>> rdg.showHTMLRelationMap(\
...               tableTitle='Robust Outranking Map',
...               rankingRule='NetFlows')
```

In Fig. 13.2 on the next page, *dark green*, respectively, *light green* marked positions show certainly, respectively, positively valid outranking situations, whereas *dark red*, respectively, *light red* marked positions show certainly, respectively, positively valid outranked situations. In the left upper corner one may verify that the five top-ranked Depts ([`ethz`, `calt`, `oxf`, `mit`, `cmel`]) are in fact mutually outranking each other and thus are all indifferent one to another. They give even robust CONDORCET winners by robustly outranking all other Depts.

Notice by the way that no certainly valid robust outranking (dark green) and no certainly valid robust outranked situations (dark red) appear in Fig. 13.3 on page 181 below, respectively, above the principal diagonal; none of these robust preferential situations are hence violated by the robust NETFLOWS ranking. The non-reflexive *white* positions in the relation map, like the one between the Georgia Institute of Technology and the MIT, mark outranking or outranked situations that are not robust with respect to the given significance weight preorder. They become, hence, doubtful and are set to the indeterminate characteristic value 0.0.

By measuring the ordinal correlation with the underlying pairwise global and marginal robust outranking situations, the quality of the robust NETFLOWS ranking result can be formally evaluated with the `computeRankingCorrelation()` method and the `showCorrelation()` method (see Chap. 16).

Listing 13.13 Measuring the quality of the NETFLOWS ranking result

```
>>> corrnf = rdg.computeRankingCorrelation(nfRanking)
>>> rdg.showCorrelation(corrnf)
```

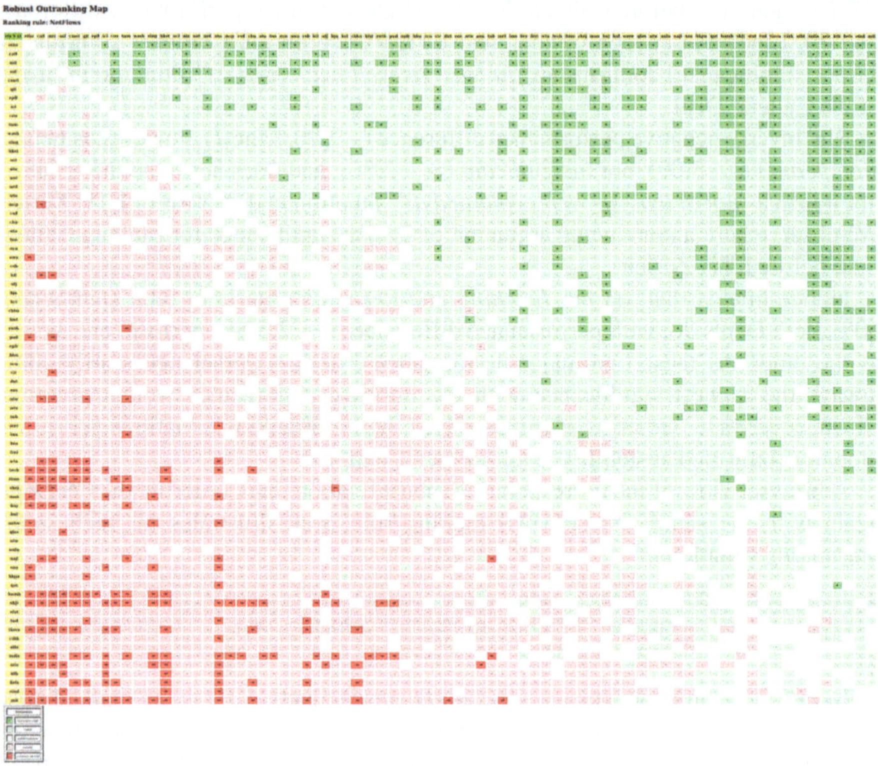

Fig. 13.2 Relation map of the robust outranking relation

```
3  Correlation indexes:
4    Crisp ordinal correlation    : +0.901
5    Epistemic determination      :  0.563
6    Bipolar-valued equivalence : +0.507
```

Listing 13.13 on the previous page Line 4 indicates that the NETFLOWS ranking result is indeed highly correlated (+0.901, in KENDALL's tau index sense) with the pairwise global robust outranking relation. Their bipolar-valued *relational equivalence* index (+0.51, Line 6) indicates a more than 75% criteria significance support.

With the `showRankingConsensusQuality()` method, we can also check how the NETFLOWS ranking rule is actually balancing the five ranking objectives.

Listing 13.14 Measuring the consensus quality of the NETFLOWS ranking result

```
>>> rdg.showRankingConsensusQuality(nfRanking)
 Criterion (weight): correlation
 -------------------------------
   gtch (0.300): +0.660
   gres (0.300): +0.638
```

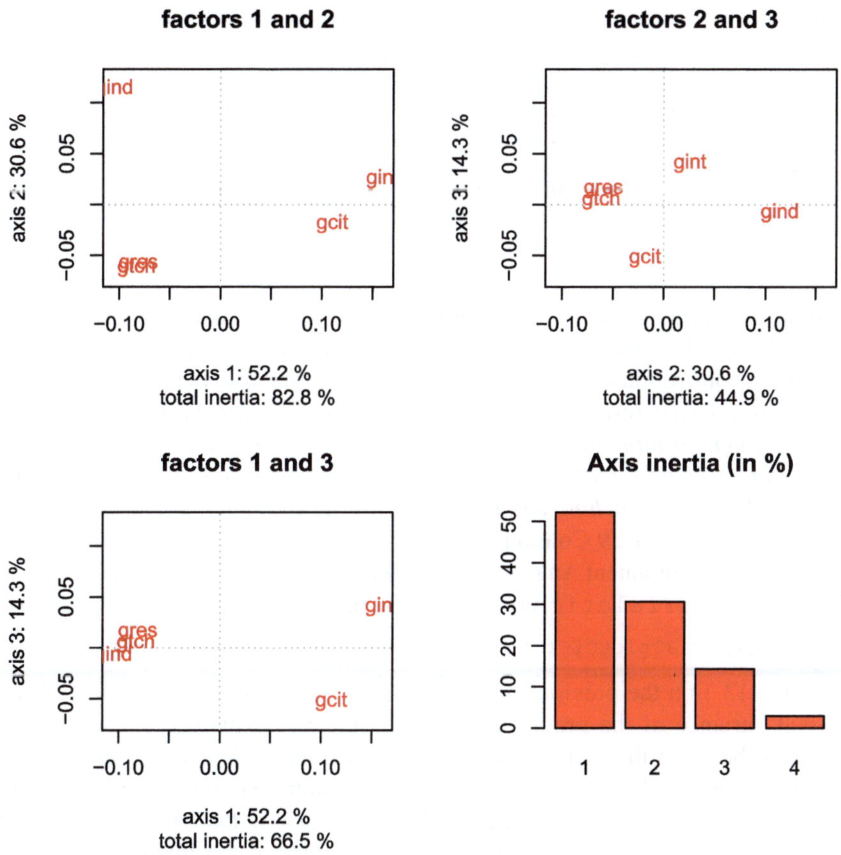

Fig. 13.3 3D PCA plot of the pairwise criteria correlation table

```
6     gcit (0.275): +0.370
7     gint (0.075): +0.155
8     gind (0.050): +0.101
9    Summary:
10    Weighted mean marginal correlation (a): +0.508
11    Standard deviation (b)                : +0.187
12    Ranking fairness (a)-(b)              : +0.321
```

The ordinal correlation indexes with the marginal performance criterion rankings are nearly respecting the given significance weights preorder: `gtch` ≈ `gres` > `gcit` > `gint` > `gind` (see Lines 4–8 above). The mean marginal ordinal correlation index is quite high (+0.51). Coupled with a low standard deviation (0.187), we obtain a quite fairly balanced ranking result (Lines 10–12).

We can furthermore inspect with the `showCriteriaCorrelationTable()` method the mutual correlation indexes observed between the individual criterion based marginal robust outranking relations.

Listing 13.15 Showing the ordinal correlation between the marginal criterion relations

```
1 >>> rdg.showCriteriaCorrelationTable()
2     Criteria ordinal correlation index
3           |  gcit    gind    gint    gres    gtch
4     -----|---------------------------------------
5     gcit |  +1.00   -0.11   +0.24   +0.13   +0.17
6     gind |          +1.00   -0.18   +0.15   +0.15
7     gint |                  +1.00   +0.04   -0.00
8     gres |                          +1.00   +0.67
9     gtch |                                  +1.00
```

Slightly contradictory (−0.11) appear the *Citations* and *Industrial income* criteria (Line 5 Column 3 in Listing 13.15 on the preceding page). Due perhaps to potential confidentiality clauses, it seems perhaps not always possible to publish industrially relevant research results in highly ranked journals. However, criteria *Citations* and *International outlook* show a slightly positive correlation (+0.24, Column 4), whereas the *International outlook* criterion shows no apparent correlation with both the major *Teaching* and *Research* criteria. The latter are, however, highly correlated (+0.67. Line 9 Column 6).

A Principal Component Analysis may well illustrate with the `export3DplotOfCriteriaCorrelation()` method the previous findings (Bisdorff 2020).

```
>>> rdg.export3DplotOfCriteriaCorrelation(graphType='pdf')
```

In Fig. 13.3 on the previous page, one may notice first that more than 80% of the total variance of the previous correlation table is explained by the apparent opposition between the marginal outrankings of criteria: *Teaching*, *Research*, and *Industry income* on the left side, and the marginal outrankings of criteria: *Citations* and *International outlook* on the right side. Notice also in the left lower corner the nearly identical positions of the marginal outrankings of the major *Teaching* and *Research* criteria. In the factors 2 and 3 plot, about 30% of the total variance is captured by the opposition between the marginal outrankings of the *Teaching* and *Research* criteria and the marginal outrankings of the *Industrial income* criterion. Finally, in the factors 1 and 3 plot, nearly 15% of the total variance is explained by the opposition between the marginal outrankings of the *International outlook* criterion and the marginal outrankings of the *Citations* criterion.

It is, finally, interesting to similarly assess the ordinal correlation between the THE average scores-based ranking and the robust outranking digraph.

Listing 13.16 Computing the ordinal quality of the THE ranking

```
>>> # theScores = [(xScore_1,x_1), (xScore_2,x_2),... ]
>>> # is sorted in decreasing order of xscores
>>> theRanking = [item[1] for item in theScores]
>>> corrthe = rdg.computeRankingCorrelation(theRanking)
>>> rdg.showCorrelation(corrthe)
    Correlation indexes:
     Crisp ordinal correlation    : +0.907
     Epistemic determination      :  0.563
     Bipolar-valued equivalence : +0.511
```

```
10 >>> rdg.showRankingConsensusQuality(theRanking)
11     Criterion (weight): correlation
12     -------------------------------
13      gtch (0.300): +0.683
14      gres (0.300): +0.670
15      gcit (0.275): +0.319
16      gint (0.075): +0.161
17      gind (0.050): +0.106
18     Summary:
19      Weighted mean marginal correlation (a): +0.511
20      Standard deviation (b)                 : +0.210
21      Ranking fairness (a)-(b)               : +0.302
```

The THE ranking result is similarly correlated (+0.907, Line 7 in Listing 13.16) with the pairwise global robust outranking situations. By its overall weighted scoring rule, the THE ranking naturally induces marginal criterion correlations that are compatible with the given significance weight preorder (Lines 13–17). Notice that the mean marginal correlation is of a similar value (+0.51, Line 19) as the robust NETFLOWS ranking. Yet, its standard deviation is slightly higher, which leads to a less fairer balancing of the three major ranking criteria.

To conclude, let us emphasise, that, without any commensurability hypothesis and by taking, furthermore, into account, first, the always present more or less imprecision of any performance evaluation and, secondly, solely ordinal criteria significance weights, we may obtain here with our robust outranking approach a very similar ranking result with a slightly better preference modelling quality. A convincing heatmap view of the 25 first-ranked Institutions may eventually be generated in the default system browser with following command (Fig. 13.4 on the following page).

```
>>> rdg.showHTMLPerformanceHeatmap(
...             WithActionNames=True,\
...             outrankingModel='this',\
...             rankingRule='NetFlows',\
...             ndigits=1,\
...             Correlations=True,\
...             fromIndex=0,toIndex=25)
```

As an exercise, the reader is invited to try out other robust outranking based ranking heuristics. Notice also that we have not challenged in this case study the THE provided criteria significance preorder. It would be very interesting to consider the five ranking objectives as equally important and, consequently, consider the ranking criteria to be equi-significant. Curious to see the ranking results under such settings.

The next case study in Chap. 14 is tackling the absolute quantile rating of the student enrolment quality of higher education institutions.

Heatmap of Performance Tableau 'robust_the_cs_2016'

criteria	gtch	gres	gcit	gint	gind
weights	+30.00	+30.00	+27.50	+7.50	+5.00
tau(*)	+0.66	+0.64	+0.37	+0.15	+0.10
Swiss Federal Institute of Technology Zürich (ethz)	89.2	97.3	97.1	93.6	64.1
Califormia Institute of Technology (calt)	91.5	96.0	99.8	59.1	85.9
Massachusetts Institute of Technology (mit)	87.3	95.4	99.4	73.9	87.5
University of Oxford (oxf)	94.0	92.0	98.8	93.6	44.3
Carnegie Mellon University (cmel)	88.1	92.3	99.4	58.9	71.1
Georgia Institute of Technology (git)	87.2	99.7	91.3	63.0	79.5
Swiss Federal Institute of Technology Lausanne (epfl)	86.3	91.6	94.8	97.2	42.7
Imperial College London (icl)	90.1	87.5	95.1	94.3	49.9
Cornell University (cou)	81.6	94.1	99.7	55.7	45.7
Technical University of München (tum)	87.6	95.1	87.9	52.9	95.1
University of Washington (wash)	84.4	88.7	99.3	57.4	41.2
National University of Singapore (sing)	89.9	91.3	83.0	95.3	50.6
Hong Kong University of Science and Technology (hkst)	74.3	92.0	96.2	84.4	55.8
University College London (ucl)	85.5	90.3	87.6	94.7	42.4
University of Illinois at Urbana-Champagne (uiu)	85.0	83.1	99.2	51.4	42.2
University of Toronto (unt)	79.9	84.4	99.6	77.6	38.4
University of Edinburgh (ued)	85.7	85.3	89.7	95.0	38.8
Nanyang Technological University of Singapore (ntu)	76.6	87.7	90.4	92.9	86.9
University of Maryland College Park (mcp)	79.7	89.3	94.6	29.8	51.7
University Of California at San Diego (csd)	75.2	81.6	99.8	39.7	59.8
Columbia University (cbu)	81.2	78.5	94.7	66.9	45.7
University of Texas at Austin (uta)	72.6	85.3	99.6	31.6	49.7
Tsinghua University (tsu)	88.1	90.2	76.7	27.1	85.9
New York University (nyu)	71.1	77.4	99.4	78.0	39.8
University of Waterloo (uwa)	75.3	82.6	91.3	72.9	41.5

Color legend:

quantile	14.29%	28.57%	42.86%	57.14%	71.43%	85.71%	100.00%

(*) **tau:** *Ordinal (Kendall) correlation between marginal criterion and global ranking relation*
Outranking model: **this**, *Ranking rule:* **NetFlows**
Ordinal (Kendall) correlation between global ranking and global outranking relation: **+0.901**
Mean marginal correlation (a) : **+0.508**
Standard marginal correlation deviation (b) : **+0.187**
Ranking fairness (a) - (b) : **+0.321**

Fig. 13.4 Extract of a heatmap browser view on the NETFLOWS ranking result

References

Bisdorff R (2004) Concordant outranking with multiple criteria of ordinal significance. 4OR 2(4):293–308. http://hdl.handle.net/10993/23721

Bisdorff R (2010) Enumerating chordless circuits in directed graphs. In: ORBEL24-2010, 24th annual conference of the Belgian operational research society (ORBEL aka Sogesci-B.V.W.B.), January 28-29, Liège (BE), Université de Liège (BE), pp 1–12. http://hdl.handle.net/10993/23926

Bisdorff R (2013) On polarizing outranking relations with large performance differences. J Multi-Crit Decis Anal Wiley 20:3–12. http://hdl.handle.net/10993/245

Bisdorff R (2020) Lecture 2: Introduction to statistical computing. In: Lectures of the Computational Statistics Course, University of Luxembourg. http://hdl.handle.net/10993/37870

Bisdorff R (2021) Documentation of the Digraph3 collection of Python modules for Algorithmic Decision Theory. https://digraph3.readthedocs.io/en/latest/

Times Higher Education (2016) World University Rankings by Computer Science Subject. http://www.timeshighereducation.com/world-university-rankings/methodology-world-university-rankings-2016-2017

Chapter 14
The Best Students, Where Do They Study? A Rating Case Study

Contents

Abstract In 2004, the German magazine *Der Spiegel*, with the help of *McKinsey & Company* and *AOL*, conducted an extensive online survey, assessing the apparent quality of German University students. The eventually published results by the *Spiegel* magazine concerned nearly 50,000 students, enrolled in one of fifteen popular academic subjects, like *German Studies*, *Life Sciences*, *Psychology*, *Law* or *CS*. Based on this published data, we present and discuss in this chapter, how to *rate* with the help of our DIGRAPH3 software resources the apparent global *enrolment quality* of new performance records.

14.1 The Rating Problem

In the 2004 DER SPIEGEL survey, more than 80,000 students, by participating, were questioned on their '*Abitur*' and university exams' marks, time of studies and age, grants, awards and publications, IT proficiency, linguistic skills, practical work experience, foreign mobility, and civil engagement. Each student received in return a quality score through a specific weighing of the collected data which depended on the subject the student is mainly studying. Publishing only those subject-university combinations, where at least 18 students had correctly filled in the questionnaire, left 41 German universities where, for at least eight out of the fifteen subjects, an average enrolment quality score could be determined (DER SPIEGEL 2004; Friedmann et al. 2004).

R. Bisdorff, *Algorithmic Decision Making with Python Resources*,
International Series in Operations Research & Management Science 324,
https://doi.org/10.1007/978-3-030-90928-4_14

Table 14.1 Enrolment quality scores per academic scores

Subject	U1	U2	U3	U4	U5
`bio`	NA	53.10	49.70	52.20	55.20
`med`	NA	NA	NA	49.50	55.50
`math`	56.80	54.70	56.30	58.60	61.30
`phys`	58.90	59.80	53.90	59.10	60.90
`chem`	52.00	50.10	54.20	53.60	56.70
`germ`	51.40	53.50	51.40	53.30	61.40
`pol`	NA	54.00	NA	50.80	59.60
`soc`	59.10	51.50	55.60	51.00	52.20
`psy`	57.70	NA	54.40	62.70	59.80
`law`	NA	NA	41.90	NA	51.10
`eco`	49.60	NA	NA	NA	54.40
`mgt`	54.00	53.40	50.70	49.60	NA
`info`	55.40	52.60	55.80	54.60	NA
`elec`	56.10	54.50	NA	57.20	NA
`mec`	54.30	55.20	NA	54.40	NA

We suppose in this rating case study that five German universities: `U1`, `U2`, `U3`, `U4`, and `U5` conducted in 2005 a similar survey among their enrolled students which gave the following enrolment quality scores per academic subject.

In Table 14.1, the fifteen popular academic subjects are grouped into topical 'Faculties': *Humanities*, *Law*, *Economics & Management*, *Life Sciences & Medicine*, *Natural Sciences & Mathematics*, and *Technology*. None of the five universities have students enrolled in all the fifteen subjects. University U1 has, for instance, no students in Life Sciences & Medicine and in Law and Politology, whereas University U5 does not offer any Technology subjects.

The average enrolment quality scores of the five universities, shown in Table 14.1, are stored in a file named `ratingCaseStudy.py` of `PerformanceTableau` format.[1] The `showHTMLPerformanceTableau()` method produces in Fig. 14.1 on the next page a colourful browser view of these performance records.

```
>>> from perfTabs import PerformanceTableau
>>> pt = PerformanceTableau('ratingCaseStudy')
>>> pt.showHTMLPerformanceTableau(Transposed=True,\
...                         title='Average enrolment scores')
```

With the best score in nine out of fifteen subjects, university `U5` presents the highest global enrolment quality of the five, whereas University `U3`, with five lowest scores, shows the lowest student enrolment quality of the five.

The university administrations would like to know now how their respective enrolment quality scores are to be appreciated in view the results of the 2004 DER

[1] The file may be found in the `examples` directory of the DIGRAPH3 resources.

Fig. 14.1 *Student enrolment quality scores per subject.* The light green, respectively, light red, figures indicate highest, respectively, lowest, score among the five universities

Average enrolment scores

criterion	U1	U2	U3	U4	U5
bio	NA	53.10	49.70	52.20	55.20
chem	52.00	50.10	54.20	53.60	56.70
eco	49.60	NA	NA	NA	54.40
elec	56.10	54.50	NA	57.20	NA
germ	51.40	53.50	51.40	53.30	61.40
info	55.40	52.60	55.80	54.60	NA
law	NA	NA	41.90	NA	51.10
math	56.80	54.70	56.30	58.60	61.30
mec	54.30	55.20	NA	54.40	NA
med	NA	NA	NA	49.50	55.50
mgt	54.00	53.40	50.70	49.60	NA
phys	58.90	59.80	53.90	59.10	60.90
pol	NA	54.00	NA	50.80	59.50
psy	57.70	NA	54.40	62.70	59.80
soc	59.10	51.50	55.60	51.00	52.20

SPIEGEL survey. Are they among the top universities, the midfield, or the bottom group?

14.2 The 2004 Performance Quintiles

The estimated lower-closed performance quintiles of the 2004 average enrolment quality per academic subject are stored in a file named `historicalQuantiles.py`,[2] which can be reloaded with the `PerformanceQuantiles` class.

Listing 14.1 Inspecting stored historical performance quantiles

```
>>> from performanceQuantiles import PerformanceQuantiles
>>> pq = PerformanceQuantiles('historicalQuintiles')
>>> pq
 *---- PerformanceQuantiles instance description ---*
  Instance class     : PerformanceQuantiles
  Instance name      : historicalQuintiles
  Objectives         : 4
  Criteria           : 15
  Quantiles          : 5
  History sizes      : {'germ': 39, 'pol': 34, 'psy': 34,
                        'soc': 32, 'law': 32, 'eco': 21,
                        'mgt': 34, 'bio': 34, 'med': 28,
                        'phys': 37, 'chem': 35, 'math': 27,
```

[2] The file may be found in the `examples` directory of the DIGRAPH3 resources.

#	Identifyer	Name	Comment	Weight	Scale			Thresholds (ax + b)		
					direction	min	max	indifference	preference	veto
1	**bio**	Life Sciences	Life Sciences & Medicine	1.00	max	45.00	65.00	0.00x + 0.10	0.00x + 0.50	
2	**chem**	Chemistry	Natural Sciences & Mathematics	1.00	max	45.00	65.00	0.00x + 0.10	0.00x + 0.50	
3	**eco**	Economics	Law, Economics & Management	1.00	max	45.00	65.00	0.00x + 0.10	0.00x + 0.50	
4	**elec**	Electrical Engineering	Technology	1.00	max	45.00	65.00	0.00x + 0.10	0.00x + 0.50	
5	**germ**	German Studies	Humanities	1.00	max	45.00	65.00	0.00x + 0.10	0.00x + 0.50	
6	**info**	Computer Science	Technology	1.00	max	45.00	65.00	0.00x + 0.10	0.00x + 0.50	
7	**law**	Law Studies	Law, Economics & Management	1.00	max	35.00	65.00	0.00x + 0.10	0.00x + 0.50	
8	**math**	Mathematics	Natural Sciences & Mathaematics	1.00	max	45.00	65.00	0.00x + 0.10	0.00x + 0.50	
9	**mec**	Mechanical Engineering	Technology	1.00	max	45.00	65.00	0.00x + 0.10	0.00x + 0.50	
10	**med**	Medicine	Life Sciences & Medicine	1.00	max	45.00	65.00	0.00x + 0.10	0.00x + 0.50	
11	**mgt**	Management	Law, Economics & Management	1.00	max	40.00	80.00	0.00x + 0.10	0.00x + 0.50	
12	**phys**	Physics	Natural Sciences & Mathematics	1.00	max	45.00	65.00	0.00x + 0.10	0.00x + 0.50	
13	**pol**	Politology	Humanities	1.00	max	50.00	70.00	0.00x + 0.10	0.00x + 0.50	
14	**psy**	Psychology	Humanities	1.00	max	50.00	70.00	0.00x + 0.10	0.00x + 0.50	
15	**soc**	Sociology	Humanities	1.00	max	45.00	65.00	0.00x + 0.10	0.00x + 0.50	

Fig. 14.2 The fifteen academic subjects taken into account for assessing the student enrolment quality

```
                        'info': 33, 'elec': 14, 'mec': 13}
   Attributes         : ['name', 'objectives', 'NA', 'criteria',
                        'quantilesFrequencies', 'historySizes',
                        'LowerClosed', 'limitingQuantiles',
                        'cdf', 'perfTabType']
```

The history sizes, reported in Listing 14.1 on the preceding page, indicate the number of universities evaluated in the 2004 survey in each one of the popular fifteen subjects. German Studies, for instance, were evaluated for 39 out of 41 universities, whereas Electrical and Mechanical Engineering were only evaluated for 14, respectively, 13 institutions. None of the fifteen subjects were evaluated in all the 41 universities.[3]

The details of the fifteen academic subjects—the performance criteria—may be consulted in a browser view (see Fig. 14.2).

```
>>> pt.showHTMLCriteria()
```

All fifteen subjects are considered equally significant (see Column Weight). The average scores in most subjects vary from 45 to 65. In some subjects, however, like Law Studies (35.0– 65.0) and Politology (50.0–70.0), a different variability is observed. The average enrolment scores per subject are hence incommensurable and global average enrolment scores over all subjects become meaningless.

To take furthermore into account the potential and very likely imprecision of the quality scores' computation, we assume that, for all subjects, an average enrolment quality score difference of 0.1 is *insignificant*, whereas a difference of 0.5 is sufficient to positively attest a *better* enrolment quality. No considerable performance difference is assumed.

[3] It would have been interesting to estimate such quantile limits from the individual quality scores of all the nearly 50,000 surveyed students. But this data was not public (Friedmann et al. 2004).

The `showLimitingQuantiles()` method prints in Listing 14.2 the estimated quintile limits of the 2004 survey.

Listing 14.2 Estimated quintile limits of the 2004 survey

```
>>> pq.showLimitingQuantiles()
 *----  performance quantiles -----*
 criteria | weights |  '0.00'  '0.20'  '0.40'  '0.60'  '0.80'  '1.00'
 ---------|-------------------------------------------------------
   'bio'  |   1.0   |   45.00   50.40   51.80   53.14   55.04   57.10
   'chem' |   1.0   |   45.00   53.30   54.20   55.80   57.40   58.80
   'eco'  |   1.0   |   49.60   52.14   53.38   54.28   56.94   60.80
   'elec' |   1.0   |   50.10   54.08   55.34   56.54   57.64   60.20
   'germ' |   1.0   |   45.00   52.14   54.02   55.84   57.48   61.40
   'info' |   1.0   |   45.00   54.40   55.44   56.68   58.10   59.80
   'law'  |   1.0   |   39.10   42.30   45.08   46.30   47.26   51.10
   'math' |   1.0   |   51.60   56.54   57.76   59.44   61.00   63.10
   'mec'  |   1.0   |   51.90   54.02   54.48   55.18   56.54   57.80
   'med'  |   1.0   |   45.00   49.20   49.84   51.10   52.42   60.10
   'mgt'  |   1.0   |   47.50   52.16   52.98   54.68   55.96   68.00
   'phys' |   1.0   |   53.90   58.78   59.90   60.96   61.96   62.80
   'pol'  |   1.0   |   50.80   54.62   56.18   57.78   59.78   65.90
   'psy'  |   1.0   |   52.50   57.58   58.46   59.80   60.94   64.10
   'soc'  |   1.0   |   45.00   51.70   53.92   55.42   56.26   59.80
```

We see confirmed again the incommensurability between the subjects, we noticed already in the apparent enrolment quality scoring, especially between Law Studies (39.1–51.1) and Politology (50.5–65.9).[4]

14.3 Rating-by-Ranking with Lower-Closed Quintile Limits

We add, now, the estimated quintile limits to the enrolment quality records of the 5 universities and rank, by using the COPELAND rule, all these records conjointly together with the help of the `LearnedQuantilesRatingDigraph` class from the `sortingDigraphs` module.

```
>>> from sortingDigraphs import\
...                     LearnedQuantilesRatingDigraph
>>> lqr = LearnedQuantilesRatingDigraph(pq,pt,\
...                     rankingRule='Copeland')
>>> lqr
 *-----  Object instance description -----------*
  Instance class       : LearnedQuantilesRatingDigraph
  Instance name        : learnedRatingDigraph
  Criteria             : 15
  Quantiles            : 5
  Lower-closed bins    : True
  New actions          : 5
  Size                 : 44
```

[4] The *Spiegel* authors opted therefore for a simple 3-tiling of the universities per evaluated academic subject, followed by an average BORDA scores-based global ranking (Friedmann et al. 2004).

Heatmap of Performance Tableau 'learnedRatingDigraph'

Ranking rule: **Copeland**; *Ranking correlation*: **0.973**

criteria	phys	math	germ	chem	bio	mgt	psy	info	pol	law	elec	med	eco	mec	soc
weights	1.0	1.0	1.0	1.0	1.0	1.0	1.0	1.0	1.0	1.0	1.0	1.0	1.0	1.0	1.0
tau(*)	+0.79	+0.77	+0.77	+0.66	+0.62	+0.59	+0.54	+0.52	+0.46	+0.41	+0.41	+0.40	+0.39	+0.38	+0.36
[0.80 -	62.0	61.0	57.5	57.4	55.0	56.0	60.9	58.1	59.8	47.3	57.6	52.4	56.9	56.5	56.3
U5	60.9	61.3	61.4	56.7	55.2	NA	59.8	NA	59.5	51.1	NA	55.5	54.4	NA	52.2
[0.60 -	61.0	59.4	55.8	55.8	53.1	54.7	59.8	56.7	57.8	46.3	56.5	51.1	54.3	55.2	55.4
[0.40 -	59.9	57.8	54.0	54.2	51.8	53.0	58.5	55.4	56.2	45.1	55.3	49.8	53.4	54.5	53.9
U1	58.9	56.8	51.4	52.0	NA	54.0	57.7	55.4	NA	NA	56.1	NA	49.6	54.3	59.1
U4	59.1	58.6	53.3	53.6	52.2	49.6	62.7	54.6	50.8	NA	57.2	49.5	NA	54.4	51.0
U2	59.8	54.7	53.5	50.1	53.1	53.4	NA	52.6	54.0	NA	54.5	NA	NA	55.2	51.5
[0.20 -	58.8	56.5	52.1	53.3	50.4	52.2	57.6	54.4	54.6	42.3	54.1	49.2	52.1	54.0	51.7
U3	53.9	56.3	51.4	54.2	49.7	50.7	54.4	55.8	NA	41.9	NA	NA	NA	NA	55.6
[0.00 -	53.9	51.6	1.0	2.0	1.0	47.5	52.5	3.0	50.8	39.1	50.1	3.0	49.6	51.9	1.0

Color legend:

quantile	20.00%	40.00%	60.00%	80.00%	100.00%

(*) tau: Ordinal (Kendall) correlation between marginal criterion and global ranking relation.

Fig. 14.3 Heatmap view of the quintiles rating-by-ranking result

```
   Determinateness (%) : 77.6
   Ranking rule        : Copeland
   Ordinal correlation : +0.97
```

The resulting ranking of the 5 universities including the lower-closed quintile score limits may be well illustrated with the `showHTMLRatingHeatmap()` method.

```
>>> lqr.showHTMLRatingHeatmap(colorLevels=5,\
...                 Correlations=True,\
...                 ndigits=1,rankingRule='Copeland')
```

The ordinal correlation (+0.97) of the COPELAND ranking with the underlying bipolar-valued outranking digraph is very high (see Fig. 14.3 Row 1). Most correlated subjects with this *rating-by-ranking* result appear to be Physics (+0.79), Mathematics (+0.77), and German Studies (+0.77). Sociology (+0.36) is the less correlated subject (see Row 4).

From the actual ranking position of the lower 5-tiling limits, we may now immediately deduce the quintile enrolment quality equivalence classes. No university reaches the highest quintile ([0.80 – [), whereas `U3` is rated into the lowest quintile ([0.00 – 0.20[). The other three universities U1, U2, and U4 are rated in the second quintile ([0.20 – 0.40[). The rating result may be easily printed out with the `showQuantilesRating()` method.

Listing 14.3 Showing the quintiling of the enrolment quality of the 5 universities

```
>>> lqr.showQuantilesRating()
```

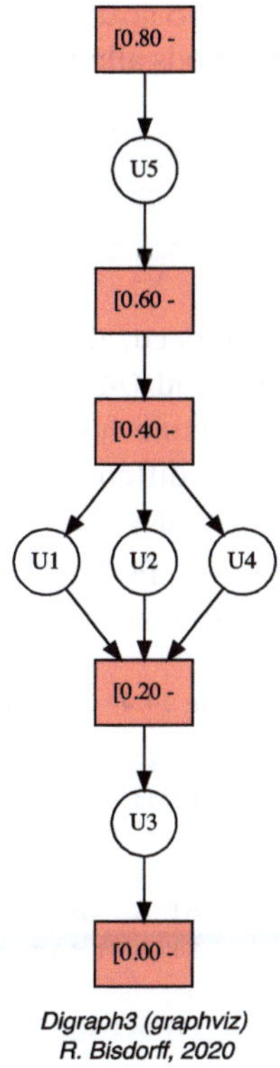

Fig. 14.4 Drawing of the quintiles rating-by-ranking result

```
 *---- Quantiles rating result ----*
  [0.60 - 0.80[ ['U5']
  [0.20 - 0.40[ ['U1', 'U4', 'U2']
  [0.00 - 0.20[ ['U3']
```

A corresponding *graphviz* drawing with the special `exportRatingByRankingGraphViz()` method may well illustrate in Fig. 14.4 these enrolment quality equivalence classes.

```
>>> lqr.exportRatingByRankingGraphViz('ratingResult')
 *---- exporting a dot file for GraphViz tools ---------*
  Exporting to ratingResult.dot
  dot -Grankdir=TB -Tpdf dot -o ratingResult.png
```

We have noticed in Chap. 8 that there is not a unique optimal rule for ranking from a given outranking digraph, like digraph `lqr` here. The COPELAND rule, for instance, has the advantage of being CONDORCET consistent, i.e. when the outranking digraph models a linear ranking, this ranking will necessarily be the result of the COPELAND rule. When this is not the case, and especially when the outranking digraph shows chordless circuits, all potential ranking rules may give very divergent ranking results and sometimes substantially divergent rating-by-ranking results.

The `computeChordlessCircuits()` and `showChordlessCircuits()` methods allow to check this issue.[5]

```
>>> lqr.computeChordlessCircuits()
>>> lqr.showChordlessCircuits()
*---- Chordless circuits ----*
1 circuits.
1:  ['U1', 'U2', 'U4'] , credibility : 0.200
```

Indeed, there appears an outranking circuit among the three universities `U1`, `U2`, and `U4` rated into the 2nd quintile. It is, hence, interesting, to verify if the epistemic fusion of the rating-by-ranking results, one may obtain when applying three different ranking rules, like the KEMENY, COPELAND, and NETFLOWS rules, does actually confirm our rating-by-ranking result shown in Fig. 14.4 on the previous page. For this purpose, we make use in Listing 14.4 of the `RankingsFusionDigraph` class.

Listing 14.4 Computing the epistemic fusion of three rating-by-ranking results

```
1 >>> # lqr.actionsRanking is Copeland ranked
2 >>> from linearOrders import KemenyOrder,\
3 ...                           NetFlowsOrder
4 >>> ke = KemenyOrder(lqr,orderLimit=10)
5 >>> nf = NetFlowsOrder(lqr)
6 >>> from transitiveDigraphs import\
7 ...                    RankingsFusionDigraph
8 >>> rankings = [lqr.actionsRanking,\
9 ...                 nf.netFlowsRanking,\
10 ...                ke.kemenyRanking]
11 >>> rankings
12  [['m5','u5','m4','m3','u1','u2','u4','m2','u3','m1'],
13   ['m5','u5','m4','m3','u1','u4','u2','u3','m2','m1'],
14   ['m5','u5','m4','m3','u1','u4','u2','m2','u3','m1']]
15 >>> rf = RankingsFusionDigraph(lqr,rankings)
16 >>> rf.exportGraphViz(fileName='fusionResult',\
17 ...              WithRatingDecoration=True)
18 exporting a dot file for GraphViz tools
19 Exporting to fusionResult.dot
20 dot -Grankdir=TB -Tpng fusionResult.dot\
21                  -o fusionResult.png
```

The fusion of the three rating-by-ranking results is shown in Fig. 14.5 on the next page where we see that they diverge solely in their rating-by-ranking of Universities `U2`, `U3` and `U4`. In Listing 14.4, Lines 12–14, one may more precisely notice that the KEMENY and COPELAND rules rate `U3` in the first quintile ([0.00 – 0.20[), whereas the NETFLOWS rule puts `U3` together with `U1`, `U2`, and `U4` into the second quintile ([0.20 – 0.40[). And, the KEMENY rule inverts the position of `U2` and `U4`.

[5] The `computeChordlessCircuits()` and `showChordlessCircuits()` methods are separate because there are various methods available for enumerating the chordless circuits in a digraph (Bisdorff 2010).

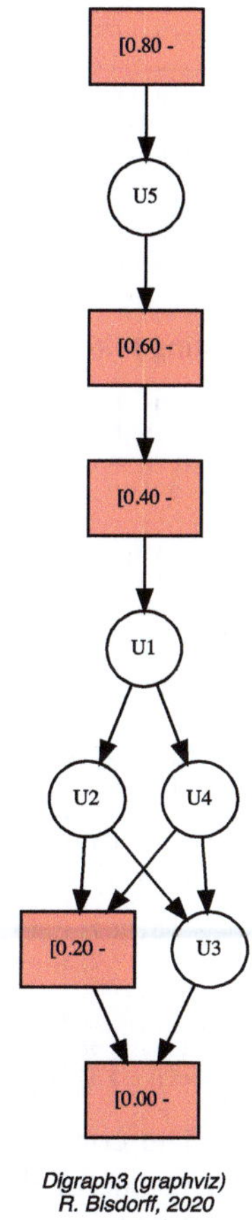

Fig. 14.5 *Disjunctive fusion of the* KEMENY, COPELAND *and* NETFLOWS *rankings.* They diverge in their rating-by-ranking of universities U2, U3, and U4

In Listing 14.5, the `showRankingConsensusQuality()` method reveals finally how fairly KEMENY and COPELAND rules actually balance the fifteen academic subjects.

Listing 14.5 Checking the consensus quality of the KEMENY ranking

```
>>> lqr.showRankingConsensusQuality(ke.kemenyRanking)
Consensus quality of ranking:
['m5', 'u5', 'm4', 'm3', 'u1', 'u2', 'u4', 'm2', 'u3', 'm1']
criterion (weight): correlation
-------------------------------
phys (0.067): +0.833
germ (0.067): +0.789
math (0.067): +0.722
bio (0.067): +0.667
mgt (0.067): +0.633
chem (0.067): +0.611
psy (0.067): +0.544
pol (0.067): +0.500
info (0.067): +0.478
mec (0.067): +0.422
law (0.067): +0.411
```

```
soc (0.067): +0.400
med (0.067): +0.400
eco (0.067): +0.389
elec (0.067): +0.367
Summary:
Weighted mean marginal correlation (a): +0.544
Standard deviation (b)                : +0.150
Ranking fairness (a)-(b)              : +0.394
```

Listing 14.6 Checking the consensus quality of the COPELAND ranking

```
>>> lqr.showRankingConsensusQuality(lqr.actionsRanking)
Consensus quality of ranking:
['m5', 'u5', 'm4', 'm3', 'u1', 'u4', 'u2', 'm2', 'u3', 'm1']
criterion (weight): correlation
------------------------------
phys (0.067): +0.789
math (0.067): +0.767
germ (0.067): +0.767
chem (0.067): +0.656
bio (0.067): +0.622
mgt (0.067): +0.589
psy (0.067): +0.544
info (0.067): +0.522
pol (0.067): +0.456
law (0.067): +0.411
elec (0.067): +0.411
med (0.067): +0.400
eco (0.067): +0.389
mec (0.067): +0.378
soc (0.067): +0.356
Summary:
Weighted mean marginal correlation (a): +0.537
Standard deviation (b)                : +0.148
Ranking fairness (a)-(b)              : +0.389
```

The consensus quality of the two rating-by-ranking results does not sensibly differ one of the other. They both show a similar high mean marginal correlation (+0.544, +0.537), similar standard deviations (+0.150, +0.148), and hence a similar ranking fairness (+0.394, +0.389).

To furthermore check the quality of our COPELAND rating-by-ranking result, we shall below compute a direct rating-by-sorting into the historic 2004 quintiles of the enrolment quality scores, without making use of any outranking digraph-based ranking rule.

14.4 Rating by Quintiles Sorting

In the case study here, the five universities `U1`, `U2`, `U3`, `U4`, and `U5` represent the decision actions: *where to study*. We say now that University x is sorted into the lower-closed quintile k when the performance record of x positively outranks the lower limit record $\mathbf{q}(p_{k-1})$ of quintile q and x does not positively outrank the upper limit record $\mathbf{q}(p_k)$ of quintile q, for $q = 1, \ldots, 5$ (see Listing 14.1 on page 189).

With the help of the bipolar-valued characteristic of the outranking relation $r(x \succsim y)$, we may indeed compute the bipolar-valued characteristic of the assertion: '*x belongs to the lower-closed quintile class* $\mathbf{q}_k$':

$$r(x \in \mathbf{q}_k) \;=\; \min\big[r\big(x \succsim \mathbf{q}(p_{k-1})\big),\, r\big(x \not\succsim \mathbf{q}(p_k)\big)\big], \tag{14.1}$$

where $k = 1, \ldots, 5$ and (p_k) denote the respective quintile proportions: 0.20, 0.40, 0.60, 0.80, and 1.00. As bipolar-valued outranking digraphs verify the coduality principle, $r\big(x \not\succsim \mathbf{q}(p_k)\big) = r\big(x precnsim \mathbf{q}(p_k)\big)$ (see Sect. 9.2).

The `showSortingCharacteristics()` method gives a precise look in Listing 14.7 on these quintiles sorting characteristics.

Listing 14.7 Showing quantiles sorting characteristics

```
>>> lqr.showSortingCharacteristics()
x   in   K_k             r(x >= m_k)     r(x < M_k)      r(x in K_k)
U5  in  [0.00 - 0.20[     0.73            -0.73            -0.73
U5  in  [0.20 - 0.40[     0.73            -0.60            -0.60
U5  in  [0.40 - 0.60[     0.60            -0.60            -0.60
U5  in  [0.60 - 0.80[     0.60             0.00             0.00
U5  in  [0.80 - <[        0.00             1.00             0.00

U2  in  [0.00 - 0.20[     0.73            -0.13            -0.13
U2  in  [0.20 - 0.40[     0.13             0.20             0.13
U2  in  [0.40 - 0.60[    -0.20             0.47            -0.20
U2  in  [0.60 - 0.80[    -0.47             0.73            -0.47
U2  in  [0.80 - <[       -0.73             1.00            -0.73

U4  in  [0.00 - 0.20[     0.87            -0.47            -0.47
U4  in  [0.20 - 0.40[     0.47             0.13             0.13
U4  in  [0.40 - 0.60[    -0.13             0.60            -0.13
U4  in  [0.60 - 0.80[    -0.60             0.67            -0.60
U4  in  [0.80 - <[       -0.67             1.00            -0.67

U1  in  [0.00 - 0.20[     0.73            -0.33            -0.33
U1  in  [0.20 - 0.40[     0.33             0.13             0.13
U1  in  [0.40 - 0.60[    -0.13             0.53            -0.13
U1  in  [0.60 - 0.80[    -0.53             0.60            -0.53
U1  in  [0.80 - <[       -0.60             1.00            -0.60

U3  in  [0.00 - 0.20[     0.67             0.13             0.13
U3  in  [0.20 - 0.40[    -0.13             0.27            -0.13
U3  in  [0.40 - 0.60[    -0.27             0.53            -0.27
U3  in  [0.60 - 0.80[    -0.53             0.67            -0.53
U3  in  [0.80 - <[       -0.67             1.00            -0.67
```

The performance record of University U5 cannot be positively rated into a precise quintile, but both the fourth and the fifth quintiles are not positively excluded as rating result. Otherwise, the bipolar-valued sorting characteristics verify the rating-by-ranking result we obtained previously with the KEMENY and COPELAND rating-by-ranking result. The `showActionsSortingResult()` method prints the eventual quintiles rating result:

Listing 14.8 Showing a quintiles rating-by-sorting result

```
>>> lqr.showActionsSortingResult()
 Quantiles sorting result per decision action
```

```
[0.20 - 0.40[: U1 with credibility: 0.13 = min(0.33,0.13)
[0.20 - 0.40[: U2 with credibility: 0.13 = min(0.13,0.20)
[0.00 - 0.20[: U3 with credibility: 0.13 = min(0.67,0.13)
[0.20 - 0.40[: U4 with credibility: 0.13 = min(0.47,0.13)
[0.60 - <[: U5 with credibility: 0.60 = min(0.60,1.00)
```

The quintiles rating-by-sorting result, shown in Listing 14.8 on the preceding page, confirms the previous COPELAND rating-by-ranking result. However, University U5 is here sorted conjointly into the fourth and the fifth quintiles and as such is part of the top rated institutions. A result already convincingly illustrated in the ranked heatmap is shown in Fig. 14.3 on page 192.

Let us conclude this hypothetical rating case study by saying that we prefer this latter rating-by-sorting approach. It is perhaps providing a less precise rating result, due the case given, to missing and contradictory performance data. Yet, the rating-by-sorting algorithm is better grounded in a powerful bipolar-valued logic and epistemic framework.

Chapter 15 proposes a series of decision problems suitable for exercises and exam questions.

References

Bisdorff R (2010) Enumerating chordless circuits in directed graphs. In: ORBEL24-2010, 24th annual conference of the Belgian operational research society (ORBEL aka Sogesci-B.V.W.B.), January 28-29, Liège (BE), Université de Liège (BE), pp 1–12. http://hdl.handle.net/10993/23926

DER SPIEGEL (2004) Studentenbefragung des SPIEGEL Die Methode. https://www.spiegel.de/lebenundlernen/uni/studentenbefragung-des-spiegel-die-methode-a-329082.html

Friedmann J, Hackenbroch V, Hipp D, Klawitter N, Koch J, Lakotta B, Mohr J, Schmitz C, Thimm K, Wüst C (2004) Die Elite von morgen. DER SPIEGEL 48. https://www.spiegel.de/politik/die-elite-von-morgen-a-afd0507a-0002-0001-0000-000037494731

Chapter 15
Exercises

Contents

Abstract In this chapter, we propose a series of decision problems of various difficulties which may serve as exercises and exam questions for an Algorithmic Decision Theory or Multiple-Criteria Decision Analysis course. They cover *selection*, *ranking*, and *rating* decision problems.

The difficulty of the exercises is marked as follows: *warming up* (§), *home work* (§§), and *research work* (§§§). Solutions should be supported both by computational Python code using the DIGRAPH3 programming resources and by methodological and algorithmic arguments from the *Algorithmic Decision Theory* Lectures (Bisdorff 2020, 2021).

15.1 Who Will Receive the Best Student Award? (§)

Data

In Table 15.1 on the following page, you find the actual grades obtained by four students: *Ariana* (`A`), *Bruce* (`B`), *Clare* (`C`), and *Daniel* (`D`) in five courses: `C1`, `C2`, `C3`, `C4`, and `C5` weighted by their respective `ECTS` points.[1]

R. Bisdorff, *Algorithmic Decision Making with Python Resources*,
International Series in Operations Research & Management Science 324,
https://doi.org/10.1007/978-3-030-90928-4_15

Table 15.1 Grades obtained by the students

Course	CF	C2	C3	C4	C5
ECTS	2	3	4	2	4
Ariana (A)	11	13	9	15	11
Bruce (B)	12	9	13	10	13
Clare (C)	8	11	14	12	14
Daniel (D)	15	10	12	8	13

The student grades are measured on an ordinal performance scale from 0 pts (weakest) to 20 pts (highest). Assume that the grading admits a preference discrimination threshold of 1 point. No considerable performance differences are given. The more ECTS points, the more importance a course takes in the curriculum of the students. An award is to be granted to the student showing the *best* result.

Questions

1. Edit a PerformanceTableau instance with the data shown above.
2. Who would you nominate for the grant?
3. Explain and motivate your selection algorithm.
4. Assume that the grading may actually admit an indifference discrimination threshold of 1 point and a preference discrimination threshold of 2 points. How stable is your grant recommendation with respect to this preference discrimination power of the grading scale?

15.2 How to Fairly Rank Movies? (§)

Data

File graffiti03.py[2] contains a performance tableau inspired by the star-rating of movies that could be seen in the city of Luxembourg during Spring 2003 (source: Graffiti magazine, Luxembourg February 2003). Its content is shown in Fig. 15.1 on the next page.

```
>>> from perfTabs import PerformanceTableau
>>> t = PerformanceTableau('graffiti03')
>>> t.showHTMLPerformanceHeatmap(WithActionNames=True,\
...                    pageTitle='Graffiti Star wars',
...                    rankingRule=None,colorLevels=5,
...                    ndigits=0)
```

[1] ECTS stands for *European Credit Transfer System*. It is a unit used to grade diplomas in all participating countries, https://wwwen.uni.lu/studies/ects_credits.

[2] The file is provided in the examples directory of the DIGRAPH3 resources.

Performance table graffiti03

criteria	ap	as	cf	cn	cs	dr	jh	jt	mk	mr	rr	td	vt
weight	1.00	1.00	1.00	1.00	1.00	1.00	2.00	1.00	1.00	1.00	1.00	1.00	2.00
ah	NA	NA	NA	-1	1	NA	1	1	2	NA	1	3	NA
aw	-1	NA	1	NA	2	NA	NA	-1	1	NA	1	NA	NA
bb	2	1	2	2	3	2	2	2	3	2	3	1	1
dl	-1	-1	-1	NA	-1	1	1	1	NA	1	1	1	-1
gny	2	4	2	4	2	3	3	4	2	4	3	2	3
gs	1	NA	-1	NA	1	1	NA	1	NA	-1	-1	-1	NA
hn	3	NA	3	2	2	NA	2	2	3	2	2	NA	1
la	3	2	3	3	2	2	3	3	3	4	3	NA	3
lor	2	3	3	NA	3	4	3	4	1	2	2	2	2
ma	NA	NA	NA	3	2	3	3	3	3	2	2	3	3
md	1	1	-1	NA	NA	-1	NA	1	-1	NA	-1	1	NA
mi	NA	-1	NA	1	-1	NA	1	2	NA	NA	-1	2	1
sa	NA	NA	NA	NA	2	NA	2	1	3	1	NA	NA	NA
sc	1	NA	1	NA	-1	NA	NA	1	1	NA	1	NA	NA
sha	2	-1	1	1	2	2	-1	2	1	1	1	NA	NA
ss	3	3	3	4	2	3	3	3	3	3	1	3	3
vf	NA	NA	NA	1	NA	NA	1	1	NA	NA	1	1	NA

Fig. 15.1 Star-rating of movies from February 2003

The critic's opinions are expressed on seven ordinal rating levels: -2 (two zeros, *I hate*), -1 (one zero, *I do not like*), 1 (one star, *maybe*), 2 (two stars, *good*), 3 (three stars, *excellent*), 4 (four stars, *not to be missed*), and 5 (five stares, *a master piece*). Notice the many missing data (NA) when a critic had not seen the respective movie. Mind also that the ratings of two movie critics (jh and vt) are given a higher significance weight.

Questions

1. The GRAFFITI magazine suggests a best rated movie by computing the average number of stars it received, ignoring the missing data and any significance weights of the critics. What movie gets the highest average?
2. By taking into account missing data and varying significance weights, how may one find the best rated movie without computing any average of stars?
3. How would one rank these movies so as to at best respect the weighted rating opinions of each movie critic?

4. In what ranking position would appear a movie not seen by any critic? Confirm computationally your answer by adding such a fictive, *not at all evaluated*, movie to the given performance tableau instance.
5. How robust are the preceding results when the significance weights of the movie critics are considered to be only of ordinal type?

15.3 What Is Your Best Choice Recommendation? (§§)

Data

A person, who wants to buy a TV set, retains after a first selection eight potential TV models.[3] To make their best choice, these eight models were evaluated with respect to three decision objectives of equal importance:

1. *Costs* of the set (to be minimised)
2. *Picture and Sound* quality of the TV (to be maximised)
3. *Maintenance contract* quality of the provider (to be maximised)

The *Costs* objective is assessed by the price of the TV set (criterion Pr to be minimised). *Picture* quality (criterion Pq), *Sound* quality (criterion Sq), and *Maintenance contract* quality (criterion Mq) are each one assessed on a four-level qualitative performance scale: -1 (*not good*), 0 (*average*), 1 (*good*), and 2 (*very good*).

The actual evaluation data are gathered in Table 15.2 on the facing page.

The *Price* criterion `Pr` supports furthermore an indifference threshold of 25.00€ and a preference threshold of 75.00€. No considerable performance differences (veto thresholds) are to be considered.

Questions

1. Edit a new `PerformanceTableau` instance with the data shown in Table 15.2 on the next page, and illustrate its content by best showing objectives, criteria, decision alternatives, and performance table. If needed, write adequate Python code.
2. What is the best TV set to recommend?
3. Illustrate your best choice recommendation with an adequate *graphviz* drawing.
4. Explain and motivate your selection algorithm.
5. Assume that the qualitative criteria: *Picture* quality (`Pq`), *Sound* quality (`Sq`), and *Maintenance contract* quality (`Mq`) are all three considered to be equi-significant and that the significance of the *Price* criterion (`Pr`) equals the significance of these three quality criteria taken together. How stable is your best choice recommendation with respect to thus reviewing the criteria significance weights?

[3] A similar didactic decision problem is discussed in Vincke (1992, pp.33–35).

Table 15.2 Performance evaluations of the potential TV sets

Criteria	`Pr` (€)	`Pq`	`Sq`	`Mq`
Significance	2	1	1	1
Model `T1`	−1300	2	2	0
Model `T2`	−1200	2	2	1
Model `T3`	−1150	2	1	1
Model `T4`	−1000	1	1	−1
Model `T5`	−950	1	1	0
Model `T6`	−950	0	1	−1
Model `T7`	−900	1	0	−1
Model `T8`	−900	0	0	0

Table 15.3 The set of potential holiday activities

Identifier	Name	Comment
`ant`	*Antequerra*	An afternoon excursion to Antequerra and surroundings
`ard`	*Ardales*	An afternoon excursion to Ardales and El Chorro
`be`	*Beach*	Sun, sea, fun, and more
`crd`	*Cordoba*	A whole day visit by car to Cordoba
`dn`	*Fa niente*	Do nothing
`lw`	*Long walk*	A whole day hiking
`mal`	*Malaga*	A whole day visit to Malaga
`sev`	*Sevilla*	A whole day visit to Sevilla
`sw`	*Short walk*	Less than a half-day hiking

15.4 Planning the Next Holiday Activity (§§)

Data

A family, staying during their holidays in Ronda (Andalusia), is planning the next day's activity (Bisdorff 2008). The alternatives shown in Table 15.3 are considered as potential actions.

The family members agree to measure their preferences with respect to a set of seven performance criteria such as the time to attend the place (to be minimised), the required physical investment, the expected quality of the food, touristic interest, relaxation, sun fun, and more (see Table 15.4 on the next page)

The evaluation, agreed with all family members, of the nine alternatives on all the criteria resulted in the performance tableau shown in Table 15.5 on the following page.

All performances on the qualitative criteria are marked on a same ordinal scale going from 0 (lowest) to 10 (highest). On the quantitative Distance criterion (to be minimised), the required travel time to go to and return from the activity is marked in negative minutes. In order to model only effective preferences, an indifference threshold of 1 point and a preference threshold of 2 points are put on the qualitative performance measures. On the Distance criterion, an indifference threshold of

Table 15.4 The set of performance criteria

Identifier	Name	Comment
`cult`	*Cultural interest*	Andalusian heritage
`dis`	*Distance*	Minutes by car to go to and return from the activity
`food`	*Food*	Quality of the expected food opportunities
`sun`	*Sun, fun, and more*	No comment
`phy`	*Physical investment*	Contribution to physical health care
`rel`	*Relaxation*	Anti-stress support
`tour`	*Tourist attraction*	How many stars in the guide?

Table 15.5 The performance tableau

Criteria	Weight	`ant`	`ard`	`be`	`crd`	`dn`	`lw`	`mal`	`sev`	`sw`
`cult`	1	7	3	0	10	0	0	5	10	0
`dis`	1	−120	−100	−30	−360	0	−90	−240	−240	0
`phy`	1	3	7	0	5	0	10	5	5	5
`rel`	1	1	5	8	3	10	5	3	3	6
`food`	1	8	10	4	8	10	1	8	10	1
`sun`	1	0	3	10	3	1	3	8	5	5
`tour`	1	5	7	3	10	0	8	10	10	5

20 minute and a preference threshold of 45 min are considered. Furthermore, a difference of more than two hours to go to and return from the activity's place is considered to raise a veto.

The individual criteria each reflect one or the other family member's preferential point of view. Therefore, they are judged equi-significant for the best action to be eventually chosen.

Questions

1. Edit a new `PerformanceTableau` instance with the data shown above and illustrate its content.
2. How do the criteria express their preferential view point on the set of alternatives? For instance, the Tourist Attraction criterion appears to be in its preferential judgments somehow positively correlated with both the cultural interest and the food criteria. It is also apparent that the distance criterion is somehow negatively correlated to these latter criteria. How to explore and illustrate these intuitions?
3. What is the preference modelling effect of the considerable performance difference of 120 minutes assumed for the distance criterion?
4. How would you rank the potential holiday activities with the fairest balance of the performance criteria?
5. What is the first choice you would recommend to the family members?

15.5 What Is the Best Public Policy? (§§)

Data Files

- File `perfTab_1.py`[4] contains a 3-objectives performance tableau with 100 performance records concerning public policies evaluated with respect to an *economic*, a *societal*, and an *environmental* decision objective.
- File `historicalData_1.py` contains a performance tableau of the same kind with 2000 historical performance records.

Questions

1. Illustrate the content of the given `perfTab_1.py` performance tableau by best showing *objectives*, *criteria*, *decision alternatives* (public policies), and *performance evaluations*. If needed, write adequate Python code.
2. Construct the corresponding bipolar-valued outranking digraph. How *confident* and/or *robust* are the apparent outranking situations?
3. What are apparently the 5 best-ranked decision alternatives in your decision problem from the different decision objectives point of views and from a global fair compromise view? Justify your ranking approach from a methodological point of view.
4. How would you rate your 100 public policies into relative deciles classes?
5. Using the given historical records in file `historicalData_1.py`, how would you rate your 100 public policies into absolute deciles classes?
6. Explain the differences you may observe between the absolute and the previous relative rating results.
7. Select among your 100 potential public policies a shortlist of up to 15 potential best policies, all reaching an absolute performance quantile of at least 66.67%.
8. Based on the previous best policies shortlist (see Question 7), what is your eventual best choice recommendation? Is it perhaps a best choice unopposed by all three objectives?

15.6 A Fair Diploma Validation Decision (§§§)

Data

Use the `RandomAcademicPerformanceTableau` class from the `randomPerfTab` module for generating realistic random students' performance tableaux concerning a curriculum of nine `ECTS` weighted courses (Bisdorff 2021). Assume that all the gradings are done on an integer scale from 0 (weakest) to 20 (best). It is known that grading procedures are inevitably somehow imprecise; therefore assume

[4] Files `perfTab_1.py` and `historicalData_1.py` are provided in the `examples` directory of the DIGRAPH3 resources (Bisdorff 2021).

an indifference discrimination threshold of 1 point and a preference discrimination threshold of 2 points. Furthermore, a performance difference of more than 12 points is considerable and will trigger a polarisation situation. To validate eventually their curriculum, the students are required to obtain more or less 10 points in each course.

Questions

1. Design and implement a fair diploma validation decision rule based on the grades obtained in the nine courses.
2. Run simulation tests with random students' performance tableaux for validating your design and implementation.

References

Bisdorff R (2008) On clustering the criteria in an outranking based decision aid approach. In: Thi HAL, Bouvry P, Pham D (eds) Computation and Optimization in Information Systems and Management Sciences, Springer, CCIS, pp 409–418. http://hdl.handle.net/10993/23718

Bisdorff R (2020) Lectures of the algorithmic decision theory course, University of Luxembourg. http://hdl.handle.net/10993/37933

Bisdorff R (2021) Technical documentation of the Digraph3 collection of Python modules. https://digraph3.readthedocs.io/en/latest/techDoc.html

Graffiti magazine, Luxembourg (February 2003) Star wars

Vincke P (1992) Multicriteria Decision-Aid. John Wiley & Sons Ltd, Chichester

Part IV
Advanced Topics

The fourth part gathers five chapters introducing and discussing further going algorithmic developments. In Chap. 16, KENDALL's ordinal correlation index is consistently extended to bipolar-valued digraphs. Chapter 17 explains the important concept of digraph kernel and illustrates the DIGRAPH3 algorithmic approach to their computation. In Chap. 18, criteria significance weights are considered to be uncertain and become random variates. This idea opens the way to compute a bipolar-valued likelihood of outranking and outranked situations leading to $\alpha\%$-confident outranking digraphs. In Chap. 19, the criteria significance weights are considered to be only of ordinal type. This working hypothesis induces the need to put into place a specific robustness analysis of the corresponding outranking digraphs. The last chapter, Chap. 20, makes use of the preceding robustness analysis for introducing and discussing ideas, like two-stage elections with multipartisan primary selection or bipolar voting systems, for tempering plurality tyranny effects in social choice problems.

Chapter 16
On Measuring the Fitness of a Multiple-Criteria Ranking

Contents

Abstract Starting from a motivating decision problem about how to list, from the best to the worst, a set of movies that are star-rated by journalists and movie critics, the chapter shows that KENDALL's ordinal correlation index tau can be extended to a bipolar-valued relational equivalence measure of bipolar-valued digraphs. This finding gives way, on the one hand, to measure the fitness and fairness of multiple-criteria ranking rules. On the other hand, it provides a tool for illustrating preference divergences between decision objectives and criteria.

16.1 Listing Movies from Best Star-Rated to Worst

In a stubborn keeping with a two-valued logic, where every argument can only be true or false, there is no place for efficiently taking into account missing data or logical indeterminateness. These cases are seen as problematic and at best are simply ignored. Worst, in modern data science, missing data get often replaced with *fictive* values, potentially falsifying hence all subsequent computations.

In social choice problems, voting abstentions are, however, frequently observed and represent a social expression that may be significant for revealing non-represented social preferences. And, in marketing studies, interviewees will not always respond to all the submitted questions. Again, such abstentions do sometimes contain nevertheless valid information concerning consumer preferences.

R. Bisdorff, *Algorithmic Decision Making with Python Resources*,
International Series in Operations Research & Management Science 324,
https://doi.org/10.1007/978-3-030-90928-4_16

Graffiti Star wars

criteria	AP	AS	CF	CS	DR	FG	GS	JH	JPT	MR	RR	SF	SJ	TD	VT
weight	1.00	1.00	1.00	1.00	1.00	1.00	1.00	3.00	1.00	1.00	1.00	1.00	1.00	1.00	3.00
mv_AL	3	-1	2	-1	NA	NA	3	NA	2	NA	NA	NA	2	NA	2
mv_BI	1	-1	1	1	-1	NA	NA	NA	2	NA	2	NA	NA	NA	NA
mv_CM	NA	3	3	2	NA	NA	3	2	3	2	1	2	2	NA	2
mv_DF	NA	4	3	1	2	NA	1	3	2	2	2	3	-1	NA	1
mv_DG	3	2	2	3	3	1	4	3	3	3	NA	NA	-1	NA	3
mv_DI	1	2	NA	1	2	2	NA	1	2	2	NA	1	1	NA	NA
mv_DJ	3	1	3	NA	NA	NA	3	NA	2	2	NA	2	4	3	-1
mv_FC	3	2	2	1	3	NA	1	3	3	3	2	NA	2	1	3
mv_FF	2	3	2	1	2	2	NA	1	2	2	1	2	1	3	1
mv_GG	2	3	3	1	NA	NA	1	-1	NA	-1	2	NA	-1	NA	1
mv_GH	1	NA	1	-1	2	2	1	2	1	3	4	NA	NA	NA	NA
mv_HN	3	1	3	1	3	NA	4	3	2	NA	1	NA	3	3	3
mv_HP	1	NA	3	1	NA	NA	NA	1	1	2	3	NA	1	2	NA
mv_HS	2	4	2	3	2	2	2	3	4	2	3	NA	1	3	NA
mv_JB	3	4	3	NA	3	2	3	2	3	2	NA	2	2	NA	3
mv_MB	NA	2	NA	NA	1	NA	1	2	2	1	2	NA	2	2	NA
mv_NP	NA	1	3	1	NA	3	2	3	3	3	2	3	NA	NA	3
mv_PE	3	4	2	NA	3	1	NA	2	4	2	NA	3	3	NA	3
mv_QS	NA	3	2	NA	4	3	NA	4	4	3	3	4	NA	4	4
mv_RG	2	2	2	2	NA	NA	3	1	2	2	1	NA	NA	3	NA
mv_RR	3	2	NA	1	4	NA	4	3	3	4	3	3	NA	4	2
mv_SM	3	3	2	2	2	2	NA	3	2	2	3	NA	2	2	3
mv_TF	-1	NA	1	1	1	NA	NA	1	2	NA	-1	NA	-1	1	NA
mv_TM	2	1	2	2	NA	NA	2	2	2	3	NA	2	NA	4	NA
mv_TP	2	3	3	1	2	NA	NA	3	2	NA	2	NA	1	NA	2

Fig. 16.1 Star-ratings of movies from September 2007

Such a case is given with a list of star-rated movies that could be seen in town in September 2007.[1] The underlying performance tableau data, stored in a file named `graffiti07.py`,[2] is shown below with the `showHTMLPerformanceTableau()` method:

```
>>> from outrankingDigraphs import\
...                          PerformanceTableau
>>> gt07 = PerformanceTableau('graffiti07')
>>> gt07.showHTMLPerformanceTableau(\
...                 title='Graffiti Star wars',\
...                 ndigits=0)
```

Figure 16.1 shows the star-ratings of 25 movies by 15 journalists and movie critics: 5 stars (*masterpiece*), 4 stars (*must be seen*), 3 stars (*excellent*), 2 stars (*good*), 1 star (*could be seen*), −1 star (*I do not like*), −2 stars (*I hate*), and `NA`

[1] *Graffiti Star wars*, Edition Revue Luxembourg, September 2007, p. 30.

[2] To be found in the `examples` directory of the DIGRAPH3 resources.

(*not seen*). Notice in the second row the higher significance (3.00) that is granted to two locally renowned movie critics, namely JH and VT. Their opinion counts for three times the opinion of the other critics. With six times a 4 stars (*must be seen*) mark, mv_QS is best-rated, followed by mv_RR with four times a 4 stars mark. Fewest stars obtain movie mv_TF with three times a −1 star (*do not like*) mark and five times a 1 star mark. Notice that many movies, like movie mv_BI, are only rated by some of the critics.

To aggregate all the critics' star-ratings, the *Graffiti* magazine computes for each movie a global score—the average weighted number of stars it obtained—just ignoring the *not seen* movies. Listing 16.1 illustrates below how to compute these global scores using the data stored in the gt07 performance tableau. Attribute gt07.actions, respectively, gt07.criteria, contains the description of the 25 movies, respectively, the 15 critics. The actual star-ratings are to be found in the gt07.evaluation attribute.

Listing 16.1 Computing the average weighted number of stars per movie

```
>>> # gt07 = PerformanceTableau('graffiti07')
>>> globalScores = {}
>>> for mv in gt07.actions:
>>>     globalScores[mv] = Decimal('0')
>>>     sumWeights = Decimal('0')
>>>     for critic in gt07.criteria:
>>>         stars = gt07.evaluation[critic][mv]
>>>         if stars != gt07.NA:
>>>             weight = gt07.criteria[critic]['weight']
>>>             globalScores[mv] += (stars * weight)
>>>             sumWeights += weight
>>>     globalScores[mv] /= sumWeights
>>> graffitiList = [(globalScores[mv],mv) for mv in globalScores]
>>> graffitiList.sort(reverse=True)
>>> for item in graffitiList:
>>>     print('%s: %.2f' % (item[1],item[0]) )
  mv_QS: 3.60
  mv_RR: 2.88
  mv_PE: 2.67
  mv_JB: 2.62
  mv_HN: 2.62
  mv_NP: 2.60
  mv_HS: 2.60
  mv_DG: 2.56
  mv_SM: 2.53
  mv_FC: 2.41
  mv_TP: 2.21
  mv_CM: 2.20
  mv_TM: 2.17
  mv_DF: 1.94
  mv_RG: 1.83
  mv_MB: 1.73
  mv_GH: 1.67
  mv_DJ: 1.67
  mv_FF: 1.61
  mv_AL: 1.60
  mv_HP: 1.55
  mv_DI: 1.42
  mv_GG: 0.71
  mv_BI: 0.71
  mv_TF: 0.55
```

The global scores ranking confirms in Lines 17–18 both leading movies—`mv_QS` (3.60) and `mv_RR` (2.88)—as well as in Line 41 the weakest rated one—`mv_TF` (0.55). Mind however that these global averages, due to the numerous missing ratings, are not computed with commensurable denominators; some critics do indeed use a more or less extended range of stars. The movies not seen, for instance, by critic `SJ` are favoured, as this critic is severer than others in her rating. Dropping the movies that were not rated by all the critics is not possible either, as none of the 25 movies was actually rated by all the 15 critics. Providing a fictive value for the many not seen situations will as well always somehow falsify the global scores. What to do?

A better approach is to rank the movies on the basis of pairwise bipolar-valued "*rated at least as well as*" statements. Under this epistemic argumentation approach, missing evaluations are naturally treated as opinion abstentions and hence do not falsify the logical computations. Such a NetFlows ranking from best-rated to weakest-rated is provided by the **heatmap** browser view generated with the `showHTMLPerformanceHeatmap()` method (see Listing 16.2).

Listing 16.2 Showing the movie from best- to worst-rated in a heatmap view

```
>>> gt07.showHTMLPerformanceHeatmap(\
...                 pageTitle='Ranking the movies',\
...                 rankingRule='NetFlows',
...                 Correlations=True,\
...                 ndigits=0)
```

The NetFlows ranking shown in Fig. 16.2 on the facing page confirms again that movie `mv_QS`, with 6 *must be seen* marks, is correctly first-ranked and the movie `mv_TF` is last-ranked with five *do not like* marks.

It is fair, however, to eventually mention here that the *Graffiti* magazine's average stars ranking rule is actually showing a very similar result. Indeed, average scores usually confirm well all evident pairwise comparisons, yet *enforce* comparability for all less evident ones. How to judge now the fitness of a given ranking rule?

This is the purpose of the ordinal correlation *tau* indexes shown in Fig. 16.2 on the next page, third row. Computing these ordinal correlation indexes is the subject of the next section.

16.2 Kendall's Ordinal Correlation Tau Index

M.G. Kendall defined his ordinal correlation τ (*tau*) index for linear orders of dimension n as a balancing of the number Co of correctly oriented pairs against the number In of incorrectly oriented pairs (Kendall 1938). The total number of irreflexive pairs being $n(n-1)$, in the case of linear orders, $Co + In = n(n-1)$. Hence $\tau = \left(\frac{Co}{n(n-1)}\right) - \left(\frac{In}{n(n-1)}\right)$. In case In is zero, $\tau = +1$ (all pairs are equivalently oriented); inversely, in case Co is zero, $\tau = -1$ (all pairs are differently oriented).

Ranking the movies

criteria	JH	JPT	AP	DR	MR	VT	GS	CS	SJ	RR	TD	CF	SF	AS	FG
weights	+3.00	+1.00	+1.00	+1.00	+1.00	+3.00	+1.00	+1.00	+1.00	+1.00	+1.00	+1.00	+1.00	+1.00	+1.00
tau(*)	+0.50	+0.43	+0.32	+0.26	+0.25	+0.23	+0.16	+0.14	+0.14	+0.13	+0.11	+0.11	+0.10	+0.08	+0.03
mv_QS	4	4	NA	4	3	4	NA	NA	NA	3	4	2	4	3	3
mv_RR	3	3	3	4	4	2	4	1	NA	3	4	NA	3	2	NA
mv_DG	3	3	3	3	3	3	4	3	-1	NA	NA	2	NA	2	1
mv_NP	3	3	NA	NA	3	3	2	1	NA	2	NA	3	3	1	3
mv_HN	3	2	3	3	NA	3	4	1	3	1	3	3	NA	1	NA
mv_HS	3	4	2	2	2	NA	2	3	1	3	3	2	NA	4	2
mv_SM	3	2	3	2	2	3	NA	2	2	3	2	2	NA	3	2
mv_JB	2	3	3	3	2	3	3	NA	2	NA	NA	3	2	4	2
mv_PE	2	4	3	3	2	3	NA	NA	3	NA	NA	2	3	4	1
mv_FC	3	3	3	3	3	3	1	1	2	2	1	2	NA	2	NA
mv_TP	3	2	2	2	NA	2	NA	1	1	2	NA	3	NA	3	NA
mv_CM	2	3	NA	NA	2	2	3	2	2	1	NA	3	2	3	NA
mv_DF	3	2	NA	2	2	1	1	1	-1	2	NA	3	3	4	NA
mv_TM	2	2	2	NA	3	NA	2	2	NA	NA	4	2	2	1	NA
mv_DJ	NA	2	3	NA	2	-1	3	NA	4	NA	3	3	2	1	NA
mv_AL	NA	2	3	NA	NA	2	3	-1	2	NA	NA	2	NA	-1	NA
mv_RG	1	2	2	NA	2	NA	3	2	NA	1	3	2	NA	2	NA
mv_MB	2	2	NA	1	1	NA	1	NA	2	2	2	NA	NA	2	NA
mv_GH	2	1	1	2	3	NA	1	-1	NA	4	NA	1	NA	NA	2
mv_HP	1	1	1	NA	2	NA	NA	1	1	3	2	3	NA	NA	NA
mv_BI	NA	2	1	-1	NA	NA	NA	1	NA	2	NA	1	NA	-1	NA
mv_DI	1	2	1	2	2	NA	NA	1	1	NA	NA	NA	1	2	2
mv_FF	1	2	2	2	2	1	NA	1	1	1	3	2	2	3	2
mv_GG	-1	NA	2	NA	-1	1	1	1	-1	2	NA	3	NA	3	NA
mv_TF	1	2	-1	1	NA	NA	NA	1	-1	-1	1	1	NA	NA	NA

Color legend:

quantile	20.00%	40.00%	60.00%	80.00%	100.00%

(*) **tau:** *Ordinal (Kendall) correlation between marginal criterion and global ranking relation*
Outranking model: **standard**, *Ranking rule:* **NetFlows**
Ordinal (Kendall) correlation between global ranking and global outranking relation: **+0.780**
Mean marginal correlation (a) : **+0.234**
Standard marginal correlation deviation (b) : **+0.147**
Ranking fairness (a) - (b) : **+0.087**

Fig. 16.2 Star-ratings of movies ranked with the NETFLOWS rule

Noticing that $\frac{Co}{n(n-1)} = 1 - \frac{In}{n(n-1)}$, and recalling that the bipolar-valued negation is operated by changing the sign of the characteristic value,

$$\tau = 1 - 2\frac{In}{n(n-1)} = -\left(2\frac{In}{n(n-1)} - 1\right) = 2\frac{Co}{n(n-1)} - 1. \tag{16.1}$$

KENDALL's original *tau* definition implemented in fact the bipolar-valued negation of the non-equivalence of two linear orders, i.e. the normalised majority margin of equivalently oriented irreflexive pairs.

Let `r1` and `r2` be two random crisp relations defined on a same set of 5 alternatives. Computing KENDALL's *tau* index may be done as shown in Listing 16.3.

Listing 16.3 Computing a relational equivalence digraph

```
1  >>> from randomDigraphs import RandomDigraph
2  >>> r1 = RandomDigraph(order=5,Bipolar=True)
3  >>> r2 = RandomDigraph(order=5,Bipolar=True)
4  >>> from digraphs import EquivalenceDigraph
5  >>> eqd = EquivalenceDigraph(r1,r2)
6  >>> eqd.showRelationTable(ReflexiveTerms=False)
7     * ---- Relation Table -----
8    r(<=>)|  'a1'   'a2'    'a3'    'a4'    'a5'
9    ------|------------------------------------
10    'a1' |    -    -1.00   1.00   -1.00    1.00
11    'a2' |  -1.00    -    -1.00    1.00   -1.00
12    'a3' |  -1.00 -1.00     -      1.00    1.00
13    'a4' |  -1.00  1.00   -1.00     -      1.00
14    'a5' |  -1.00  1.00   -1.00    1.00     -
15    Valuation domain: [-1.00;1.00]
16 >>> eqd.correlation
17   {'correlation': -0.1, 'determination': 1.0}
```

In the table of the equivalence relation (r1 ⇔ r2) above (see Listing 16.1 on page 211, Lines 10–14), we observe that the normalised majority margin of equivalent versus non-equivalent irreflexive pairs amounts to $(9 - 11)/20 = -0.1$, i.e. the value of KENDALL's *tau* index in this plainly determined crisp case (see Line 17).

What happens now with more or less epistemically determined and even partially indeterminate relations? May we proceed in a similar way?

16.3 Bipolar-Valued Relational Equivalence

Two random bipolar-valued digraphs d1 and d2 of order five, generated in Listing 16.4, will help exploring this idea.

Listing 16.4 Two random bipolar-valued digraphs

```
>>> from randomDigraphs import RandomValuationDigraph
>>> d1 = RandomValuationDigraph(order=5,seed=1)
>>> d1.showRelationTable(ReflexiveTerms=False)
  * ---- Relation Table -----
  r(d1)|   'a1'    'a2'    'a3'    'a4'    'a5'
 ------|------------------------------------
  'a1' |    -    -0.66    0.44    0.94   -0.84
  'a2' |  -0.36    -     -0.70    0.26    0.94
  'a3' |   0.14   0.20     -      0.66   -0.04
  'a4' |  -0.48 - 0.76    0.24     -     -0.94
  'a5' |  -0.02   0.10    0.54    0.94     -
  Valuation domain: [-1.00;1.00]
>>> d2 = RandomValuationDigraph(order=5,seed=2)
>>> d2.showRelationTable(ReflexiveTerms=False)
  * ---- Relation Table -----
```

```
16    r(d2) |  'a1'    'a2'    'a3'    'a4'    'a5'
17    ------|-----------------------------------
18    'a1'  |   -     -0.86   -0.78   -0.80   -0.08
19    'a2'  |  -0.58    -      0.88    0.70   -0.22
20    'a3'  |  -0.36   0.54     -     -0.46    0.54
21    'a4'  |  -0.92   0.48    0.74     -     -0.60
22    'a5'  |   0.10   0.62    0.00    0.84     -
23    Valuation domain: [-1.00;1.00]
```

In the generated random digraphs d1 and d2, 9 pairs, like (a1,a2) or (a3,a2), for instance, appear equivalently oriented (see Lines 7 and 18 or Lines 9 and 20). The EquivalenceDigraph class computes this bipolar-valued relational equivalence between digraphs d1 and d2 (see Listing 16.5).

Listing 16.5 Bipolar-valued equivalence digraph

```
1  >>> from digraphs import EquivalenceDigraph
2  >>> eqd = EquivalenceDigraph(d1,d2)
3  >>> eqd.showRelationTable(ReflexiveTerms=False)
4    * ---- Relation Table -----
5     r(<=>)|  'a1'   'a2'    'a3'    'a4'    'a5'
6     ------|----------------------------------
7     'a1'  |   -     0.66   -0.44   -0.80    0.08
8      'a2' |  0.36    -     -0.70    0.26   -0.22
9      'a3' | -0.14   0.20     -     -0.46   -0.04
10     'a4' |  0.48  -0.48    0.24     -      0.60
11     'a5' | -0.02   0.10    0.00    0.84     -
12    Valuation domain: [-1.00;1.00]
```

In our bipolar-valued epistemic logic, logical disjunctions and conjunctions are implemented as max and, respectively, min operations. Notice also that the logical equivalence (d1 $\Leftrightarrow$ d2) corresponds to a double implication (d1 $\Rightarrow$ d2) $\wedge$ (d2 $\Rightarrow$ d1) and that the implication (d1 $\Rightarrow$ d2) is logically equivalent to the disjunction ($\neg$d1 $\vee$ d2).

When $r(x \texttt{ d1 } y)$ and $r(x \texttt{ d2 } y)$ denote the bipolar-valued characteristic values of relation d1 and, respectively, d2, we may hence compute as follows a majority margin $M(\texttt{d1} \Leftrightarrow \texttt{d2})$ between equivalently and not equivalently oriented irreflexive pairs (x, y):

$$M(\texttt{d1} \Leftrightarrow \texttt{d2}) = \sum_{(x \neq y)} \Big[\min\Big(\max\big(-r(x\,\texttt{d1}\,y), r(x\,\texttt{d2}\,y)\big), \max\big(-r(x\,\texttt{d2}\,y), r(x\,\texttt{d1}\,y)\big)\Big)\Big]. \tag{16.2}$$

$M(\texttt{d1} \Leftrightarrow \texttt{d2})$ is thus given by the sum of the non-reflexive terms of the relation table of *eqd*, the relational equivalence digraph computed above (see Listing 16.5). In the crisp case, $M(\texttt{d1} \Leftrightarrow \texttt{d2})$ is normalised with the maximum number of possible irreflexive pairs, namely $n(n-1)$. In the extended r-valued case, the maximal possible equivalence majority margin M corresponds to the sum D of the conjoint determinations of $(x\ \mathtt{d1}\ y)$ and $(x\ d2\ y)$ (see Bisdorff 2012):

$$D = \sum_{x \neq y} \min \Big[abs\big(r(x\ \mathtt{d1}\ y)\big), abs\big(r(x\ \mathtt{d2}\ y)\big) \Big], \tag{16.3}$$

and we obtain hence in the general r-valued case:

$$\tau(\mathtt{d1}, \mathtt{d2}) = \frac{M(\mathtt{d1} \Leftrightarrow \mathtt{d2})}{D}. \tag{16.4}$$

$\tau(\mathtt{d1}, \mathtt{d2})$ corresponds so to the classical ordinal correlation index, yet restricted to the conjointly determined parts of the given digraphs d1 and d2. In the limit case of two crisp linear orders, D equals $n(n-1)$, i.e. the number of irreflexive pairs, and we recover KENDALL's original *tau* index definition.

It is worthwhile noticing that the ordinal correlation index $\tau(\mathtt{d1}, \mathtt{d2})$ one obtains above corresponds in fact to the ratio of:

- $r(\mathtt{d1} \Leftrightarrow \mathtt{d2}) = \frac{M(\mathtt{d1} \Leftrightarrow \mathtt{d2})}{n(n-1)}$:
  the normalised majority margin of the pairwise *relational* equivalence statements, also called *valued ordinal correlation.*
- $d = \frac{D}{n(n-1)}$:
  the normalised epistemic determination of the corresponding pairwise relational equivalence statements, in fact the *determinateness* of the relational equivalence digraph.

The epistemic determination effect is thus successfully *out-factored* from the ordinal correlation effect. With completely determined relations, $\tau(\mathtt{d1}, \mathtt{d2}) = r(\mathtt{d1} \Leftrightarrow \mathtt{d2})$. The ordinal correlation with a completely indeterminate digraph, i.e. when $D = 0$, is set by convention to the indeterminate correlation value 0.0. With uniformly chosen random r-valued digraphs, the expected τ index is 0.0, denoting in fact an indeterminate relational equivalence. The corresponding expected normalised determination d is about 0.333 (see Bisdorff (2012)).

We verify Eq. 16.4 below with the help of the equivalence digraph eqd computed in Listing 16.5 on the previous page.

Listing 16.6 Computing the ordinal correlation index from the equivalence digraph

```
>>> # eqd = EquivalenceDigraph(d1,d2)
>>> M = Decimal('0'); D = Decimal('0')
>>> n2 = eqd.order*(eqd.order - 1)
>>> for x in eqd.actions:
...     for y in eqd.actions:
...         M += eqd.relation[x][y]
...         D += abs(eqd.relation[x][y])
>>> print('r(rd1<=>rd2) = %+.3f, d = %.3f, tau = %+.3f' %\
          (M/n2,D/n2,M/D))
  r(rd1<=>rd2) = +0.026, d = 0.356, tau = +0.073
```

The DIGRAPH3 resources directly provide for the preceding computations the `computeOrdinalCorrelation()` method which renders a dictionary with a `correlation` (τ) and a `determination` (d) attribute. r(`d1` $\Leftrightarrow$ `d2`) is recovered by multiplying τ with d (see Listing 16.7, Line 4).

Listing 16.7 Computing the valued ordinal correlation index

```
1 >>> corrd1d2 = d1.computeOrdinalCorrelation(d2)
2 >>> tau = corrd1d2['correlation']
3 >>> d = corrd1d2['determination']
4 >>> r = tau * d
5 >>> print('tau(d1,d2) = %+.3f, d = %.3f,\
6 ...             r(d1<=>d2) = %+.3f' % (tau, d, r))
7   tau(d1,d2) = +0.073, d = 0.356, r(d1<=>d2) = +0.026
```

The DIGRAPH3 resources provide for convenience a direct `showCorrelation()` method:

```
>>> d1.showCorrelation(\
...          d1.computeOrdinalCorrelation(d2) )
  Correlation indexes:
    Extended Kendall tau         : +0.073
    Epistemic determination      :  0.356
    Bipolar-valued equivalence : +0.026
```

We are now ready for assessing the quality of the NETFLOWS ranking of the movies shown in the heatmap view of Fig. 16.2 on page 213.

16.4 Fitness of Ranking Heuristics

The NETFLOWS ranking of the movies shown in the heatmap view in Fig. 16.2 on page 213 is based on the bipolar-valued outranking digraph modelling the pairwise global "*rated at least as well as*" relation among the 25 movies from the performance tableau instance `gt07`.

Listing 16.8 The bipolar-valued outranking digraph of the star-rated movies

```
>>> bod = BipolarOutrankingDigraph(gt07)
  *------- Object instance description ------*
   Instance class     : BipolarOutrankingDigraph
   Instance name      : rel_grafittiPerfTab.xml
   Actions            : 25
   Criteria           : 15
   Size               : 390
   Determinateness    : 65%
   Valuation domain : {'min': Decimal('-1.0'),
                        'med': Decimal('0.0'),
                        'max': Decimal('1.0'),}
>>> g.computeCoSize()
  188
```

Listing 16.8 on the previous page reveals that the outranking digraph `bod` contains 390 positively validated (Line 7), 188 positively invalidated (Line 13), and 22 indeterminate outranking situations from the potential $25 \times 24 = 600$ irreflexive movie pairs.

Listing 16.9 Line 2 illustrates first how to compute, with the `NetFlowsOrder` class from the `linearOrders` module, the global NETFLOWS ranking `nf`, as shown in the ordered heatmap of Fig. 16.2 on page 213. In Lines 8–27, we compute subsequently the bipolar-valued relational equivalence index between each one of the individual critic's star-ratings and the `nf` ranking.

Listing 16.9 Computing marginal criterion correlations with global NETFLOWS ranking

```
1 >>> from linearOrders import NetFlowsOrder
2 >>> nf = NetFlowsOrder(bod)
3 >>> nf.netFlowsRanking
4   ['mv_QS', 'mv_RR', 'mv_DG', 'mv_NP', 'mv_HN', 'mv_HS', 'mv_SM',
5    'mv_JB', 'mv_PE', 'mv_FC', 'mv_TP', 'mv_CM', 'mv_DF', 'mv_TM',
6    'mv_DJ', 'mv_AL', 'mv_RG', 'mv_MB', 'mv_GH', 'mv_HP', 'mv_BI',
7    'mv_DI', 'mv_FF', 'mv_GG', 'mv_TF']
8 >>> for i,item in enumerate(\
9 ...         bod.computeMarginalVersusGlobalRankingCorrelations(\
10 ...                  nf.netFlowsRanking,ValuedCorrelation=True) ):\
11 ...      print('r(%s<=>nf) = %+.3f' % (item[1],item[0]) )
12
13  r(JH<=>nf)  = +0.500
14  r(JPT<=>nf) = +0.430
15  r(AP<=>nf)  = +0.323
16  r(DR<=>nf)  = +0.263
17  r(MR<=>nf)  = +0.247
18  r(VT<=>nf)  = +0.227
19  r(GS<=>nf)  = +0.160
20  r(CS<=>nf)  = +0.140
21  r(SJ<=>nf)  = +0.137
22  r(RR<=>nf)  = +0.133
23  r(TD<=>nf)  = +0.110
24  r(CF<=>nf)  = +0.110
25  r(SF<=>nf)  = +0.103
26  r(AS<=>nf)  = +0.080
27  r(FG<=>nf)  = +0.027
```

In Listing 16.9 (see Lines 13–27), we obtain the relational equivalence characteristic values shown in the third row of the ranked heatmap (see Fig. 16.2 on page 213). The global NETFLOWS ranking `nf` represents obviously a rather balanced compromise with respect to each movie critic's star-ratings, as there appears no negative correlation with anyone of them. The ranking `nf` apparently takes also correctly into account that the journalist JH, a locally renowned movie critic, shows a higher significance weight (see Line 13).

The ordinal correlation between the global NETFLOWS ranking and the outranking digraph `bod` may be furthermore computed as illustrated in Listing 16.10:

Listing 16.10 Computing the ordinal correlation between NETFLOWS and global outranking digraph

```
1 >>> corrgnf = bod.computeOrdinalCorrelation(nf)
2 >>> bod.showCorrelation(corrgnf)
3  Correlation indexes:
4    Extended Kendall tau        : +0.780
5    Epistemic determination     :  0.300
6    Bipolar-valued equivalence  : +0.234
```

One may notice in Line 4 above that the ordinal correlation τ(bod, nf) index between the NETFLOWS ranking nf and the determined part of the outranking digraph bod is quite high (+0.780). Due to the rather high number of missing data, the r-valued relational equivalence between the nf and the bod digraph, with a characteristic value of only +0.234, may be misleading. Yet, +0.234 still corresponds to a 62% majority support of the movie critics' star-ratings.

It would be interesting to compare similarly the correlations one may obtain with other global ranking heuristics, like the COPELAND ranking rule.

16.5 Illustrating Preference Divergences

The bipolar-valued relational equivalence indexes give us, via the showCriteriaCorrelationTable(ValuedCorrelation=True) method, a further measure for studying how *divergent* may appear the rating opinions expressed by the movie critics.

It is remarkable to notice in Fig. 16.3 that, due to the quite numerous missing data, all pairwise valued ordinal correlation indexes $r(x \Leftrightarrow y)$ appear to be of low value, except the diagonal ones. These reflexive indexes $r(x \Leftrightarrow x)$ would trivially all amount to +1.0 in a plainly determined case. Here they indicate a reflexive normalised determination score d, i.e. the proportion of pairs of movies each critic

```
>>> g = BipolarOutrankingDigraph(t,Normalized=True)
>>> g.showCriteriaCorrelationTable(ValuedCorrelation=True)
Criteria valued ordinal correlation index
     |   AP     AS     CF     CS     DR     FG     GS     JH    JPT     MR     RR     SF     SJ     TD     VT
-----|------------------------------------------------------------------------------------------------------------
  AP | +0.63  +0.04  +0.19  +0.09  +0.22  -0.01  +0.11  +0.23  +0.25  +0.08  +0.02  +0.04  +0.19  +0.04  +0.12
  AS |        +0.77  +0.12  +0.12  +0.04  -0.02  -0.06  +0.02  +0.24  -0.08  +0.07  +0.04  -0.07  -0.01  +0.02
  CF |               +0.77  +0.07  +0.11  +0.03  +0.05  +0.07  +0.10  -0.03  +0.01  +0.00  +0.06  +0.03  -0.04
  CS |                      +0.63  +0.04  -0.02  +0.07  +0.13  +0.25  +0.01  +0.03  +0.00  +0.02  +0.03  +0.07
  DR |                             +0.45  +0.03  +0.07  +0.17  +0.23  +0.16  +0.06  +0.03  +0.10  +0.07  +0.10
  FG |                                    +0.15  -0.01  +0.04  +0.01  +0.06  -0.00  +0.02  +0.01  +0.01  +0.02
  GS |                                           +0.40  +0.07  +0.07  +0.09  -0.02  +0.00  +0.06  +0.04  +0.04
  JH |                                                  +0.77  +0.28  +0.26  +0.15  +0.12  +0.10  +0.05  +0.14
 JPT |                                                         +0.92  +0.15  +0.06  +0.09  +0.08  +0.08  +0.17
  MR |                                                                +0.63  +0.10  +0.08  +0.03  +0.09  +0.10
  RR |                                                                       +0.51  +0.04  +0.01  +0.05  +0.05
  SF |                                                                              +0.18  +0.01  +0.02  +0.05
  SJ |                                                                                     +0.51  +0.03  +0.07
  TD |                                                                                            +0.26  +0.00
  VT |                                                                                                   +0.40
```

Fig. 16.3 Pairwise valued correlation of the movie critics

did evaluate. Critic JPT (the editor of the Graffiti magazine), for instance, evaluated all but one ($d = 24 \times 23/600 = 0.92$), whereas critic FG evaluated only 10 movies among the 25 in discussion ($d = 10 \times 9/600 = 0.15$).

To get a picture of the actual divergence of rating opinions concerning jointly seen pairs of movies, we may develop a Principal Component Analysis of the corresponding τ correlation matrix.[3] The 3D plot of the first 3 principal axes is shown in Fig. 16.2 on page 213.

```
>>> bod.export3DplotOfCriteriaCorrelation(\
...                           ValuedCorrelation=False)
```

The first 3 principal axes support together about 70% of the total inertia. Most eccentric and opposed in their respective rating opinions appear, on the first principal axis with 27.2% inertia, the conservative daily press against labour and public press. On the second principal axis with 23.7.7% inertia, it is the people press versus the cultural critical press. And, on the third axis with still 19.3% inertia, the written media appear most opposed to the radio media (Fig. 16.5 on the next page).

16.6 Exploring the "*better rated*" and the "*as well as rated*" Opinions

In order to furthermore study the quality of a ranking result, it may be interesting to have a separate view on the asymmetric and symmetric parts of the "*at least as well rated as*" opinions (see Sect. 2.3).

Let us first inspect the pairwise asymmetric part, namely the "*better rated than*" and "*less well rated than*" opinions of the movie critics.

```
>>> from digraphs import AsymmetricPartialDigraph
>>> ag = AsymmetricPartialDigraph(bod)
>>> ag.showHTMLRelationTable(\
...     actionsList=g.computeNetFlowsRanking(),ndigits=0)
```

We notice in Fig. 16.4 on the facing page that the NETFLOWS ranking rule inverts in fact just three "*less well rated than*" opinions and four "*better rated than*" ones. A similar look in Fig. 16.6 on page 222 at the symmetric part—the pairwise "*as well rated as*" opinions—suggests a preordered preference structure in several equivalently rated classes.

```
>>> from digraphs import SymmetricPartialDigraph
>>> sg = SymmetricPartialDigraph(bod)
>>> sg.showHTMLRelationTable(\
...             actionsList=g.computeNetFlowsRanking(),\
...             ndigits=0)
```

[3] The 3D PCA plot method requires a running R statistics software (https://www.r-project.org/) installation and the Calmat matrix calculator (see the calmat directory in the DIGRAPH3 resources).

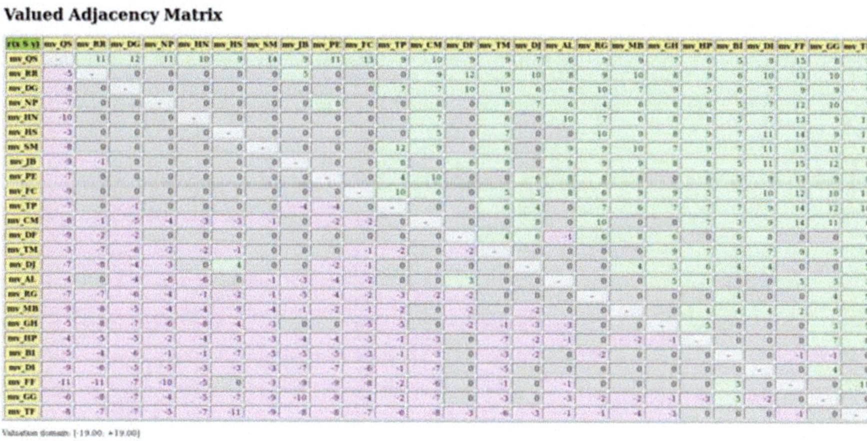

Fig. 16.4 Asymmetric part of the "*at least as well star-rated as*" statements

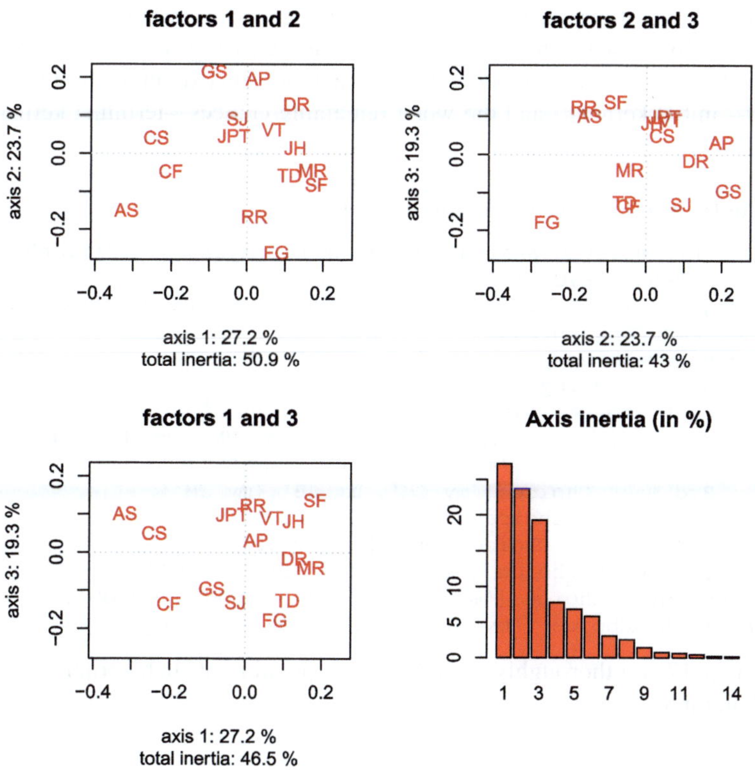

Fig. 16.5 3D PCA plot of the criteria ordinal correlation matrix

Valued Adjacency Matrix

r(x S y)	mv_QS	mv_RR	mv_DG	mv_NP	mv_HN	mv_HS	mv_SM	mv_JB	mv_PE	mv_FC	mv_TP	mv_CM
mv_QS	-	0	0	0	0	0	0	0	0	0	0	0
mv_RR	0	-	5	7	8	6	4	0	2	9	10	0
mv_DG	0	9	-	9	10	5	9	7	8	13	0	0
mv_NP	0	5	7	-	10	3	7	9	8	11	9	0
mv_HN	0	4	8	8	-	5	9	9	8	10	10	0
mv_HS	0	2	5	5	3	-	10	2	5	8	9	0
mv_SM	0	6	7	5	7	6	-	6	8	10	12	0
mv_JB	0	0	5	1	3	4	8	-	9	6	0	13
mv_PE	0	2	6	0	4	3	6	11	-	5	0	0
mv_FC	0	5	11	9	8	2	8	8	7	-	10	0
mv_TP	0	4	0	3	2	3	0	0	0	0	-	6
mv_CM	0	0	0	0	0	0	0	3	0	0	4	-
mv_DF	0	0	0	2	2	2	2	0	1	1	5	1
mv_TM	0	0	0	0	0	0	0	0	1	0	0	2
mv_DJ	0	0	0	0	2	4	2	1	0	0	0	0
mv_AL	0	0	0	0	0	1	0	0	0	0	3	1
mv_RG	0	0	0	0	0	0	0	0	0	0	0	0
mv_MB	0	0	0	0	0	0	0	0	0	0	0	1
mv_GH	0	0	0	0	0	0	0	0	1	0	0	1
mv_HP	0	0	0	0	0	0	0	0	0	0	0	0
mv_BI	0	0	0	0	0	0	0	0	0	0	0	0
mv_DI	0	0	0	0	0	0	0	0	0	0	0	0
mv_FF	0	0	0	0	0	0	0	0	0	0	0	0
mv_GG	0	0	0	0	0	0	0	0	0	0	0	0
mv_TF	0	0	0	0	0	0	0	0	0	0	0	0

r(x S y)	mv_DF	mv_TM	mv_DJ	mv_AL	mv_RG	mv_MB	mv_GH	mv_HP	mv_BI	mv_DI	mv_FF	mv_GG	mv_TF
mv_QS	0	0	0	0	0	0	0	0	0	0	0	0	0
mv_RR	0	0	0	8	0	0	0	0	0	0	0	0	0
mv_DG	0	0	0	0	0	0	0	0	0	0	0	0	0
mv_NP	12	0	0	0	0	0	0	0	0	0	0	0	0
mv_HN	10	0	8	0	0	0	0	0	0	0	0	0	0
mv_HS	10	0	0	1	0	0	0	0	0	0	14	0	0
mv_SM	10	6	4	0	0	0	0	0	0	0	0	0	0
mv_JB	6	8	5	0	0	0	8	0	0	0	0	0	0
mv_PE	5	7	0	0	0	0	5	0	0	0	0	0	0
mv_FC	11	0	0	0	0	0	0	0	0	0	0	0	0
mv_TP	11	0	4	5	0	0	0	0	0	0	0	0	0
mv_CM	3	6	8	9	0	7	3	0	0	0	0	0	0
mv_DF	-	0	6	0	0	0	0	5	6	6	13	13	0
mv_TM	0	-	2	2	7	6	0	0	0	0	0	0	0
mv_DJ	0	4	-	3	5	0	0	0	0	4	3	1	0
mv_AL	0	2	3	-	2	2	0	0	3	1	0	0	0
mv_RG	0	1	3	4	-	1	0	0	0	8	9	0	0
mv_MB	0	2	0	2	3	-	2	0	4	4	2	0	0
mv_GH	0	0	0	0	0	0	-	0	2	3	3	0	0
mv_HP	1	0	0	0	2	0	0	-	3	6	5	0	0
mv_BI	0	0	0	1	0	0	2	1	-	1	0	0	4
mv_DI	0	0	0	1	4	0	1	8	5	-	6	0	8
mv_FF	1	0	5	0	9	0	1	7	0	12	-	9	0
mv_GG	3	0	1	0	0	0	0	0	0	0	3	-	2
mv_TF	0	0	0	0	0	0	0	0	2	2	0	2	-

Valuation domain: [-19.00; +19.00]

Fig. 16.6 Symmetric part of the "*at least as well star-rated as*" statements

Such a kind of preordering of the movies can, for instance, be computed in Listing 16.11 with the `RankingByChoosingDigraph` class from the `transitiveDigraphs` module, where we iteratively extract the best remaining choices—initial kernels—and the worst remaining choices—terminal kernels (see Chap. 17).

Listing 16.11 Bipolar ranking-by-choosing the movies

```
>>> from transitiveDigraphs import RankingByChoosingDigraph
>>> rbc = RankingByChoosingDigraph(bod)
Threading ... # if multiple processing cores are detected
Exiting computing threads
>>> rbc.showRankingByChoosing()
 Ranking by Choosing and Rejecting
   1st Best Choice ['mv_QS']
     2nd Best Choice ['mv_DG','mv_FC','mv_HN','mv_HS','mv_NP',
                      'mv_PE','mv_RR','mv_SM']
       3rd Best Choice ['mv_CM','mv_JB','mv_TM']
         4th Best Choice ['mv_AL','mv_TP']
         4th Worst Choice ['mv_AL','mv_TP']
       3rd Worst Choice ['mv_GH','mv_MB','mv_RG']
     2nd Worst Choice ['mv_DF','mv_DJ','mv_FF','mv_GG']
   1st Worst Choice ['mv_BI','mv_DI','mv_HP','mv_TF']
```

In Chap. 17, we thoroughly discuss the computation of such kernels in bipolar-valued digraphs.

References

Bisdorff R (2012) On measuring and testing the ordinal correlation between bipolar outranking relations. In: Mousseau V, Pirlot M (eds) DAP'2012 From Multiple Criteria Decision Aid to Preference Learning, University of Mons (Belgium), pp 91–100. http://hdl.handle.net/10993/23909

Kendall MG (1938) A new measure of rank correlation. Biometrica 30:81–93

Chapter 17
On Computing Digraph Kernels

Contents

Abstract We illustrate in this chapter, first, the concept of graph kernel, i.e. maximal independent set of vertices. In non-symmetric digraphs, the kernel concept becomes richer and separates into initial and terminal kernels. In, furthermore, lateralised outranking digraphs, initial and terminal kernels become separate and may deliver suitable first and, respectively, last choice recommendations. After commenting the tractability of kernel computations, we close the chapter with the solving of bipolar-valued kernel equation systems.

17.1 What Is a Graph Kernel?

We call *choice* in a graph, respectively, a digraph, a subset of its vertices, respectively, of its nodes or actions. A choice Y is called *internally stable* or *independent* when there exist no links—(edges) or relations (arcs)—between its members. Furthermore, a choice Y is called *externally stable* when for each vertex, node or action x not in Y, there exists at least a member y of Y such that x is linked or related to y. Now, an internally *and* externally stable choice is called a *kernel*.

A first trivial example is immediately given by the maximal independent vertices sets (MISs) of the n-cycle graph (see Fig. 21.5 on page 317). Indeed, each MIS in the n-cycle graph is by definition independent, i.e. *internally stable*, and each non-selected vertex in the n-cycle graph is in relation with either one or even two members of the MIS.

R. Bisdorff, *Algorithmic Decision Making with Python Resources*,
International Series in Operations Research & Management Science 324,
https://doi.org/10.1007/978-3-030-90928-4_17

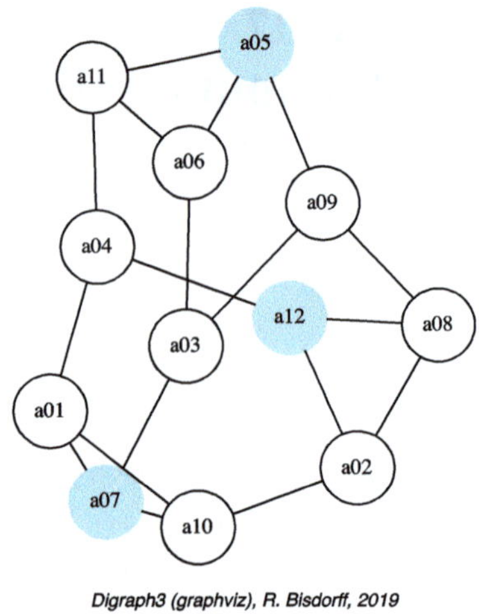

Fig. 17.1 *A random MIS coloured in the random 3-regular graph* `r3g12`. All non-blue vertices are covered by a blue vertex

In all graphs or symmetric digraphs, the *maximality condition* imposed on the internal stability is equivalent to the *external stability* condition. Indeed, if there would exist a vertex or node not related to any of the elements of a choice, we may safely add this vertex or node to the given choice without violating its internal stability. All kernels must hence be maximal independent choices. In fact, in a topological sense, they correspond to maximal *holes* in the given graph.

Figure 17.1 illustrates this coincidence between MISs and kernels in graphs and symmetric digraphs with a random 3-regular graph instance generated in Listing 17.1. A random MIS in this graph may be computed by using the `MISModel` class (see Line 5 below).

Listing 17.1 Generating a random 3-regular graph of order 12

```
1  >>> from graphs import RandomRegularGraph
2  >>> r3g12 = RandomRegularGraph(order=12,\
3  ...                             degree=3,seed=4)
4  >>> from graphs import MISModel
5  >>> mg = MISModel(r3g12)
6    Iteration:  1
7     Running a Gibbs Sampler for 660 step !
8     {'a05', 'a07', 'a12'}  is maximal !
9  >>> mg.exportGraphViz('random3RegularGraph-mis')
10   *---- exporting a dot file for GraphViz tools ---*
11    Exporting to random3RegularGraph-mis.dot
12    fdp -Tpng random3RegularGraph-mis.dot\
13                 -o random3RegularGraph-mis.png
```

It is easily verified in Fig. 17.1 that the computed MIS renders indeed a valid kernel of the given graph. The complete set of kernels of this 3-regular graph instance coincides hence with the set of its MISs.

Listing 17.2 Printing out all maximal independent sets of the random 3-regular graph

```
>>> g.showMIS()
  *---  Maximal Independent Sets ---*
   ['a05', 'a07', 'a12']
   ['a01', 'a06', 'a08']
   ['a07', 'a08', 'a11']
   ['a01', 'a09', 'a11', 'a12']
   ['a01', 'a02', 'a09', 'a11']
   ['a01', 'a06', 'a09', 'a12']
   ['a01', 'a02', 'a06', 'a09']
   ['a07', 'a09', 'a11', 'a12']
   ['a02', 'a07', 'a09', 'a11']
   ['a06', 'a07', 'a09', 'a12']
   ['a04', 'a06', 'a07', 'a08']
   ['a02', 'a04', 'a05', 'a07']
   ['a04', 'a05', 'a07', 'a08']
   ['a09', 'a10', 'a11', 'a12']
   ['a06', 'a09', 'a10', 'a12']
   ['a04', 'a06', 'a08', 'a10']
   ['a04', 'a06', 'a09', 'a10']
   ['a01', 'a03', 'a11', 'a12']
   ['a01', 'a02', 'a03', 'a11']
   ['a01', 'a03', 'a08', 'a11']
   ['a01', 'a03', 'a05', 'a12']
   ['a01', 'a02', 'a03', 'a05']
   ['a01', 'a03', 'a05', 'a08']
   ['a02', 'a03', 'a04', 'a05']
   ['a03', 'a10', 'a11', 'a12']
   ['a03', 'a08', 'a10', 'a11']
   ['a03', 'a05', 'a10', 'a12']
   ['a02', 'a04', 'a06', 'a07', 'a09']
   ['a03', 'a04', 'a05', 'a08', 'a10']
  number of solutions:  29
  cardinality distribution
  card.:  [0, 1, 2, 3, 4, 5, 6, 7, 8, 9, 10, 11, 12]
  freq.:  [0, 0, 0, 3, 24, 2, 0, 0, 0, 0, 0, 0, 0]
  stability number :  5
  execution time: 0.00056 sec.
  Results in self.misset
>>> g.misset
 [frozenset({'a05', 'a07', 'a12'}),
  frozenset({'a11', 'a01', 'a12', 'a09'}),
  frozenset({'a11', 'a01', 'a02', 'a09'}),
   ...
   ...
  frozenset({'a03', 'a10', 'a08', 'a11'}),
  frozenset({'a03', 'a10', 'a12', 'a05'}),
  frozenset({'a03', 'a10', 'a04', 'a05', 'a08'})]
```

All graphs and symmetric digraphs admit MISs, hence also kernels. In the context of digraphs, i.e. *oriented* graphs, the kernel concept gets much richer and separates from the symmetric MIS concept.

17.2 Initial and Terminal Kernels

In an oriented graph context, the internal stability condition of the kernel concept remains untouched; however, the external stability condition gets indeed split up by the orientation into two lateral cases:

1. A *dominant* stability condition, where each non selected node is dominated by at least one member of the kernel
2. An *absorbent* stability condition, where each non-selected node is absorbed by at least one member of the kernel

A both internally stable **and** dominant, respectively, absorbent choice is called a dominant or *initial*, respectively, an absorbent or *terminal* kernel. From a topological perspective, the initial kernel concept looks from the outside of the digraph into its interior, whereas the terminal kernel looks from the interior of a digraph towards its outside. From an algebraic perspective, the initial kernel is a prefix operand, and the terminal kernel is a postfix operand in the kernel equation systems (see Sect. 17.6 on page 242).

Furthermore, as the kernel concept involves conjointly a positive logical refutation (the internal stability) and a positive logical affirmation (the external stability), it appeared rather quickly necessary in our operational developments to adopt a bipolar characteristic $[-1.0, 1.0]$ valuation domain, modelling logical negation by a change of numerical sign and including explicitly a third median value (0.0), expressing logical *indeterminateness*—neither positive nor negative (Bisdorff 2000, 2002, 2004).

In such a bipolar-valued context, we call *prekernel* a choice which is *externally stable* and for which the internal stability condition is *valid or indeterminate*. We say that the independence condition is in this case only *weakly validated*. Notice that all kernels are hence prekernels, but not vice versa.

In graphs or symmetric digraphs, where there is essentially no apparent *laterality*, all prekernels are initial and terminal at the same time. A universal example is given by the *complete* digraph .

Listing 17.3 The prekernels of a complete digraph

```
>>> from digraphs import CompleteDigraph
>>> u = CompleteDigraph(order=5)
>>> u
  *------- Digraph instance description ------*
    Instance class   : CompleteDigraph
    Instance name    : complete
    Digraph Order      : 5
    Digraph Size       : 20
    Valuation domain : [-1.00 ; 1.00]
    --------------------------------
>>> u.showPreKernels()
  *--- Computing preKernels ---*
   Dominant kernels :
```

```
   ['1'] ind. : 1.0; dom. : 1.0; abs. : 1.0
   ['2'] ind. : 1.0; dom. : 1.0; abs. : 1.0
   ['3'] ind. : 1.0; dom. : 1.0; abs. : 1.0
   ['4'] ind. : 1.0; dom. : 1.0; abs. : 1.0
   ['5'] ind. : 1.0; dom. : 1.0; abs. : 1.0
    Absorbent kernels :
   ['1'] ind. : 1.0; dom. : 1.0; abs. : 1.0
   ['2'] ind. : 1.0; dom. : 1.0; abs. : 1.0
   ['3'] ind. : 1.0; dom. : 1.0; abs. : 1.0
   ['4'] ind. : 1.0; dom. : 1.0; abs. : 1.0
   ['5'] ind. : 1.0; dom. : 1.0; abs. : 1.0
 *----- statistics -----
    graph name:  complete
    number of solutions
    dominant kernels :  5
    absorbent kernels:  5
    cardinality frequency distributions
    cardinality      :  [0, 1, 2, 3, 4, 5]
    dominant kernel :   [0, 5, 0, 0, 0, 0]
    absorbent kernel:   [0, 5, 0, 0, 0, 0]
    Execution time   : 0.00004 sec.
    Results in sets: dompreKernels
    and abspreKernels.
```

In a complete digraph, each single node is indeed both an initial and a terminal prekernel candidate and there is no definite begin or end of the digraph to be detected. Laterality is here entirely relative to a specific singleton chosen as reference point of view.

The same absence of laterality is apparent (see Listing 17.4) in two other universal digraph models, the *empty* and the *indeterminate* digraph.

Listing 17.4 The prekernels of the empty or indeterminate digraph

```
>>> from digraphs import EmptyDigraph
>>> ed = EmptyDigraph(order=5)
>>> ed.showPreKernels()
 *--- Computing preKernels ---*
  Dominant kernel :
    ['1', '2', '3', '4', '5']
       independence :  1.0
       dominance    :  1.0
       absorbency   :  1.0
  Absorbent kernel :
    ['1', '2', '3', '4', '5']
       independence :  1.0
       dominance    :  1.0
       absorbency   :  1.0
>>> from digraphs import IndeterminateDigraph
>>> id = IndeterminateDigraph(order=5)
>>> id.showPreKernels()
 *--- Computing preKernels ---*
   Dominant prekernel :
    ['1', '2', '3', '4', '5']
```

```
21       independence :  0.0
22       dominance    :  1.0
23       absorbency   :  1.0
24    Absorbent prekernel :
25    ['1', '2', '3', '4', '5']
26       independence :  0.0
27       dominance    :  1.0
28       absorbency   :  1.0
```

In the empty digraph, the whole set of nodes gives indeed at the same time the unique initial and terminal prekernel (see Lines 6 and 11), similarly, for the *indeterminate* digraph (see Lines 20 and 25).

Both these results make sense, as in a completely empty or indeterminate digraph, there is no *interior* of the digraph defined, only a *border* which is hence at the same time an initial and terminal prekernel (see Sect. 2.4). Notice, however, that in the latter indeterminate case, the complete set of nodes verifies only weakly the internal stability condition (see Lines 21 and 26).

Other common digraph models, although being clearly oriented, may show nevertheless no apparent laterality, like *chordless circuits*, i.e. holes surrounded by an oriented cycle—a circuit—of odd length. They do not admit in fact any initial or terminal prekernel.

Listing 17.5 The prekernels of the 5-circuit digraph

```
>>> from digraphs import CirculantDigraph
>>> c5 = CirculantDigraph(order=5,circulants=[1])
>>> c5.showPreKernels()
  *----- statistics -----
   digraph name:  c5
   number of solutions
    dominant prekernels :  0
    absorbent prekernels:  0
```

Chordless circuits of *even* length $2 \times k$, with $k > 1$, contain however two isomorphic prekernels of cardinality k which qualify conjointly as initial and terminal candidates.

Listing 17.6 The prekernels of the 6-circuit digraph

```
>>> c6 = CirculantDigraph(order=6,circulants=[1])
>>> c6.showPreKernels()
  *--- Computing preKernels ---*
    Dominant preKernels :
    ['1', '3', '5'] ind. : 1.0, dom. : 1.0, abs. : 1.0
    ['2', '4', '6'] ind. : 1.0, dom. : 1.0, abs. : 1.0
    Absorbent preKernels :
    ['1', '3', '5'] ind. : 1.0, dom. : 1.0, abs. : 1.0
    ['2', '4', '6'] ind. : 1.0, dom. : 1.0, abs. : 1.0
```

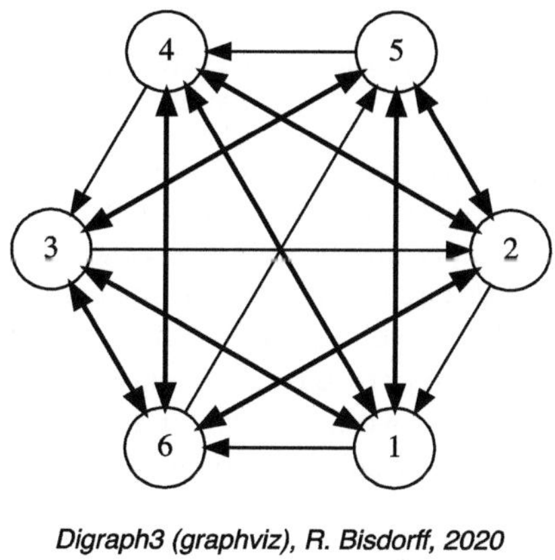

Fig. 17.2 *The dual of the chordless 6-circuit.* No initial or terminal prekernel—weakly independent and dominant, respectively, absorbent subset of nodes—may be found in this kind of digraphs

Dual of weak 6-circuit

Converse of weak 6-circuit

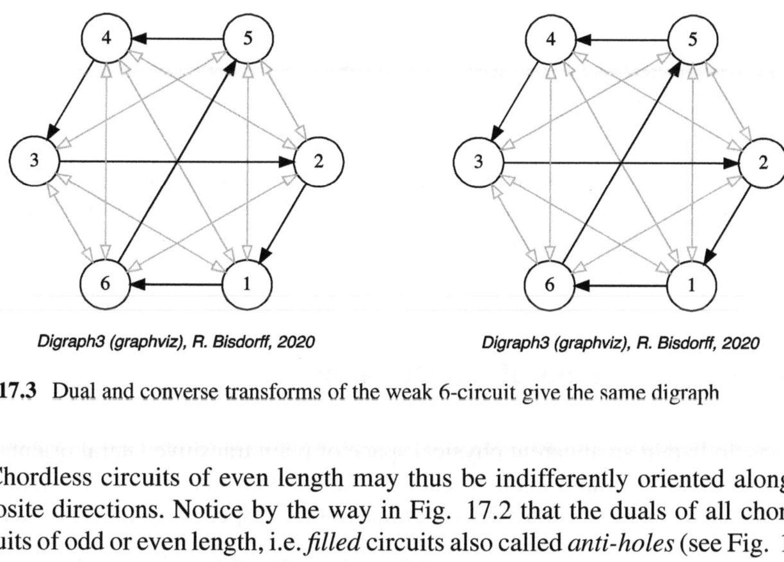

Fig. 17.3 Dual and converse transforms of the weak 6-circuit give the same digraph

Chordless circuits of even length may thus be indifferently oriented along two opposite directions. Notice by the way in Fig. 17.2 that the duals of all chordless circuits of odd or even length, i.e. *filled* circuits also called *anti-holes* (see Fig. 17.3), never contain any potential prekernel candidates.

Listing 17.7 The prekernels of the dual of the 6-circuit digraph

```
>>> dc6 = -c6   # dc6 = DualDigraph(c6)
>>> dc6.showPreKernels()
  *----- statistics -----
    graph name:  dual_c6
    number of solutions
     dominant prekernels :  0
```

```
     absorbent prekernels:  0
>>> dc6.exportGraphViz(fileName='dualChordlessCircuit')
  *---- exporting a dot file for GraphViz tools ----*
    Exporting to dualChordlessCircuit.dot
    circo -Tpng dualChordlessCircuit.dot\
               -o dualChordlessCircuit.png
```

We call *weak*, a chordless circuit with indeterminate inner part.

In Listing 17.8, we use the `IndeterminateInnerPart` parameter of the `CirculantDigraph` class for constructing such a weak chordless 6-circuit digraph. It is worth noticing in Fig. 17.3 on the preceding page that the *dual* version of a weak circuit corresponds to its *converse* version.[1]

Listing 17.8 The weak 6-circuit digraph

```
>>> from digraphs import CirculantDigraph
>>> c6 = CirculantDigraph(order=6,circulants=[1],\
...                       IndeterminateInnerPart=True)
>>> (-c6).exportGraphViz()
  *---- exporting a dot file for GraphViz tools ---------*
   Exporting to dual_c6.dot
   circo -Tpng dual_c6.dot -o dual-c6.png
>>> (~c6).exportGraphViz()
  *---- exporting a dot file for GraphViz tools ---------*
   Exporting to converse_c6.dot
   circo -Tpng converse_c6.dot -o converse-c6.png
```

Weak chordless circuits of length n are in fact part of the class of digraphs that are invariant under the *codual* transform, $cn = -(\sim cn) =\sim (-cn)$.[2] When digraph cn is a weak chordless n-circuit, cn, $-cn$, $\sim cn$, and $\sim (-cn)$ will all admit the same set of prekernels.

17.3 Kernels in Lateralized Digraphs

Humans do live in an apparent physical space of plain transitive lateral orientation, fully empowered in finite geometrical 3D models with linear orders, where first, respectively, last ranked, nodes deliver unique initial, respectively, terminal, kernels.

[1] Not to be confused with the dual graph of a plane graph g that has a vertex for each face of g. Here we mean the *less than* (strict converse) relation corresponding to a *greater than or equal to* relation, or the *less than or equal to* relation corresponding to a (strict converse) *greater than* relation.

[2] The class of *self-codual* bipolar-valued digraphs consists of all weakly asymmetric digraphs, i.e. digraphs containing only asymmetric and/or indeterminate links. Limit cases consist of, on the one side, full tournaments with indeterminate reflexive links and, on the other side, fully indeterminate digraphs. In this class, the converse (inverse $\sim$) operator is indeed identical to the dual (negation $-$) one.

Similarly, in finite preorders, the first, respectively, last, equivalence classes deliver the unique initial, respectively, unique terminal, kernels. More generally, in finite partial orders, i.e. asymmetric and transitive digraphs, topological sort algorithms will easily reveal on the first, respectively, last, level all unique initial, respectively, terminal, kernels.

In genuine random digraphs, however, we may need to check for each of its MISs, whether one, both, or none of the lateralized external stability conditions may be satisfied. Consider, for instance, in Listing 17.9, the following random digraph instance of order 7 and generated with an arc probability of 30%.

Listing 17.9 Generating a random digraph `rd` of order 7 and arc probability 0.3

```
>>> from randomDigraphs import RandomDigraph
>>> rd = RandomDigraph(order=7,arcProbability=0.3,seed=5)
>>> rd.exportGraphViz('randomLaterality')
  *---- exporting a dot file for GraphViz tools ---*
   Exporting to randomLaterality.dot
   dot -Grankdir=BT -Tpng randomLaterality.dot\
                    -o randomLaterality.png
```

The random digraph shown in Fig. 17.4 has no apparent special properties, except from being connected (see Line 3 in Listing 17.10 on the next page).

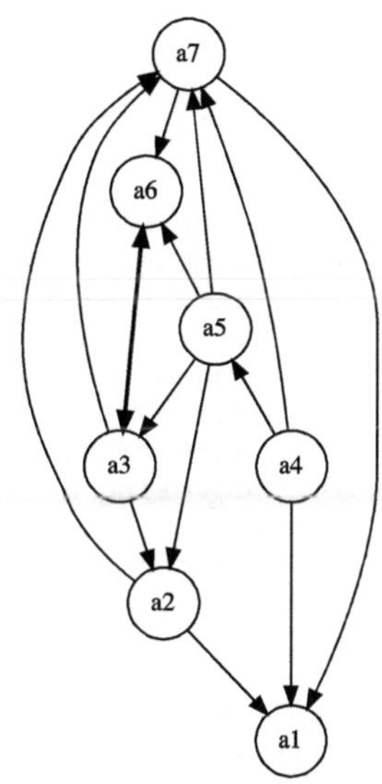

Digraph3 (graphviz), R. Bisdorff, 2020

Fig. 17.4 *A random digraph instance of order 7 and arc probability* 0.3. The digraph instance is neither asymmetric (`a3` $\Leftrightarrow$ `a6`) nor symmetric (`a2` $\rightarrow$ `a1`, `a1` $\not\rightarrow$ `a2`), and the digraph is not transitive (`a5` $\rightarrow$ `a2` $\rightarrow$ `a1`, but `a5` $\not\rightarrow$ `a1`)

Listing 17.10 Inspecting the properties of random digraph `rd`

```
1  >>> rd.showComponents()
2    *--- Connected Components ---*
3    1: ['a1', 'a2', 'a3', 'a4', 'a5', 'a6', 'a7']
4  >>> rd.computeSymmetryDegree(Comments=True)
5    Symmetry degree (%) of digraph <randomDigraph>:
6     arcs x>y: 14, symmetric: 1, asymmetric: 13
7     symmetric/arcs =  0.071
8  >>> rd.computeChordlessCircuits()
9    [] # no chordless circuits detected
10 >>> rd.computeTransitivityDegree(Comments=True)
11   Transitivity degree (%) of graph <randomDigraph>:
12    triples x>y>z: 23, closed: 11, open: 12
13    closed/triples =  0.478
```

There are no chordless circuits (see Line 9 above), and more than half of the required transitive closure is missing (see Line 13 above).

Now, we know that its potential prekernels must be among its set of maximal independent choices.

Listing 17.11 Inspecting the prekernels of random digraph `rd`

```
>>> rd.showMIS()
  *---  Maximal independent choices ---*
    ['a2', 'a4', 'a6']
    ['a6', 'a1']
    ['a5', 'a1']
    ['a3', 'a1']
    ['a4', 'a3']
    ['a7']
>>> rd.showPreKernels()
  *--- Computing preKernels ---*
   Dominant preKernels :
    ['a2', 'a4', 'a6']
       independence :  1.0
       dominance    :  1.0
       absorbency   :  -1.0
       covering     :  0.500
    ['a4', 'a3']
       independence :  1.0
       dominance    :  1.0
       absorbency   :  -1.0
       covering     :  0.600
   Absorbent preKernels :
    ['a3', 'a1']
       independence :  1.0
       dominance    :  -1.0
       absorbency   :  1.0
       covering     :  0.500
    ['a6', 'a1']
       independence :  1.0
       dominance    :  -1.0
```

```
31        absorbency     :  1.0
32        covering       :  0.600
```

Among the six MISs contained in random digraph `rd`, we discover in Listing 17.11 on the facing page two initial and two terminal prekernels. Notice by the way the covering values (between 0.0 and 1.0) shown by the `showPreKernels()` method (Lines 16, 21, 27, and 32). The higher this value, the more the corresponding prekernel candidate makes apparent the digraph's laterality. In Fig. 17.5, the same digraph `rd` is redrawn by looking into its interior via the best covering initial prekernel candidate: the dominant choice {`a3`, `a4`} (coloured in yellow), and looking out of it via the best covered terminal prekernel candidate: the absorbent choice {`a1`, `a6`} (coloured in blue).

```
>>> rd.exportGraphViz(fileName='orientedLaterality',\
...                   firstChoice=set(['a3', 'a4']),\
...                   lastChoice=set(['a1', 'a6']))
  *---- exporting a dot file for GraphViz tools ----*
   Exporting to orientedLaterality.dot
   dot -Grankdir=BT -Tpng orientedLaterality.dot\
                    -o orientedLaterality.png
```

As all reasonable bipolar-valued outranking digraphs usually show some laterality—the marginal criteria preferences being not all totally contradictory—initial and terminal prekernels provide convincing first, respectively, last, choice recommendations as illustrated in Chap. 4 on page 41.

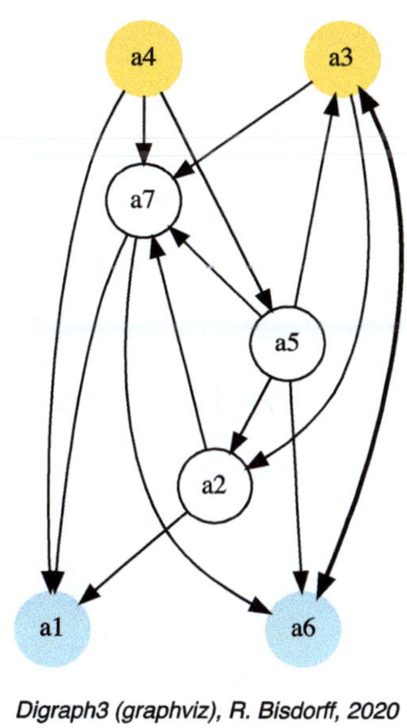

Fig. 17.5 A random digraph oriented by best covering initial and best covered terminal prekernels

17.4 Computing First and Last Choice Recommendations

To illustrate this idea, let us finally compute in Listing 17.12 first and last choice recommendations in a random bipolar-valued outranking digraph.

Listing 17.12 Generating a random bipolar-valued outranking digraph

```
>>> from outrankingDigraphs import\
...                    RandomBipolarOutrankingDigraph
>>> g = RandomBipolarOutrankingDigraph(seed=5)
>>> g
  *------- Object instance description ------*
   Instance class    : RandomBipolarOutrankingDigraph
   Instance name     : randomOutranking
   Actions           : 7
   Criteria          : 7
   Size              : 26
   Determinateness   : 34.275
   Valuation domain  : {'min': -100.0, 'med': 0.0, 'max': 100.0}
>>> g.showHTMLPerformanceTableau()
```

The associated random performance tableau shown in Fig. 17.6 reveals the performance evaluations of 7 potential decision alternatives with respect to 7 decision criteria supporting each an increasing performance scale from 0.0 to 100.0. Notice the missing performance data concerning decision alternatives a2 and a5. The resulting *strict* outranking—i.e. a weighted majority supported—*'better than without considerable counter-performance'*—digraph is shown in Fig. 17.7 on the next page.

```
>>> gcd = ~(-g)   # Codual: the converse of the negation
>>> gcd.exportGraphViz(fileName='tutOutRanking')
  *---- exporting a dot file for GraphViz tools ----*
   Exporting to tutOutranking.dot
   dot -Grankdir=BT -Tpng tutOutranking.dot\
                    -o tutOutranking.png
```

Performance table randomOutranking

criterion	g1	g2	g3	g4	g5	g6	g7
a1	64.90	1.31	13.88	98.24	94.10	14.57	31.00
a2	NA	NA	61.75	87.24	69.06	6.51	81.85
a3	11.32	27.95	12.67	28.93	96.66	30.14	48.07
a4	46.91	91.63	0.18	96.15	89.37	60.31	31.58
a5	NA	76.57	87.14	53.92	29.88	0.34	48.12
a6	54.38	15.96	20.95	67.78	36.12	67.79	70.47
a7	57.39	79.71	21.55	20.48	16.60	33.79	5.70

Fig. 17.6 The performance tableau of a random outranking digraph instance

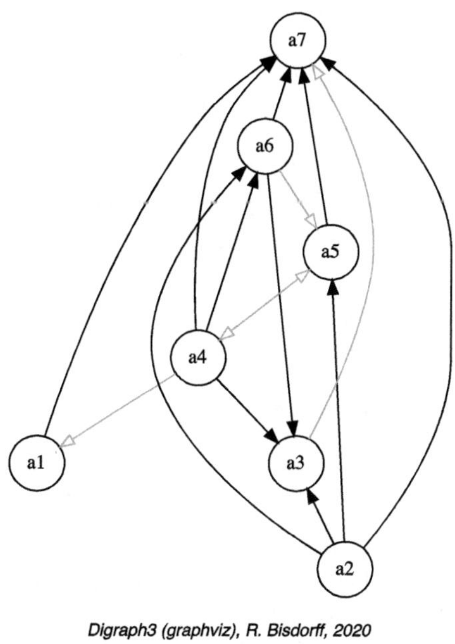

Fig. 17.7 *A random strict outranking digraph instance.* All decision alternatives appear strictly better performing than alternative `a7`. We call it a CONDORCET loser and it is an evident terminal prekernel candidate. On the other side, three alternatives: `a1`, `a2`, and `a4` are not dominated. They give together an initial prekernel candidate

Listing 17.13 Computing the prekernels of the strict outranking digraph `gcd`

```
1  >>> gcd.showPreKernels()
2    *--- Computing preKernels ---*
3     Dominant preKernels :
4      ['a1', 'a2', 'a4']
5         independence :  0.00
6         dominance    :  6.98
7         absorbency   :  -48.84
8         covering     :  0.667
9     Absorbent preKernels :
10     ['a3', 'a7']
11        independence :  0.00
12        dominance    :  -74.42
13        absorbency   :  16.28
14        covered      :  0.800
```

With such unique disjoint initial and terminal prekernels (see Lines 4 and 10 in Listing 17.13), the random digraph `gcd` is hence clearly lateralized. Indeed, these initial and terminal prekernels of the codual outranking digraph reveal *first*, respectively, *last*, choice recommendations one may formulate on the basis of a given outranking digraph instance.

Listing 17.14 Computing a first and last choice recommendation from digraph `gcd`

```
>>> g.showBestChoiceRecommendation()
 Rubis best choice recommendation(s) (BCR)
  (in decreasing order of determinateness)
```

```
4    Credibility domain: [-100.00,100.00]
5    === >> potential first choice(s)
6     * choice              : ['a1', 'a2', 'a4']
7       independence        : 0.00
8       dominance           : 6.98
9       absorbency          : -48.84
10      covering (%)        : 66.67
11      determinateness (%) : 57.97
12      - most credible action(s) =
13                          {'a4': 20.93,'a2': 20.93}
14   === >> potential last choice(s)
15    * choice              : ['a3', 'a7']
16      independence        : 0.00
17      dominance           : -74.42
18      absorbency          : 16.28
19      covered (%)         : 80.00
20      determinateness (%) : 64.62
21      - most credible action(s) = { 'a7': 48.84, }
```

Notice in Lines 13 and 21 in Listing 17.14 on the previous page that solving the valued kernel equation system provides furthermore a positive characterisation of the most credible decision alternatives in each respective choice recommendation. Alternatives `a2` and `a4` give equivalent candidates for a unique *first* choice, and alternative `a7` is clearly confirmed as the *last* choice.

In Fig. 17.8, we orient the drawing of the strict outranking digraph instance with the help of these first and last choice recommendations.

```
>>> gcd.exportGraphViz(fileName='firstLastOrientation',\
...                      firstChoice=['a2','a4'],\
...                      lastChoice=['a7'])
 *---- exporting a dot file for GraphViz tools -----*
  Exporting to firstLastOrientation.dot
  dot -Grankdir=BT -Tpng firstLastOrientation.dot\
                    -o firstLastOrientation.png
```

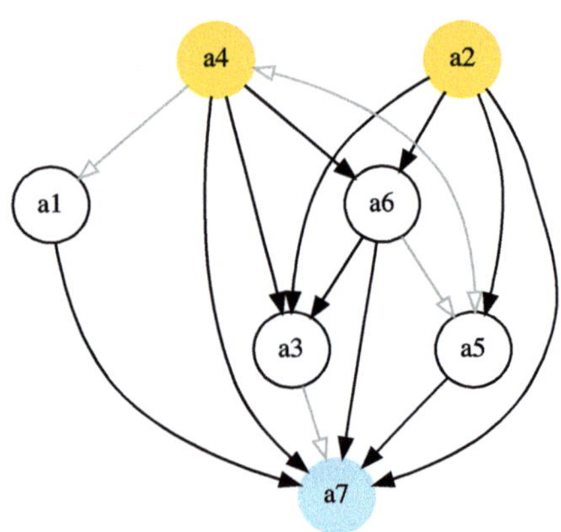

Fig. 17.8 *The strict outranking digraph oriented by its first and last choice recommendations*. The grey arrows, like the one between alternatives `a4` and `a1`, represent indeterminate preferential situations. Alternative `a1` appears hence to be rather incomparable to all the other, except alternative `a7`

Heatmap of Performance Tableau 'randomOutranking'

criteria	g4	g7	g5	g6	g1	g2	g3
weights	9	10	6	5	4	8	1
tau(*)	+0.64	+0.40	+0.29	+0.17	+0.02	-0.05	-0.10
a4	96	32	89	60	47	92	0
a2	87	82	69	7	NA	NA	62
a6	68	70	36	68	54	10	21
a1	98	31	94	15	65	1	14
a5	54	48	30	0	NA	77	87
a3	29	48	97	30	11	28	13
a7	20	6	17	34	57	80	22

Color legend:

quantile	20.00%	40.00%	60.00%	80.00%	100.00%

(*) tau: *Ordinal (Kendall) correlation between marginal criterion and global ranking relation*
Ranking rule: **Copeland**
Ordinal (Kendall) correlation between global ranking and global outranking relation: **+0.848**

Fig. 17.9 Heatmap with Copeland ranking of the performance tableau

It may be interesting to compare this result with a COPELAND ranking of the underlying performance tableau (see Sect. 8.2 on ranking with incommensurable criteria).

```
>>> g.showHTMLPerformanceHeatmap(colorLevels=5, ndigits=0,\
...                 Correlations=True, rankingRule='Copeland')
```

In the resulting linear ranking (see Fig. 17.9), alternative a4 is set at first rank, followed by alternative a2. This makes sense as alternative a4 shows three performances in the first quintile, whereas a2 is only partially evaluated and shows only two such excellent performances. But a4 also shows a very weak performance in the first quintile. Both decision actions, hence, do not show eventually a performance profile that would make apparent a clear preference situation in favour of one or the other. In this sense, the prekernels based best choice recommendations may appear more faithful with respect to the actually definite strict outranking relation than any *'forced'* linear ranking result as shown in Fig. 17.9.

17.5 Tractability of Kernel Computation

Let us give some hints on the *tractability* of prekernel computations. Detecting all prekernels in a digraph is a computationally difficult problem. Checking external stability conditions for an independent choice is equivalent to checking its maximality and may be done in the linear complexity of the order of the digraph. However, checking all independent choices contained in a digraph may get hard already for tiny sparse digraphs of order $n > 20$ (Bisdorff 2006). Indeed, the worst case is given by an empty or indeterminate digraph where the set of all potential independent choices to check is in fact the power set of the vertices.

```
1 >>> from digraphs import EmptyDigraph
2 >>> e = EmptyDigraph(order=20)
3 >>> e.showMIS()    # by visiting all 2^20 independent choices
4   *---  Maximal independent choices ---*
5     [ '1', '2', '3', '4', '5', '6', '7', '8', '9','10',
6      '11','12','13','14','15','16','17','18','19','20']
7     number of solutions:  1
8     execution time: 1.47640 sec.   # <-----
9 >>> 2**20
10  1048576
```

Now, there exist more efficient specialised algorithms for directly enumerating MISs and dominant or absorbent kernels contained in specific digraph models without visiting all independent choices (Bisdorff 2006). *Alain Hertz* provided kindly such a MISs enumeration algorithm for the DIGRAPH3 project (the `showMIS_AH()` method). When the number of independent choices is big compared to the actual number of MISs, like in very sparse or empty digraphs, the performance difference may be dramatic (see Line 8 above and Line 16 below).

Listing 17.15 Enumerating MISs by visiting only maximal independent choices (*A. Hertz*)

```
1 >>> e.showMIS_AH()
2   # by visiting only maximal independent choices
3   *-----------------------------------*
4   * Python implementation of Hertz's  *
5   * algorithm for generating all MISs *
6   * R.B. version 7(6)-25-Apr-2006     *
7   *-----------------------------------*
8   ===>>> Initial solution :
9   [ '1','2','3','4','5','6','7','8','9','10','11',
10     '12','13','14','15','16','17','18','19','20']
11  *---- results ----*
12  [ '1','2','3','4','5','6','7','8','9','10','11',
13     '12','13','14','15','16','17','18','19','20']
14  *---- statistics ----*
15  MIS solutions     :  1
16  execution time    : 0.00026 sec. # <----
17  iteration history:  1
```

For more or less dense strict outranking digraphs of modest order, as facing usually in MCDA applications, enumerating all independent choices remains however in most cases tractable, especially by using a very efficient iterator generator with the `independentChoices()` method shown in Listing 17.16.

Listing 17.16 Generating all independent choices in a digraph

```
def independentChoices(self,U):
    """
    Generator for all independent choices with associated
    dominated, absorbed and independent neighborhoods
    of digraph instance self.
```

```
    Initiate with U = self.singletons().
    Yields [(independent choice, domnb, absnb, indnb)].
    """
    if U == []:
        yield [(frozenset(),set(),set(),set(self.actions))]
    else:
        x = list(U.pop())
        for S in self.independentChoices(U):
            yield S
            if x[0] <=  S[0][3]:
                Sxgamdom = S[0][1]  | x[1]
                Sxgamabs = S[0][2]  | x[2]
                Sxindep = S[0][3] &  x[3]
                Sxchoice = S[0][0]  | x[0]
                Sx =  [(Sxchoice,Sxgamdom,Sxgamabs,Sxindep)]
                yield Sx
```

And, checking maximality of independent choices via the external stability conditions during their enumeration with the `computePreKernels()` method shown in Listing 17.17 provides the effective advantage of computing all initial and terminal prekernels in a single loop (see Line 10 and Bisdorff 2006).

Listing 17.17 Computing dominant and absorbent prekernels

```
1  def computePreKernels(self):
2      """
3      computing dominant and absorbent preKernels:
4      Result in self.dompreKernels and self.abspreKernels
5      """
6      actions = set(self.actions)
7      n = len(actions)
8      dompreKernels = set()
9      abspreKernels = set()
10     for choice in self.independentChoices(self.singletons()):
11         restactions = actions - choice[0][0]
12         if restactions <= choice[0][1]:
13             dompreKernels.add(choice[0][0])
14         if restactions <= choice[0][2]:
15             abspreKernels.add(choice[0][0])
16     self.dompreKernels = dompreKernels
17     self.abspreKernels = abspreKernels
```

Finally, we use our bipolar-valued epistemic logic framework for computing the credibility that an individual node of the digraph fits with being a member of an initial or terminal prekernel. For this purpose, we use kernel equation systems (Schmidt and Ströhlein 1985).

17.6 Solving Kernel Equation Systems

Let $G(X, R)$ be a crisp irreflexive digraph defined on a finite set X of nodes and where R is the corresponding $\{-1, +1\}$-valued adjacency matrix. Let Y be the $\{-1, +1\}$-valued membership characteristic (row) vector of a choice in X.

When Y satisfies the following equation system:

$$Y \circ R \; = \; -Y \, , \tag{17.1}$$

where for all x in X,

$$(Y \circ R)(x) \; = \; \max_{y \in X, x \neq y} \big(\min(Y(x), R(x, y))\big) \, , \tag{17.2}$$

then Y characterises an *initial kernel* (Bisdorff 2006; Bisdorff et al. 2006).

When transposing now the membership characteristic vector Y into a column vector Y^t, the following equation system: $R \circ Y^t \; = \; -Y^t$ makes Y^t similarly characterise a *terminal kernel*.

Let us verify this result in Listing 17.18 on a tiny random digraph.

Listing 17.18 Verifying the kernel equation system on a tiny random digraph

```
>>> from digraphs import RandomDigraph
>>> g = RandomDigraph(order=3,seed=1)
>>> g.showRelationTable()
  * ---- Relation Table -----
     R   | 'a1'   'a2'   'a3'
  ------|---------------------
   'a1' |  -1     +1     -1
   'a2' |  -1     -1     +1
   'a3' |  +1     +1     -1
>>> g.showPreKernels()
  *--- Computing preKernels ---*
   Dominant preKernels :
     ['a3']
   Absorbent preKernels :
     ['a2']
```

It is easy to verify by hand that the characteristic vector $[-1, -1, +1]$ satisfies the initial kernel equation system; node `a3` gives an *initial* kernel. Similarly, the characteristic vector $[-1, +1, -1]$ verifies indeed the terminal kernel equation system and node `a2` gives hence a *terminal* kernel.

We succeeded now in generalising crisp kernel equation systems to genuine bipolar-valued digraphs (Bisdorff 2006; Bisdorff et al. 2006). The constructive proof, found by *Marc Pirlot*, is based on the following *fixpoint equation* that may be used for computing bipolar-valued kernel membership vectors:

$$T(Y) \; := \; -(Y \circ R) = Y \, . \tag{17.3}$$

John von Neumann showed indeed that, when a digraph $G(X, R)$ is acyclic with a unique initial kernel K characterised by its membership characteristics vector Y_k, then the following double fixpoint equation:

$$T^2(Y) \;:=\; -\big(-(Y \circ R) \circ R\big) \;=\; Y \tag{17.4}$$

will admit a stable *high* and a stable *low* fixpoint solutions that converge both to Y_k (Schmidt and Ströhlein 1985).

Inspired by the crisp double fixpoint equation 17.4, we observed that for a given bipolar-valued digraph $G(X, R)$, each one of its dominant or absorbent prekernels K_i in X determines an induced partial digraph $G(X, R/K_i)$ which is acyclic and admits K_i as unique prekernel (Bisdorff 1997).

Following the *von Neumann* fixpoint algorithm, a similar bipolar-valued extended double fixpoint algorithm, applied to $G(X, R/K_i)$, allows us to compute the associated bipolar-valued kernel characteristic vectors Y_i in polynomial complexity.

Algorithm 17.1 Computing bipolar-valued kernel characteristic vectors

in : bipolar-valued digraph $G(X, R)$,
out : set $\{Y_1, Y_2, ..\}$ of bipolar-valued kernel membership characteristic vectors.

1. Enumerate all initial and terminal prekernels $K_1, K_2, ...$ in the given bipolar-valued digraph (see Listing 17.17 on page 241);
2. For each crisp initial kernel K_i:
   a. Construct a partially determined subgraph $G(X, R/K_i)$ supporting exactly this unique initial kernel K_i;
   b. Use the double fixpoint equation T^2 (17.4) with the partially determined adjacency matrix R/K_i for computing a stable low and a stable high fixpoint;
   c. Determine the bipolar-valued K_i-membership characteristic vector Y_i with an epistemic disjunction of the previous low and high fixpoints;
3. Repeat step (2) for each terminal kernel K_j by using the double fixpoint equation T^2 with the transpose of the adjacency matrix R/K_j.

Time for a Practical Illustration

We reconsider the random digraph `g` generated in Listing 17.12 on page 236. Digraph `g` models the pairwise outranking situations between seven decision alternatives evaluated on seven incommensurable performance criteria. We recompute its corresponding bipolar-valued prekernels on the associated codual digraph `gcd`.

```
>>> gcd = ~(-g) # strict outranking digraph
>>> gcd.showPreKernels()
 *--- Computing prekernels ---*
  Dominant prekernels :
   ['a1', 'a4', 'a2']
      independence :   +0.000
```

```
7        dominance     :  +0.070
8        absorbency    :  -0.488
9        covering      :  +0.667
10   Absorbent prekernels :
11     ['a7', 'a3']
12       independence :  +0.000
13       dominance     :  -0.744
14       absorbency    :  +0.163
15       covered       :  +0.800
16  *----- statistics -----
17   graph name:  converse-dual_rel_randomperftab
18   number of solutions
19    dominant kernels :  1
20    absorbent kernels:  1
21   cardinality frequency distributions
22    cardinality      :  [0, 1, 2, 3, 4, 5, 6, 7]
23    dominant kernel :  [0, 0, 0, 1, 0, 0, 0, 0]
24    absorbent kernel:  [0, 0, 1, 0, 0, 0, 0, 0]
25   Execution time   : 0.00022 sec.
```

The codual outranking digraph `gcd`, modelling a strict outranking relation, admits an initial prekernel {`a1`, `a2`, `a4`} and a terminal one {`a3`, `a7`} (see Lines 5 and 11 above).

In Listing 17.19, we now compute, with the `domkernelrestrict()` method, the initial prekernel-restricted adjacency table (see Fig. 17.10).

Listing 17.19 Computing a dominant prekernel restricted adjacency table

```
>>> k1Relation = gcd.domkernelrestrict(['a1','a2','a4'])
>>> gcd.showHTMLRelationTable(\
...         actionsList=['a1','a2','a4','a3','a5','a6','a7'],\
...         relation=k1Relation,\
...         tableTitle='K1 restricted adjacency table')
```

The corresponding initial prekernel membership characteristic vector may be computed with the `computeKernelVector()` method.

Fig. 17.10 *Initial kernel* `a1, a2, a4`} *restricted adjacency table*
The outranking situation between alternatives `a4` and `a1` being *indeterminate*, this initial prekernel is only weakly independent

K1 restricted adjacency table

r(x S y)	a1	a2	a4	a3	a5	a6	a7
a1	-	-0.23	-1.00	0.00	0.00	0.00	0.16
a2	-0.21	-	-0.21	0.21	0.44	0.05	0.49
a4	0.00	-0.21	-	0.21	0.00	0.07	0.58
a3	-0.28	-0.21	-0.74	-	0.00	0.00	0.00
a5	-0.26	-0.67	0.00	0.00	-	0.00	0.00
a6	-0.12	-0.49	-0.49	0.00	0.00	-	0.00
a7	-0.51	-0.49	-0.86	0.00	0.00	0.00	-

Valuation domain: [-1.00; +1.00]

Listing 17.20 Fixpoint iterations for initial prekernel {a1, a2, a4}

```
>>> gcd.computeKernelVector(['a1','a2','a4'],\
...                    Initial=True,Comments=True)
 --> Initial prekernel: {'a1', 'a2', 'a4'}
 initial low vector : [-1.00,-1.00,-1.00,-1.00,-1.00,-1.00,-1.00]
 initial high vector: [+1.00,+1.00,+1.00,+1.00,+1.00,+1.00,+1.00]
 1st low vector     : [ 0.00,+0.21,-0.21, 0.00,-0.44,-0.07,-0.58]
 1st high vector    : [+1.00,+1.00,+1.00,+1.00,+1.00,+1.00,+1.00]
 2nd low vector     : [ 0.00,+0.21,-0.21, 0.00,-0.44,-0.07,-0.58]
 2nd high vector    : [ 0.00,+0.21,-0.21,+0.21,-0.21,-0.05,-0.21]
 3rd low vector     : [ 0.00,+0.21,-0.21,+0.21,-0.21,-0.07,-0.21]
 3rd high vector    : [ 0.00,+0.21,-0.21,+0.21,-0.21,-0.05,-0.21]
 4th low vector     : [ 0.00,+0.21,-0.21,+0.21,-0.21,-0.07,-0.21]
 4th high vector    : [ 0.00,+0.21,-0.21,+0.21,-0.21,-0.07,-0.21]
 Iterations         : 4
 low & high fusion  : [ 0.00,+0.21,-0.21,+0.21,-0.21,-0.07,-0.21]
 Choice vector for initial prekernel: {'a1', 'a2', 'a4'}
   'a2': +0.21
   'a4': +0.21
   'a1':  0.00
   'a6': -0.07
   'a3': -0.21
   'a5': -0.21
   'a7': -0.21
```

In Listing 17.20, we start the fixpoint computation with an empty set characterisation as first low vector and a complete set X characterising high vector. After each iteration, the low vector is set to the negation of the previous high vector and the high vector is set to the negation of the previous low vector.

A unique stable prekernel characteristic vector Y_1 is here attained at the fourth iteration with positive members a2: +0.21 and a4: +0.21 (60.5% criteria significance majority), a1: 0.00 being an ambiguous potential member. Alternatives a3, a5, a6, and a7 are all negative members, i.e. positive *non members* of this outranking prekernel.

Let us also compute the restricted adjacency table for the outranked, i.e. the *terminal* prekernel {a3, a7} (Fig. 17.11).

```
>>> k2Relation = gcd.abskernelrestrict(['a3','a7'])
>>> gcd.showHTMLRelationTable(\
...   actionsList=['a3','a7','a1','a2','a4','a5','a6'],\
...   relation=k2Relation,\
...   tableTitle='K2 restricted adjacency table')
```

Fig. 17.11 *Terminal prekernel* {a3, a7} *restricted adjacency table*. Again, we notice that this terminal prekernel is only weakly independent

K2 restricted adjacency table

r(x S y)	a3	a7	a1	a2	a4	a5	a6
a3	-	0.00	-0.28	-0.21	-0.74	-0.40	-0.53
a7	-1.00	-	-0.51	-0.49	-0.86	-0.67	-0.63
a1	0.00	0.16	-	0.00	0.00	0.00	0.00
a2	0.21	0.49	0.00	-	0.00	0.00	0.00
a4	0.21	0.58	0.00	0.00	-	0.00	0.00
a5	0.00	0.16	0.00	0.00	0.00	-	0.00
a6	0.30	0.26	0.00	0.00	0.00	0.00	-

Valuation domain: [-1.00; +1.00]

The corresponding bipolar-valued characteristic vector Y_2 may be computed as follows:

```
1  >>> gcd.computeKernelVector(['a3','a7'],\
2  ...                          Initial=False,Comments=True)
3   --> Terminal prekernel: {'a3', 'a7'}
4   initial low vector  : [-1.00,-1.00,-1.00,-1.00,-1.00,-1.00,-1.00]
5   initial high vector : [+1.00,+1.00,+1.00,+1.00,+1.00,+1.00,+1.00]
6   1st low vector      : [-0.16,-0.49, 0.00,-0.58,-0.16,-0.30,+0.49]
7   1st high vector     : [+1.00,+1.00,+1.00,+1.00,+1.00,+1.00,+1.00]
8   2nd low vector      : [-0.16,-0.49, 0.00,-0.58,-0.16,-0.30,+0.49]
9   2nd high vector     : [-0.16,-0.49, 0.00,-0.49,-0.16,-0.26,+0.49]
10  3rd low vector      : [-0.16,-0.49, 0.00,-0.49,-0.16,-0.26,+0.49]
11  3rd high vector     : [-0.16,-0.49, 0.00,-0.49,-0.16,-0.26,+0.49]
12  Iterations          : 3
13  high & low fusion   : [-0.16,-0.49, 0.00,-0.49,-0.16,-0.26,+0.49]
14  Choice vector for terminal prekernel: {'a3','a7'}
15     'a7': +0.49
16     'a3':  0.00
17     'a1': -0.16
18     'a5': -0.16
19     'a6': -0.26
20     'a2': -0.49
21     'a4': -0.49
```

A unique stable bipolar-valued high and low fixpoint is attained at the third iteration with alternative a7 positively confirmed (about $(1.0 + 0.49)/2 = 0.75\%$ criteria significance majority, see Line 15 above) as member of this terminal prekernel, whereas the membership of alternative a3 in this prekernel appears indeterminate. All the remaining nodes have negative membership characteristic values and are hence positively excluded from this prekernel.

When we reconsider the graphviz drawing of this outranking digraph in Fig. 17.12, it becomes obvious here why alternative a1 is *neither included nor excluded* from the initial prekernel. The same observation is applicable to alternative a3 that can *neither be included nor excluded* from the terminal prekernel. It may even happen, in case of more indeterminate outranking situations, that no alternative is positively included or excluded from a weakly independent prekernel, the corresponding bipolar-valued membership characteristic vector being completely indeterminate.

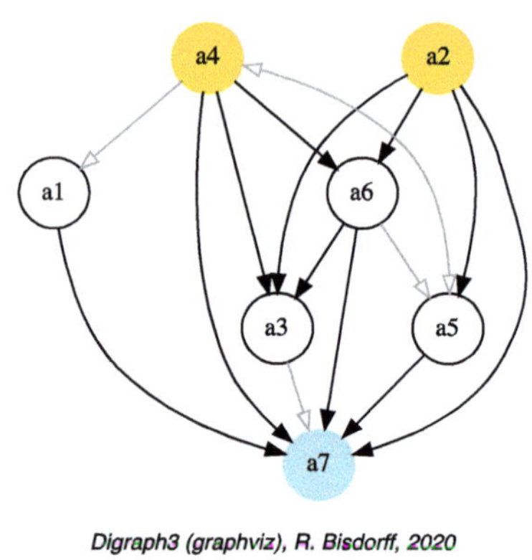

Fig. 17.12 *The strict outranking digraph oriented by its initial and terminal prekernels.* Notice the indeterminate outranking situation between alternatives a4 and a1 and the same indeterminate outranking situation between alternatives a3 and a7

To illustrate finally why sometimes it is necessary to operate an epistemic disjunctive fusion of unequal stable low and high membership characteristic vectors (see Algorithm 17.1 on page 243, Step 2c), let us consider, for instance, the following crisp 7-cycle graph:

```
>>> from digraphs import CirculantDigraph
>>> c7 = CirculantDigraph(order=7,circulants=[-1,1])
>>> c7
  *------- Digraph instance description ------*
   Instance class        : CirculantDigraph
   Instance name         : c7
   Digraph Order         : 7
   Digraph Size          : 14
   Valuation domain      : [-1.00;1.00]
   Determinateness (%)   : 100.00
   Attributes            : ['name','order','circulants',
                             'actions','valuationdomain',
                             'relation','gamma','notGamma']
```

Digraph `c7` is a symmetric crisp digraph showing, among others, the maximal independent set {2,5,7}, i.e. an initial as well as terminal kernel. The `computeKernelVector()` method computes the corresponding initial kernel characteristic vector.

```
1  >>> c7.computeKernelVector(['2','5','7'],\
2  ...                        Initial=True,Comments=True)
3   --> Initial kernel: {'2', '5', '7'}
4   initial low vector  : [-1.0, -1.0, -1.0, -1.0, -1.0, -1.0, -1.0]
5   initial high vector : [+1.0, +1.0, +1.0, +1.0, +1.0, +1.0, +1.0]
6   1 st low vector     : [-1.0,  0.0, -1.0, -1.0,  0.0, -1.0,  0.0]
7   1 st high vector    : [+1.0, +1.0, +1.0, +1.0, +1.0, +1.0, +1.0]
8   2 nd low vector     : [-1.0,  0.0, -1.0, -1.0,  0.0, -1.0,  0.0]
9   2 nd high vector    : [ 0.0, +1.0,  0.0,  0.0, +1.0,  0.0, +1.0]
10  stable low vector   : [-1.0,  0.0, -1.0, -1.0,  0.0, -1.0,  0.0]
11  stable high vector  : [ 0.0, +1.0,  0.0,  0.0, +1.0,  0.0, +1.0]
12  Iterations          : 3
13  low & high fusion   : [-1.0, +1.0, -1.0, -1.0, +1.0, -1.0, +1.0]
14  Choice vector for initial prekernel: {'2', '5', '7'}
15     '7': +1.00
16     '5': +1.00
17     '2': +1.00
18     '6': -1.00
19     '4': -1.00
20     '3': -1.00
21     '1': -1.00
```

Notice that the stable low vector characterises the *negative membership* part, whereas the stable high vector characterises the *positive membership* part (see Lines 10–11 above).The bipolar epistemic fusion assembles eventually both stable parts into the correct prekernel characteristic vector (Line 13).

The adjacency matrix of a symmetric digraph staying *unchanged* by the transposition operator, the previous computations, when qualifying the same kernel as a terminal instance, will hence produce exactly the same result.

It is worthwhile noticing the essential computational role, the logical indeterminate value 0.0 is playing in this double fixpoint algorithm. To implement such kind of algorithms without a logical *neutral* term would be like implementing numerical algorithms without a possible usage of the number 0. Infinitely many trivial impossibility theorems and dubious logical results come up.

Notes

Following the observation that an independent absorbent choice in an acyclic digraph corresponds to the zero values of the associated GRUNDY function, Riguet (1948) introduced the name "noyau" (kernel) for such a choice. Terminal kernels where in the sequel studied by Berge (1962) in the context of Combinatorial Game Theory. Initial kernels—independent and dominating choices—were introduced under the name game solutions by von Neumann and Morgenstern (1944). The absorbent version of the crisp kernel equation system 17.1 on page 242 was first introduced by Schmidt and Ströhlein (1985) in the context of their thorough exploration of relational algebra.

The fuzzy version of kernel equation systems was first investigated by Kitainik (1993). Commenting on this work at a meeting in Spring 1995 of the EURO Working Group on Multicriteria Decision Aiding in Lausanne (Switzerland), *Marc Roubens* feared that solving such fuzzy kernel equation systems could be computationally difficult. Triggered by his pessimistic remark and knowing about kernel equation systems and the VON NEUMANN fixpoint theorem (Schmidt and Ströhlein 1985; von Neumann and Morgenstern 1944), I immediately started to implement in Prolog a solver for the valued version of Eq. 17.4 on page 243, the equation system serving as constraints for a discrete labelling of all possible rational solution vectors. And in Summer 1995, we luckily obtained with a commercial finite domain solver the very first valued initial and terminal kernels from a didactic outranking digraph of order 8, well known in the multiple-criteria decision aiding community. The computation took several seconds on a CRAY 6412 superserver with 12 processors operating in a nowadays ridiculous CPU speed of 90 MHz. The labelled solution vectors, obtained in the sequel for any outranking digraph with a single initial or terminal kernel, were structured in a way that suggested the converging stages of the VON NEUMANN fixpoint algorithm and so gave the initial hint for Algorithm 17.1 (Bisdorff 1996, 1997; Bisdorff and Roubens 2004).

In our present Python 3.9 implementation, such a tiny problem is solved in less than a thousandth of a second on a common laptop. And this remains practically the same for any relevant example of outranking digraph observed in a real decision-aiding problem. Several times we wrote in our personal journal that there is certainly now no more potential for any substantial improvement of this computational efficiency; Only to discover, shortly later, that following a new theoretical idea or choosing a more efficient implementation—using, for instance, the amazing instrument of iterator generators in Python—execution times could well be divided by 20.

This nowadays available computational efficiency confers the bipolar-valued kernel concept a methodological premium for solving specific decision problems on the basis of the bipolar-valued outranking digraph (see Chaps. 4 and 12). But it also opens new opportunities for verifying and implementing kernel extraction algorithms for more graph theoretical purposes. New results, like enumerating the non-isomorphic maximal independent sets—the kernels—of known difficult graph instances like the n-cycle, could be obtained (see Sect. 21.7 and Bisdorff and Marichal 2008).

References

Berge C (1962) The theory of graphs (original title: The theory of graphs and its applications). Dover 2001 (originally published by Wiley 1962). Translated from a French edition by Dunod, 1958

Bisdorff R (1996) On computing kernels on fuzzy simple graphs by combinatorial enumeration using a CPL(FD) system. In: 8th Benelux workshop on logic programming, Louvain-la-Neuve (BE), 9 September 1996, Université catholique de Louvain. http://hdl.handle.net/10993/46933

Bisdorff R (1997) On computing kernels on l-valued simple graphs. In: Proceedings of EUFIT'97, 5th European congress on fuzzy and intelligent technologies, Aachen, September 8–11, 1997, pp 97–103

Bisdorff R (2000) Logical foundation of fuzzy preferential systems with application to the electre decision aid methods. Comput Oper Res 27:673–687. http://hdl.handle.net/10993/23724

Bisdorff R (2002) Logical foundation of multicriteria preference aggregation. In: Bouyssou D, Jacquet-Lagrèze E, Perny P, Słowiński R, Vanderpooten D, Vincke P (eds) Aiding decisions with multiple criteria. Kluwer Academic Publishers, pp 379–403. http://hdl.handle.net/10993/45313

Bisdorff R (2004) On a natural fuzzification of Boolean logic. In: Klement EP, Pap E (eds) Linz seminar on fuzzy set theory, mathematics of fuzzy systems. Bildungszentrum St. Magdalena, Linz (Austria), pp 20–26. http://hdl.handle.net/10993/23721

Bisdorff R (2006) On enumerating the kernels in a bipolar-valued digraph. Annales du Lamsade 6:1–38. http://hdl.handle.net/10993/38741

Bisdorff R, Marichal J (2008) Counting non-isomorphic maximal independent sets of the n-cycle graph. J Integer Seq 11(Art. 08.5.7):1–16. https://cs.uwaterloo.ca/journals/JIS/VOL11/Marichal/marichal.html

Bisdorff R, Roubens M (2004) Choice procedures in pairwise comparison multiple-attribute decision making methods. In: Berghammer R, Möller B, Struth G (eds) Relational and Kleene-algebraic methods in computer science: 7th international seminar on relational methods in computer science and 2nd international workshop on applications of Kleene algebra, Bad Malente, May 12–17, 2003. Lecture notes in computer science, vol 3051. Springer, Heidelberg, pp 1–7

Bisdorff R, Pirlot M, Roubens M (2006) Choices and kernels from bipolar valued digraphs. Eur J Oper Res 175:155–170. http://hdl.handle.net/10993/23720

Kitainik L (1993) Fuzzy decision procedures with binary relations: towards a unified theory. Kluwer Academic Publisher, Boston

Riguet J (1948) Relations binaires, fermetures, correspondances de galois. Bull Soc Math France 76:114–155

Schmidt G, Ströhlein T (1985) On kernels of graphs and solutions of games: a synopsis based on relations and fixpoints. SIAM, J Algebraic Discrete Methods 6:54–65

von Neumann J, Morgenstern O (1944) Theory of games and economic behaviour. Princeton University Press, Princeton

Chapter 18
On Confident Outrankings with Uncertain Criteria Significance Weights

Contents

Abstract When modelling preferences following the outranking approach, the signs of the majority margins do sharply distribute validation and invalidation of pairwise outranking situations. How can we be confident in the resulting outranking digraph, when we acknowledge the usual imprecise knowledge of criteria significance weights coupled with small majority margins? In this chapter we propose to link the qualifying significance majority with a required $\alpha\%$-confidence level. We model therefore the significance weights as random variables following more or less widespread distributions around an average significance value that corresponds to the given deterministic weight. As the bipolar-valued random credibility of an outranking statement hence results from the simple sum of positive or negative independent random variables, we may apply the Central Limit Theorem (CLT) for computing the bipolar likelihood that the expected majority margin will indeed be positive and, respectively, negative.

18.1 Modelling Uncertain Criteria Significance Weights

Let us consider the significance weights of a family F of m criteria to be independent random variables W_j, distributing the potential significance weight of each criterion $j = 1, ..., m$ around a mean value $E(W_j)$ with variance $V(W_j)$.

Choosing a specific stochastic model of uncertainty is usually application specific. In the limited scope of this chapter, we will illustrate the consequence of this design decision on the resulting outranking modelling with four slightly different models for taking into account the uncertainty with which we know the

R. Bisdorff, *Algorithmic Decision Making with Python Resources*,
International Series in Operations Research & Management Science 324,
https://doi.org/10.1007/978-3-030-90928-4_18

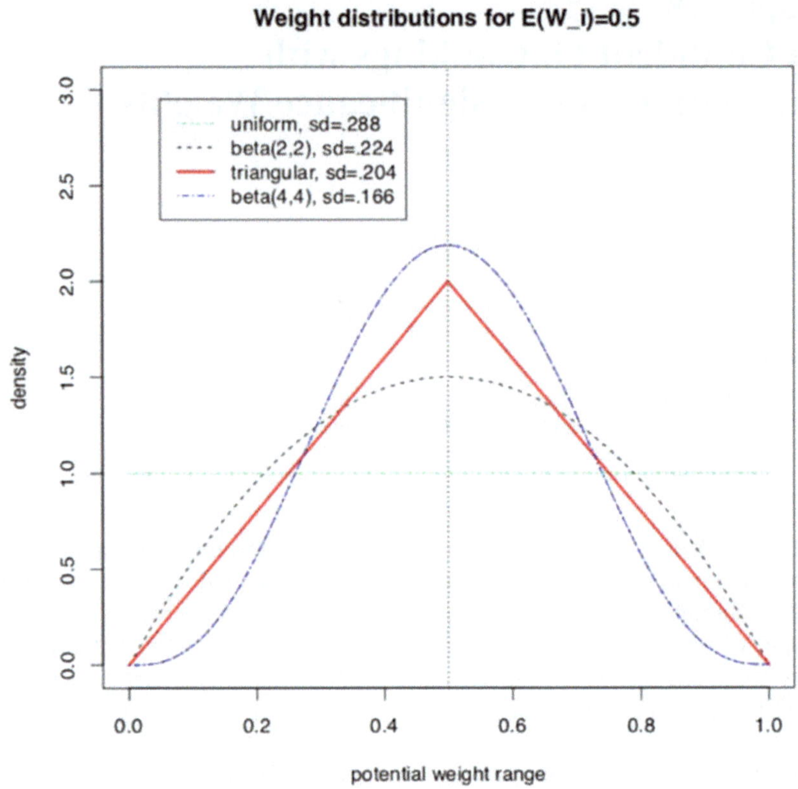

Fig. 18.1 Four models of uncertain criteria significance weights

numerical significance weights: *uniform*, *triangular*, and two models of *Beta laws*, one more widespread and, the other, more concentrated.

When considering, for instance, that the potential range of a significance weight is distributed between 0 and two times its mean value, we obtain the following random variates (Bisdorff 2020):

- A continuous *uniform* distribution on the range 0 to $2E(W_j)$. Thus, $W_j \rightsquigarrow U(0, 2E(W_j))$ and $V(W_j) = 1/3(E(W_j))^2$.
- A *symmetric beta* distribution with, for instance, parameters $\alpha = 2$ and $\beta = 2$. Thus, $W_i \rightsquigarrow Beta(2, 2) \times 2E(W_j)$ and $V(W_j) = 1/5(E(Wj))^2$.
- A *symmetric triangular* distribution on the same range with mode $E(W_j)$. Thus $W_j \rightsquigarrow Tr(0, 2E(W_j), E(Wj))$ with $V(W_j) = 1/6(E(Wj))^2$.
- A *narrower beta* distribution with, for instance, parameters $\alpha = 4$ and $\beta = 4$. Thus $W_j \rightsquigarrow Beta(4, 4) \times 2E(W_j)$ and $V(W_j) = 1/9(E(W_j))^2$.

It is worthwhile noticing in Fig. 18.1 that these four uncertainty models all admit the same expected value, $E(W_j)$, however, with a respective variance which goes decreasing from $1/3$ to $1/9$ of the square of $E(W_j)$ (Bisdorff 2014).

18.2 Bipolar-Valued Likelihood of Outranking Situations

Let $A = \{x, y, z, ...\}$ be a finite set of n potential decision actions, evaluated on $F = \{1, \dots, m\}$—a finite and coherent family of m performance criteria. On each criterion $j \in F$, the decision actions are evaluated on a real performance scale $[0; M_j]$, supporting an upper-closed indifference threshold ind_j and a lower-closed preference threshold pr_j such that $0 \leq ind_j < pr_j \leqslant M_j$. The marginal performance of object x on criterion j is denoted x_j. Each criterion j is thus characterising a marginal double threshold order $\geqslant_j$ on A:

$$r(x \geq_j y) = \begin{cases} +1 & \text{if} \quad x_j - y_j \geqslant -ind_j, \\ -1 & \text{if} \quad x_j - y_j \leqslant -pr_j, \\ 0 & \text{otherwise.} \end{cases} \tag{18.1}$$

Semantics of the marginal bipolar-valued characteristic function:

- $+1$ signifies x is at least as well evaluated as y on criterion j.
- -1 signifies that x is not at least as well evaluated as y on criterion j.
- 0 signifies that it is unclear whether, on criterion j, x is at least as well evaluated as y (Fig. 18.2).

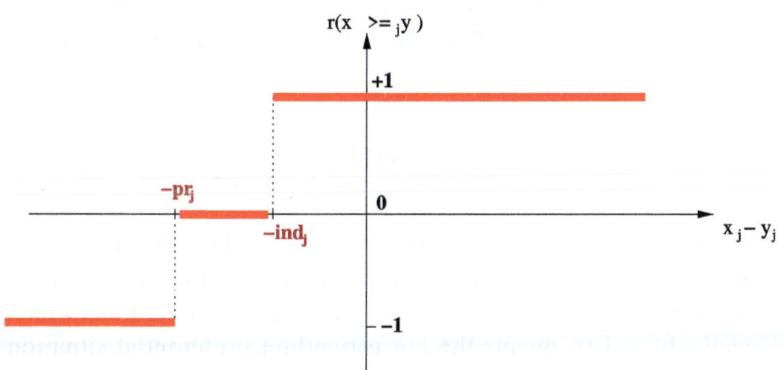

Fig. 18.2 Bipolar-valued outranking characteristic function

Each criterion j in F contributes the random significance W_j of his '*at least as well evaluated as*' characteristic $r(x \geqslant_j y)$ to the global characteristic $\tilde{r}(x \geqslant y)$ in the following way:

$$\tilde{r}(x \geqslant y) = \sum_{j \in F} W_j \times r(x \geqslant_j y)). \tag{18.2}$$

Thus, $\tilde{r}(x \geqslant y)$ becomes a simple sum of positive or negative independent random variables with known means and variances where $\tilde{r}(x \geqslant y) > 0$ signifies x is globally at least as well evaluated as y, $\tilde{r}(x \geqslant y) < 0$ signifies that x is not globally at least as well evaluated as y, and $\tilde{r}(x \geqslant y) = 0$ signifies that it is unclear whether x is globally at least as well evaluated as y.

From the *Central Limit Theorem* (CLT), we know that such a sum of random variables leads, with m getting large, to a Gaussian distribution Y with

$$E(Y) = \textstyle\sum_{j \in F} \big(E(W_j) \times r(x \geqslant_j y) \big), \text{ and} \tag{18.3}$$

$$V(Y) = \textstyle\sum_{j \in F} \big(V(W_j) \times |r(x \geqslant_j y)| \big). \tag{18.4}$$

And the *likelihood of validation*, respectively, *invalidation*, of an '*at least as well evaluated as*' situation, denoted $lh(x \geqslant y)$, may hence be assessed by the probability $P(Y > 0) = 1.0 - P(Y \leqslant 0)$ that Y takes a positive value, respectively, $P(Y < 0)$ takes a negative value. In the bipolar-valued case here, we can judiciously make usage of the standard Gaussian *error function* , i.e. the bipolar $2P(Z) - 1.0$ version of the standard Gaussian $P(Z)$ probability distribution function (Press et al. 2007):

$$lh(x \geqslant y) = -\text{erf}\big(\frac{1}{\sqrt{2}} \frac{-E(Y)}{\sqrt{V(Y)}}\big). \tag{18.5}$$

The range of the bipolar-valued $lh(x \geqslant y)$ hence becomes $[-1.0; +1.0]$, and $-lh(x \geqslant y) = lh(x \not\geqslant y)$, i.e. a *negative likelihood* represents the likelihood of the correspondent *negated* '*at least as well evaluated as*' situation. A likelihood of $+1.0$ (respectively, -1.0) means the corresponding preferential situation appears *certainly* validated (respectively, invalidated).

Example 18.1 Let x and y be evaluated with respect to 7 equi-significant criteria; four criteria positively support that x is *at least as well assessed as* y and three criteria support that x is *not at least as well evaluated as* y. Suppose $E(W_j) = w$ for $j = 1, \ldots, 7$ and $W_j \leadsto Tr(0, 2w, w)$ for $j = 1, \ldots, 7$. The expected value of the global '*at least as well evaluated as*' characteristic value becomes $E\big(\tilde{r}(x \geqslant y)\big) = 4w - 3w = w$ with a variance $V\big(\tilde{r}(x \geqslant y)\big) = 7\frac{1}{6}w^2$.

If $w = 1$, $E\big(\tilde{r}(x \geqslant y)\big) = 1$ and $sd\big(\tilde{r}(x \geqslant y)\big) = 1.08$. By the CLT, the bipolar likelihood of the *at least as well evaluated as* situation becomes $lh(x \geqslant y) = 0.66$,

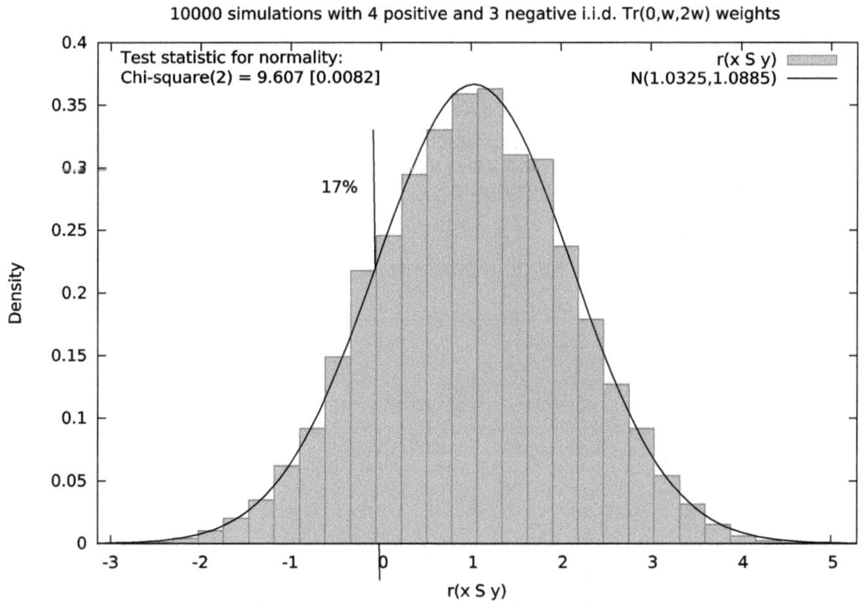

Fig. 18.3 Distribution of 10 000 random outranking characteristic values

which corresponds to a global support of $(0.66 + 1.0)/2 = 83\%$ of the criteria significance weights.

A *Monte Carlo* simulation with 10 000 runs empirically confirms the effective convergence to a Gaussian (see Fig. 18.3 realised with *gretl.*[1])

Indeed, $\tilde{r}(x \geqslant y) \rightsquigarrow Y = \mathcal{N}(1.03, 1.089)$, with an empirical probability of observing a negative majority margin of about 17%.

18.3 Confidence level Of Outranking Digraphs

Definition 18.1 (Confident Outranking Situation) Following the classical outranking approach (see (Bisdorff 2013)), we say, from an epistemic perspective, that decision action x *outranks* decision action y at *confidence* level $\alpha\%$, when:

- An expected majority of criteria validates, at confidence level $\alpha\%$ or higher, a global '*at least as well evaluated as*' situation between x and y.
- No considerably less performing is observed on a discordant criterion.

[1] The GNU Regression, Econometrics and Time-Series Library, http://gretl.sourceforge.net/.

Dually, decision action x *does not outrank* decision action y at confidence level $\alpha\%$, when:

- An expected minority only of criteria at confidence level $-\alpha\%$ or lesser validates a global '*at least as well evaluated as*' situation between x and y.
- No considerably better performing situation is observed on a discordant criterion.

Otherwise, the situation is indeterminate.

Time for a coded example
Let us consider the following random performance tableau:

```
>>> from randomPerfTabs import RandomPerformanceTableau
>>> t = RandomPerformanceTableau(\
...             numberOfActions=7,\
...             numberOfCriteria=7,seed=100)
>>> t.showPerformanceTableau(Transposed=True)
  *----  performance tableau -----*
  criteria | weights |  'a1'   'a2'   'a3'   'a4'   'a5'   'a6'   'a7'
  ---------|-----------------------------------------------------------
    'g1'   |    1    | 15.17  44.51  57.87  58.00  24.22  29.10  96.58
    'g2'   |    1    | 82.29  43.90    NA   35.84  29.12  34.79  62.22
    'g3'   |    1    | 44.23  19.10  27.73  41.46  22.41  21.52  56.90
    'g4'   |    1    | 46.37  16.22  21.53  51.16  77.01  39.35  32.06
    'g5'   |    1    | 47.67  14.81  79.70  67.48    NA   90.72  80.16
    'g6'   |    1    | 69.62  45.49  22.03  33.83  31.83    NA   48.80
    'g7'   |    1    | 82.88  41.66  12.82  21.92  75.74  15.45   6.05
```

For the corresponding confident outranking digraph, we assume uncertain criteria significance weights of *triangular* distribution and require a confidence level of $\alpha = 90\%$. The `ConfidentBipolarOutrankingDigraph` class provides such a construction.

Listing 18.1 Computing a 90% confident outranking digraph

```
>>> from outrankingDigraphs import\
...          ConfidentBipolarOutrankingDigraph
>>> g90 = ConfidentBipolarOutrankingDigraph(t,\
...          distribution ='triangular',confidence=90)
>>> g90
  *------- Object instance description ------*
   Instance class       : ConfidentBipolarOutrankingDigraph
   Instance name        : rel_randomperftab_CLT
   Actions            : 7
   Criteria           : 7
   Size                 : 15
   Uncertainty model    : triangular(a=0,b=2w)
   Likelihood domain    : [-1.0;+1.0]
   Confidence level     : 0.80 (90.0%)
   Confident majority   : 0.14 (57.1%)
   Determinateness (%)  : 62.07
   Valuation domain     : [-1.00;1.00]
   Attributes           : ['name', 'bipolarConfidenceLevel',
                'distribution', 'betaParameter', 'actions',
                'order', 'valuationdomain', 'criteria',
                'evaluation', 'concordanceRelation',
```

```
                    'vetos', 'negativeVetos',
                    'largePerformanceDifferencesCount',
                    'likelihoods', 'confidenceCutLevel',
                    'relation', 'gamma', 'notGamma']
```

The resulting 90%-confident expected outranking relation is shown below.

Listing 18.2 90%-confident outranking relation with triangular distributed significance weights

```
1  >>> g90.showRelationTable(LikelihoodDenotation=True)
2   * ---- Outranking Relation Table -----
3     r/(lh) |  'a1'    'a2'    'a3'    'a4'    'a5'    'a6'    'a7'
4     -------|--------------------------------------------------------
5      'a1'  | +0.00   +0.71   +0.29   +0.29   +0.29   +0.29   +0.00
6            | ( - )  (+1.00) (+0.95) (+0.95) (+0.95) (+0.95) (+0.65)
7      'a2'  | -0.71   +0.00   -0.29   +0.00   +0.00   +0.29   -0.57
8            |(-1.00)  ( - )  (-0.95) (-0.65) (+0.73) (+0.95) (-1.00)
9      'a3'  | -0.29   +0.29   +0.00   -0.29   +0.00   +0.00   -0.29
10           |(-0.95) (+0.95)  ( - )  (-0.95) (-0.73) (-0.00) (-0.95)
11     'a4'  | +0.00   +0.00   +0.57   +0.00   +0.29   +0.57   -0.43
12           |(-0.00) (+0.65) (+1.00)  ( - )  (+0.95) (+1.00) (-0.99)
13     'a5'  | -0.29   +0.00   +0.00   +0.00   +0.00   +0.29   -0.29
14           |(-0.95) (-0.00) (+0.73) (-0.00)  ( - )  (+0.99) (-0.95)
15     'a6'  | -0.29   +0.00   +0.00   -0.29   +0.00   +0.00   +0.00
16           |(-0.95) (-0.00) (+0.73) (-0.95) (+0.73)  ( - )  (-0.00)
17     'a7'  | +0.00   +0.71   +0.57   +0.43   +0.29   +0.00   +0.00
18           |(-0.65) (+1.00) (+1.00) (+0.99) (+0.95) (-0.00)  ( - )
19    Valuation domain    : [-1.000; +1.000]
20    Uncertainty model   : triangular(a=2.0,b=2.0)
21    Likelihood domain   : [-1.0;+1.0]
22    Confidence level    : 0.80 (90.0%)
23    Confident majority  : 0.14 (57.1%)
24    Determinateness     : 0.24 (62.1%)
```

The (lh) figures, indicated in Listing 18.2 above, correspond to bipolar likelihoods and the required bipolar confidence level equals $(0.90 + 1.0)/2 = 0.80$ (see Line 22 above). Action `a1` thus confidently outranks all other actions, except `a7` where the actual likelihood (+0.65) is lower than the required one (+0.80), and we furthermore observe a considerable counter-performance on criterion `g1`.

Notice also the lack of confidence in the outranking situations we observe between action `a2` and actions `a4` and `a5`. In the deterministic case, we would have $r(a2 \geqslant a4) = -0.143$ and $r(a2 \geqslant a5) = +0.143$. All outranking situations with a characteristic value lower than or equal to $abs(0.143)$, i.e. a majority support of $1.143/2 = 57.1\%$ and less, appear indeed to be *not confident* at α level 90% (see Line 23 above).

Figure 18.4 on the following page shows the corresponding strict 90%-confident outranking digraph, oriented by its initial and terminal strict prekernels.

```
>>> gcd90 = ~ (-g90)
>>> gcd90.showPreKernels()
 *--- Computing preKernels ---*
 Dominant preKernels :
  ['a1', 'a7']
     independence :  0.0
     dominance    :  0.2857
     absorbency   :  -0.7143
```

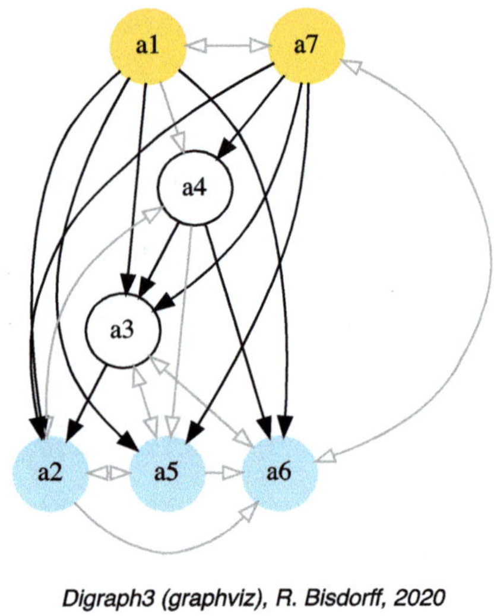

Fig. 18.4 90%-confident strict outranking digraph oriented by its initial and terminal prekernels

```
      covering      :  0.800
   Absorbent preKernels :
    ['a2', 'a5', 'a6']
      independence :  0.0
      dominance     :  -0.2857
      absorbency    :  0.2857
      covered       :  0.583
>>> gcd90.exportGraphViz(fileName='confidentOutranking',
...          firstChoice=['a1', 'a7'],\
...          lastChoice=['a2', 'a5', 'a6'])
  *---- exporting a dot file for GraphViz tools ----*
  Exporting to confidentOutranking.dot
  dot -Grankdir=BT -Tpng confidentOutranking.dot\
             -o confidentOutranking.png
```

Now, what becomes this 90%-confident outranking digraph when we require a stronger confidence level of, say 99%?

Listing 18.3 99%-confident outranking relation

```
>>> g99 = ConfidentBipolarOutrankingDigraph(t,\
...      distribution='trinagular',confidence=99)
>>> g99.showRelationTable()
  * ---- Outranking Relation Table -----
   r/(lh) |  'a1'    'a2'    'a3'    'a4'    'a5'    'a6'    'a7'
   -------|------------------------------------------------------
    'a1'  |  +0.00   +0.71   +0.00   +0.00   +0.00   +0.00   +0.00
          |  ( - )  (+1.00) (+0.95) (+0.95) (+0.95) (+0.95) (+0.65)
    'a2'  |  -0.71   +0.00   +0.00   +0.00   +0.00   +0.00   -0.57
          | (-1.00)  ( - )  (-0.95) (-0.65) (+0.73) (+0.95) (-1.00)
    'a3'  |  +0.00   +0.00   +0.00   +0.00   +0.00   +0.00   +0.00
          | (-0.95) (+0.95)  ( - )  (-0.95) (-0.73) (-0.00) (-0.95)
    'a4'  |  +0.00   +0.00   +0.57   +0.00   +0.00   +0.57   -0.43
          | (-0.00) (+0.65) (+1.00)  ( - )  (+0.95) (+1.00) (-0.99)
    'a5'  |  +0.00   +0.00   +0.00   +0.00   +0.00   +0.29   +0.00
```

```
16          | (-0.95) (-0.00) (+0.73) (-0.00)  ( - )   (+0.99) (-0.95)
17     'a6' |  +0.00   +0.00   +0.00   +0.00   +0.00   +0.00   +0.00
18          | (-0.95) (-0.00) (+0.73) (-0.95) (+0.73)  ( - )  (-0.00)
19     'a7' |  +0.00   +0.71   +0.57   +0.43   +0.00   +0.00   +0.00
20          | (-0.65) (+1.00) (+1.00) (+0.99) (+0.95) (-0.00)  ( - )
21     Valuation domain    : [-1.000; +1.000]
22     Uncertainty model   : triangular(a=0.0,b=2.0)
23     Likelihood domain   : [-1.0;+1.0]
24     Confidence level    : 0.98 (99.0%)
25     Confident majority  : 0.29 (64.3%)
26     Determinateness     : 0.13 (56.6%)
```

At 99% confidence, the minimal required significance majority support amounts to 64.3% (see Listing 18.3, Line 25 above). As a result, most outranking situations do not get anymore validated, like the outranking situations between action `a1` and actions `a3`, `a4`, `a5`, and `a6` (see Line 5 above). The overall epistemic determination of the digraph consequently drops from 62.1% to 56.6% (see Line 26).

Finally, what becomes the previous 90%-confident outranking digraph when the uncertainty concerning the criteria significance weights is modelled with a larger variance, like *uniform* variates (see Line 2 below).

Listing 18.4 90%-confident outranking digraph with uniform variates

```
1  >>> gu90 = ConfidentBipolarOutrankingDigraph(t,\
2  ...          confidence=90,distribution='uniform')
3  >>> gu90.showRelationTable()
4   * ---- Outranking Relation Table -----
5   r/(lh) |   'a1'    'a2'    'a3'    'a4'    'a5'    'a6'    'a7'
6   -------|-------------------------------------------------------
7     'a1' |  +0.00   +0.71   +0.29   +0.29   +0.29   +0.29   +0.00
8          |  ( - )  (+1.00) (+0.84) (+0.84) (+0.84) (+0.84) (+0.49)
9     'a2' |  -0.71   +0.00   -0.29   +0.00   +0.00   +0.29   -0.57
10         | (-1.00)  ( - )  (-0.84) (-0.49) (+0.56) (+0.84) (-1.00)
11    'a3' |  -0.29   +0.29   +0.00   -0.29   +0.00   +0.00   -0.29
12         | (-0.84) (+0.84)  ( - )  (-0.84) (-0.56) (-0.00) (-0.84)
13    'a4' |  +0.00   +0.00   +0.57   +0.00   +0.29   +0.57   -0.43
14         | (-0.00) (+0.49) (+1.00)  ( - )  (+0.84) (+1.00) (-0.95)
15    'a5' |  -0.29   +0.00   +0.00   +0.00   +0.00   +0.29   -0.29
16         | (-0.84) (-0.00) (+0.56) (-0.00)  ( - )  (+0.92) (-0.84)
17    'a6' |  -0.29   +0.00   +0.00   -0.29   +0.00   +0.00   +0.00
18         | (-0.84) (-0.00) (+0.56) (-0.84) (+0.56)  ( - )  (-0.00)
19    'a7' |  +0.00   +0.71   +0.57   +0.43   +0.29   +0.00   +0.00
20         | (-0.49) (+1.00) (+1.00) (+0.95) (+0.84) (-0.00)  ( - )
21     Valuation domain    : [-1.000; +1.000]
22     Uncertainty model   : uniform(a=0.0,b=2.0)
23     Likelihood domain   : [-1.0;+1.0]
24     Confidence level    : 0.80 (90.0%)
25     Confident majority  : 0.14 (57.1%)
26     Determinateness     : 0.24 (62.1%)
```

Despite lower likelihood values (see the `g90` digraph in Listing 18.2 on page 257), we keep here the same confident majority level of 57.1% (Line 25) and, hence, also the same 90%-confident outranking digraph.

For concluding, it is worthwhile noticing again that it is the *neutral* value of the bipolar-valued epistemic logic that allows to easily handle $\alpha\%$ confidence or not of outranking situations when confronted with uncertain criteria significance weights. Remarkable furthermore is the usage, the standard Gaussian error function (erf) provides by delivering *signed* likelihood values immediately concerning either a positive relational statement or when negative its negated version (Press et al. 2007).

Chapter 19 analyses the robustness of the outranking digraph when the criteria significance weights faithfully indicate solely an order of importance.

References

Bisdorff R (2013) On polarizing outranking relations with large performance differences. Journal of Multi-Criteria Decision Analysis, Wiley 20:3–12, http://hdl.handle.net/10993/245

Bisdorff R (2014) On confident outrankings with multiple criteria of uncertain significance. In: Mousseau V, Pirlot M (eds) DAP'2014 From Multiple Criteria Decision Aid to Preference Learning, Ecole Centrale de Paris, pp 1–6. http://hdl.handle.net/10993/23910

Bisdorff R (2020) Lecture 3: continuous random variables. In: Lectures of the computational statistics course. University of Luxembourg. http://hdl.handle.net/10993/37870

Press W, Teukolsky S, Vetterling W, Flannery B (2007) Numerical recipes: the art of scientific computing, 3rd edn., Special Functions, Section 6.2. Cambridge University Press, Cambridge, pp 209–214

Chapter 19
Robustness Analysis of Outranking Digraphs

Contents

Abstract The required cardinal significance weights of the performance criteria represent the '*Achilles*' heel of the outranking approach. Rarely will indeed a decision maker be cognitively competent for suggesting precise decimal-valued criteria significance weights. More often, the decision problem will involve more or less equally important decision objectives with more or less equi-significant criteria per objective. In this chapter, we study the robustness of the outranking digraph when the criteria significance weights faithfully indicate solely an order of importance.

19.1 Cardinal or Ordinal Criteria Significance Weights?

A random example of such a decision problem with equally important decision objectives and equi-significant criteria may be generated with the `Random3ObjectivesPerformanceTableau` class.

Listing 19.1 Generate a random 3-objectives performance tableau

```
>>> from randomPerfTabs import \
...              Random3ObjectivesPerformanceTableau
>>> pt = Random3ObjectivesPerformanceTableau(\
...              numberOfActions=7,\
...              numberOfCriteria=9,seed=102)
>>> pt
```


R. Bisdorff, *Algorithmic Decision Making with Python Resources*,
International Series in Operations Research & Management Science 324,
https://doi.org/10.1007/978-3-030-90928-4_19

```
7        *------- PerformanceTableau instance description ------*
8        Instance class     : Random3ObjectivesPerformanceTableau
9        Seed               : 102
10       Instance name      : random3ObjectivesPerfTab
11       Actions            : 7
12       Objectives         : 3
13       Criteria           : 9
14       NA proportion (%) : 0.0
15       Attributes         : ['name', 'valueDigits', 'BigData',
16                 'OrdinalScales', 'missingDataProbability',
17                 'negativeWeightProbability', 'randomSeed',
18                 'sumWeights', 'valuationPrecision',
19                 'commonScale', 'objectiveSupportingTypes',
20                 'actions', 'objectives', 'criteriaWeightMode',
21                 'criteria', 'evaluation', 'weightPreorder']
22   >>> pt.showObjectives()
23     *------ show objectives -------*
24      Eco: Economical aspect
25        ec1 criterion of objective Eco 8
26        ec4 criterion of objective Eco 8
27        ec8 criterion of objective Eco 8
28        Total weight: 24.00 (3 criteria)
29      Soc: Societal aspect
30        so2 criterion of objective Soc 12
31        so7 criterion of objective Soc 12
32        Total weight: 24.00 (2 criteria)
33      Env: Environmental aspect
34        en3 criterion of objective Env 6
35        en5 criterion of objective Env 6
36        en6 criterion of objective Env 6
37        en9 criterion of objective Env 6
38        Total weight: 24.00 (4 criteria)
```

In Listing 19.1 above, one may notice a performance tableau pt describing seven decision alternatives that are assessed with respect to three equally important decision objectives concerning:

1. An *economical* aspect with a coalition of three performance criteria of significance weight 8 (Line 24)
2. A *societal* aspect with a coalition of two performance criteria of significance weight 12 (Line 29)
3. An *environmental* aspect with a coalition of four performance criteria of significance weight 6 (Line 33)

The question we tackle in this chapter is the following: how *dependent* on the actual values of the significance weights appears to be the corresponding bipolar-valued outranking digraph? In the previous chapter, Chap. 18, we assumed that the criteria significance weights were random variables. Here, we shall assume that we know for sure only the preordering of the significance weights.

In the random performance tableau pt, we observe three increasing weight equivalence classes.

Listing 19.2 The significance weights preorder

```
>>> pt.showWeightPreorder()
    ['en3', 'en5', 'en6', 'en9'] (6) <
    ['ec1', 'ec4', 'ec8'] (8) <
    ['so2', 'so7'] (12)
```

How robust will appear now the validated outranking and outranked situations when allowing all possible significance weights that are compatible with the weights preorder shown above (Bisdorff et al. 2009, 2014)?

19.2 Qualifying the Stability of Outranking Situations

Let us construct the bipolar-valued outranking digraph corresponding to the random 3-objectives performance tableau `pt`.

Listing 19.3 Example bipolar outranking digraph

```
1  >>> from outrankingDigraphs import BipolarOutrankingDigraph
2  >>> g = BipolarOutrankingDigraph(pt)
3  >>> g.showRelationTable()
4   * ---- Relation Table -----
5    r(>=) |  'p1'   'p2'   'p3'   'p4'   'p5'   'p6'   'p7'
6    ------|-------------------------------------------------
7     'p1' | +1.00  -0.42  +0.00  -0.69  +0.39  +0.11  -0.06
8     'p2' | +0.58  +1.00  +0.83  +0.00  +0.58  +0.58  +0.58
9     'p3' | +0.25  -0.33  +1.00  +0.00  +0.50  +1.00  +0.25
10    'p4' | +0.78  +0.00  +0.61  +1.00  +1.00  +1.00  +0.67
11    'p5' | -0.11  -0.50  -0.25  -0.89  +1.00  +0.11  -0.14
12    'p6' | +0.22  -0.42  +0.00  -1.00  +0.17  +1.00  -0.11
13    'p7' | +0.22  -0.50  +0.17  -0.06  +0.78  +0.42  +1.00
```

In Listing 19.3, Lines 7–13, we notice on the principal diagonal of the relation table the certainly validated reflexive terms $+1.00$; they are trivially independent of any significance weights. Now, we know for sure that *unanimous* outranking situations are also completely independent of the significance weights. And, all outranking situations that are supported by a majority significance in each coalition of equi-significant criteria are as well independent of the actual importance we attach to each individual criteria coalition. We are furthermore able to effectively test if an outranking situation remains valid with all potential significance weights that are compatible with the given preordering shown in Listing 19.2 on the preceding page (Bisdorff 2014). Mind that usually one obtains more or less numerous outranking situations that are in fact *dependent* on the precise cardinal significance weights given to the criteria.

Such a *stability denotation* of outranking situations can be inspected using the `StabilityDenotation=True` flag with the `showRelationTable()` method.

Listing 19.4 Bipolar-valued outranking relation table with stability denotation

```
1  >>> g.showRelationTable(StabilityDenotation=True)
2    * ---- Relation Table -----
3    r/(stab)   |  'p1'   'p2'   'p3'   'p4'   'p5'   'p6'   'p7'
4    ----------|------------------------------------------
5      'p1'     | +1.00 -0.42 +0.00 -0.69 +0.39 +0.11 -0.06
6               |  (+4)  (-2)  (+0)  (-3)  (+2)  (+2)  (-1)
7      'p2'     | +0.58 +1.00 +0.83  0.00 +0.58 +0.58 +0.58
8               |  (+2)  (+4)  (+3)  (+2)  (+2)  (+2)  (+2)
9      'p3'     | +0.25 -0.33 +1.00  0.00 +0.50 +1.00 +0.25
10              |  (+2)  (-2)  (+4)   (0)  (+2)  (+2)  (+1)
11     'p4'     | +0.78  0.00 +0.61 +1.00 +1.00 +1.00 +0.67
12              |  (+3)  (-1)  (+3)  (+4)  (+4)  (+4)  (+2)
13     'p5'     |-0.11 -0.50 -0.25 -0.89 +1.00 +0.11 -0.14
14              |  (-2)  (-2)  (-2)  (-3)  (+4)  (+2)  (-2)
15     'p6'     | +0.22 -0.42  0.00 -1.00 +0.17 +1.00 -0.11
16              |  (+2)  (-2)  (+1)  (-2)  (+2)  (+4)  (-2)
17     'p7'     | +0.22 -0.50 +0.17 -0.06 +0.78 +0.42 +1.00
18              |  (+2)  (-2)  (+1)  (-1)  (+3)  (+2)  (+4)
```

In Listing 19.4 above, we may hence distinguish the following bipolar-valued stability levels:

- **±4**: *unanimous* outranking, respectively, outranked, situation. The pairwise trivial reflexive outrankings, for instance, all show this stability level.
- **±3**: validated outranking, respectively, outranked, situation in *each* coalition of equi-significant criteria. This is, for instance, the case for the outranking situation observed between alternatives `p1` and `p4` (see Lines 6 and 12).
- **±2**: validated outranking, respectively, outranked situation with *all* potential significance weights that are *compatible* with the given significance preorder (see Listing 19.2 on page 262). This is the case when comparing alternatives `p1` and `p2` (see Lines 6 and 8).
- **±1**: validated outranking, respectively, outranked situation with the *precisely given decimal* significance weights, a situation we may observe between alternatives `p3` and `p7` (see Lines 10 and 18).
- **0**: `indeterminate` outranking situation, like the one between alternatives `p1` and `p3` (see Lines 6 and 10).

It is worthwhile noticing that, in the one limit case where all performance criteria appear equi-significant – there is given a single equivalence class containing all the criteria significance weights – one may only distinguish stability levels ±4 and ±3. In the other limit case, when all performance criteria admit different significance weights, the significance weights may be linearly ordered without ties and no stability level ±3 can be observed.

As mentioned above, all reflexive comparisons trivially confirm an unanimous outranking situation: all decision alternatives are indeed always "*at least as well evaluated as*" themselves. But there appear also two non-reflexive unanimous outranking situations: when comparing, for instance, alternative `p4` with alternatives `p5` and `p6` (see Listing 19.4, Lines 14 and 16). Let us compare with the `showPairwiseComparison()` method the multicriteria performance records of alternatives `p4` and `p5`.

```
>>> g.showPairwiseComparison('p4','p5')
 *------------  pairwise comparison ----*
  Comparing actions : (p4, p5)
  crit. wght.  g(x)   g(y)     diff   | ind    pref     r()
  ec1    8.00  85.19  46.75  +38.44  | 5.00  10.00   +8.00
  ec4    8.00  72.26   8.96  +63.30  | 5.00  10.00   +8.00
  ec8    8.00  44.62  35.91   +8.71  | 5.00  10.00   +8.00
  en3    6.00  80.81  31.05  +49.76  | 5.00  10.00   +6.00
  en5    6.00  49.69  29.52  +20.17  | 5.00  10.00   +6.00
  en6    6.00  66.21  31.22  +34.99  | 5.00  10.00   +6.00
  en9    6.00  50.92   9.83  +41.09  | 5.00  10.00   +6.00
  so2   12.00  49.05  12.36  +36.69  | 5.00  10.00  +12.00
  so7   12.00  55.57  44.92  +10.65  | 5.00  10.00  +12.00
  Valuation in range: -72.00 to +72.00;                -------
                            global concordance:  +72.00
```

Alternative `p4` is indeed unanimously "*at least as well evaluated as*" alternative `p5` and $r(p4 \succsim p5) = 72/72 = +1.00$ (see Listing 19.4 on the facing page, Line 11). The converse comparison does not, however, reveal such an unanimous outranked situation.

```
1  >>> g.showPairwiseComparison('p5','p4')
2   *------------  pairwise comparison ----*
3    Comparing actions : (p5, p4)
4    crit. wght.  g(x)   g(y)     diff   | ind    pref     r()
5    ec1    8.00  46.75  85.19  -38.44  | 5.00  10.00   -8.00
6    ec4    8.00   8.96  72.26  -63.30  | 5.00  10.00   -8.00
7    ec8    8.00  35.91  44.62   -8.71  | 5.00  10.00   +0.00
8    en3    6.00  31.05  80.81  -49.76  | 5.00  10.00   -6.00
9    en5    6.00  29.52  49.69  -20.17  | 5.00  10.00   -6.00
10   en6    6.00  31.22  66.21  -34.99  | 5.00  10.00   -6.00
11   en9    6.00   9.83  50.92  -41.09  | 5.00  10.00   -6.00
12   so2   12.00  12.36  49.05  -36.69  | 5.00  10.00  -12.00
13   so7   12.00  44.92  55.57  -10.65  | 5.00  10.00  -12.00
14   Valuation in range: -72.00 to +72.00;                ------
15                             global concordance:  -64.00
```

The converse comparison only qualifies at stability level −3 (see Listing 19.4 on the preceding page, Line 13); $r(p5 \succsim p4) = -64/72 = -0.89$). On criterion `ec8`, we observe in fact a small negative performance difference of −8.71 (see Line 7 above) which is effectively below the supposed preference discrimination threshold of 10.00. Yet, the outranked situation is supported by a majority of criteria in each decision objective. Hence, the reported preferential situation is completely independent of any chosen significance weights.

Let us now consider a comparison, like the one between alternatives `p2` and `p1`, that is qualified at stability level +2, respectively, −2.

Listing 19.5 Comparison of alternatives p2 and p1

```
>>> g.showPairwiseOutrankings('p2','p1')
  *------------  pairwise comparison ----*
   Comparing actions : (p2, p1)
   crit. wght.  g(x)   g(y)     diff   | ind   pref     r()
   ec1    8.00  89.77  38.11  +51.66  | 5.00  10.00   +8.00
   ec4    8.00  86.00  22.65  +63.35  | 5.00  10.00   +8.00
   ec8    8.00  89.43  77.02  +12.41  | 5.00  10.00   +8.00
   en3    6.00  20.79  58.16  -37.37  | 5.00  10.00   -6.00
   en5    6.00  23.83  31.40   -7.57  | 5.00  10.00   +0.00
   en6    6.00  18.66  11.41   +7.25  | 5.00  10.00   +6.00
   en9    6.00  26.65  44.37  -17.72  | 5.00  10.00   -6.00
   so2   12.00  89.12  22.43  +66.69  | 5.00  10.00  +12.00
   so7   12.00  84.73  28.41  +56.32  | 5.00  10.00  +12.00
   Valuation in range: -72.00 to +72.00;                -------
                               global concordance:  +42.00
    *------------  pairwise comparison ----*
    Comparing actions : ('p1', 'p2')
    crit. wght.  g(x)   g(y)     diff   | ind   pref     r()
    ec1    8.00  38.11  89.77  -51.66  | 5.00  10.00   -8.00
    ec4    8.00  22.65  86.00  -63.35  | 5.00  10.00   -8.00
    ec8    8.00  77.02  89.43  -12.41  | 5.00  10.00   -8.00
    en3    6.00  58.16  20.79  +37.37  | 5.00  10.00   +6.00
    en5    6.00  31.40  23.83   +7.57  | 5.00  10.00   +6.00
    en6    6.00  11.41  18.66   -7.25  | 5.00  10.00   +0.00
    en9    6.00  44.37  26.65  +17.72  | 5.00  10.00   +6.00
    so2   12.00  22.43  89.12  -66.69  | 5.00  10.00  -12.00
    so7   12.00  28.41  84.73  -56.32  | 5.00  10.00  -12.00
    Valuation in range: -72.00 to +72.00;                -------
                                 global concordance: -30.00
```

In both comparisons, the evaluations observed with respect to the environmental decision objective are not validating with a significant majority the outranking, respectively, outranked, situation. Hence, the stability of the reported preferential situations is in fact dependent on choosing significance weights that are compatible with the given significance weights preorder (see Listing 19.2 on page 262).

Let us finally inspect a comparison that is only qualified at stability level +1, like the one between alternatives p7 and p3.

Listing 19.6 Comparison of alternatives p7 and p3

```
>>> g.showPairwiseOutrankings('p7','p3')
 *------------  pairwise comparison ----*
  Comparing actions : ('p7', 'p3')
  crit. wght.  g(x)   g(y)     diff   | ind   pref     r()
  ec1    8.00  15.33  80.19  -64.86  | 5.00  10.00   -8.00
  ec4    8.00  36.31  68.70  -32.39  | 5.00  10.00   -8.00
  ec8    8.00  38.31  91.94  -53.63  | 5.00  10.00   -8.00
  en3    6.00  30.70  46.78  -16.08  | 5.00  10.00   -6.00
  en5    6.00  35.52  27.25   +8.27  | 5.00  10.00   +6.00
  en6    6.00  69.71   1.65  +68.06  | 5.00  10.00   +6.00
  en9    6.00  13.10  14.85   -1.75  | 5.00  10.00   +6.00
```

```
 so2  12.00  68.06  58.85   +9.21 | 5.00  10.00  +12.00
 so7  12.00  58.45  15.49  +42.96 | 5.00  10.00  +12.00
 Valuation in range: -72.00 to +72.00;           -------
                             global concordance: +12.00
*-----------  pairwise comparison ----*
 Comparing actions : ('p3', 'p7')
 crit. wght.  g(x)  g(y)     diff  | ind   pref     r()
 ec1   8.00  80.19  15.33  +64.86 | 5.00  10.00   +8.00
 ec4   8.00  68.70  36.31  +32.39 | 5.00  10.00   +8.00
 ec8   8.00  91.94  38.31  +53.63 | 5.00  10.00   +8.00
 en3   6.00  46.78  30.70  +16.08 | 5.00  10.00   +6.00
 en5   6.00  27.25  35.52   -8.27 | 5.00  10.00   +0.00
 en6   6.00   1.65  69.71  -68.06 | 5.00  10.00   -6.00
 en9   6.00  14.85  13.10   +1.75 | 5.00  10.00   +6.00
 so2  12.00  58.85  68.06   -9.21 | 5.00  10.00   +0.00
 so7  12.00  15.49  58.45  -42.96 | 5.00  10.00  -12.00
 Valuation in range: -72.00 to +72.00;           -------
                             global concordance: +18.00
```

In both cases, choosing only significances that are compatible with the given weights preorder will not always result in positively validated outranking situations.

19.3 Computing the Stability Denotation of Outranking Situations

Stability levels ± 4 and ± 3 are, the case given, easy to detect. Detecting a stability level ± 2 is far less obvious. Now, it is precisely again the bipolar-valued epistemic characteristic domain that will give us a way to implement an effective test for stability level $+2$ and -2 (Bisdorff 2004a,b).

Let us consider the significance equivalence classes we observe in the given weights preorder. Here, we observe three weight classes: 6, 8, and 12, in an increasing order (see Listing 19.2 on page 262). In the pairwise comparisons, shown above, these equivalence classes may appear positively or negatively, besides the indeterminate significance of value 0.00. We thus get the following ordered bipolar list of significance weights: $W = [-12, -8, -6, 0, 6, 8, 12]$.

In all the pairwise marginal comparisons shown in the previous section, we may observe that each one of the nine criteria assigns one precise item out of this list W. Let us denote by $q[i]$ the number of criteria assigning item $W[i]$ and $Q[i]$ the cumulative sums of these $q[i]$ counts, where i is an index in the range of the length of list W.

In the comparison of alternatives p2 and p1, for instance (see Listing 19.5 on page 265), we observe the following counts:

$W[i]$	-12	-8	-6	0	6	8	12
$q[i]$	0	0	2	1	1	3	2
$Q[i]$	0	0	2	3	4	7	9

Let us denote by $-q$ and $-Q$ the reversed versions of the q and the Q lists. We thus obtain the following result:

$W[i]$	-12	-8	-6	0	6	8	12
$-q[i]$	2	3	1	1	2	0	0
$-Q[i]$	2	5	6	7	9	9	9

Now, a pairwise outranking situation will be qualified at stability level $+2$, i.e. positively validated with any significance weights that are compatible with the given weights preorder, when for all i, we observe $Q[i] \leq -Q[i]$, and there exists one i such that $Q[i] < -Q[i]$. Similarly, a pairwise outranked situation will be qualified at stability level -2, when for all i, we observe $Q[i] \geq -Q[i]$, and there exists one i such that $Q[i] > -Q[i]$ (Bisdorff 2004b).

Let us verify that the outranking situation observed, for instance, between alternatives p2 and p1 does indeed verify this *first-order distributional dominance* condition.

$W[i]$	-12	-8	-6	0	6	8	12
$Q[i]$	0	0	2	3	4	7	9
$-Q[i]$	2	5	6	7	9	9	9

Notice that outranking situations qualified at stability levels ± 4 and ± 3 evidently verify the stability level ± 2 test above. The outranking situation between alternatives p7 and p3 does not, however, verify this test (see Listing 19.6 on page 266).

$W[i]$	-12	-8	-6	0	6	8	12
$q[i]$	0	3	1	0	3	0	2
$Q[i]$	0	3	4	4	7	7	9
$-Q[i]$	2	2	5	5	6	9	9

This time, not all the $Q[i]$ are *lower than or equal to* the corresponding $-Q[i]$ terms. Hence the outranking situation between p7 and p3 is not positively validated with all potential significance weights that are compatible with the given weights preorder.

Using this stability denotation, we may, hence, define the following *robust* version of a bipolar-valued outranking digraph.

19.4 Robust Bipolar-Valued Outranking Digraphs

Definition 19.1 (Robust Outranking Situation)

- We say that decision alternative x *robustly outranks* decision alternative y when:
  - x is at least as well evaluated as y at stability level +2 or higher.
  - We do not observe any considerable counter-performance of x on a discordant criterion.
- Dually, we say that decision alternative x *does not robustly outrank* decision alternative y when:
  - x is not at least as well evaluated as y at stability level −2 or lower.
  - We do not observe any considerable better performance of x on a discordant criterion.
- Otherwise, the outranking situation is indeterminate.

The corresponding *robust* outranking digraph can be computed as follows with the `RobustOutrankingDigraph` class:

Listing 19.7 Computing a robust outranking digraph

```
>>> from outrankingDigraphs import\
...                 RobustOutrankingDigraph
>>> rg = RobustOutrankingDigraph(pt) # same pt
>>> rg
 *------- Object instance description ------*
  Instance class         : RobustOutrankingDigraph
  Instance name          : robust_random3ObjectivesPerfTab
  Actions                : 7
  Criteria               : 9
  Size                   : 22
  Determinateness (%)    : 68.45
  Valuation domain       : [-1.00;1.00]
  Attributes             : ['name', 'methodData', 'actions',
         'order', 'criteria', 'evaluation',
         'vetos', 'valuationdomain',
         'cardinalRelation', 'ordinalRelation',
         'equisignificantRelation', 'unanimousRelation',
         'relation', 'gamma', 'notGamma']
>>> rg.showRelationTable()
 * ---- Relation Table -----
```

```
21    r/(stab) | 'p1'  'p2'  'p3'  'p4'  'p5'  'p6'  'p7'
22   ---------|------------------------------------------
23      'p1'   | +1.00 -0.42 +0.00 -0.69 +0.39 +0.11 +0.00
24             |  (+4)  (-2)  (+0)  (-3)  (+2)  (+2)  (-1)
25      'p2'   | +0.58 +1.00 +0.83 +0.00 +0.58 +0.58 +0.58
26             |  (+2)  (+4)  (+3)  (+2)  (+2)  (+2)  (+2)
27      'p3'   | +0.25 -0.33 +1.00 +0.00 +0.50 +1.00 +0.00
28             |  (+2)  (-2)  (+4)  (+0)  (+2)  (+2)  (+1)
29      'p4'   | +0.78 +0.00 +0.61 +1.00 +1.00 +1.00 +0.67
30             |  (+3)  (-1)  (+3)  (+4)  (+4)  (+4)  (+2)
31      'p5'   | -0.11 -0.50 -0.25 -0.89 +1.00 +0.11 -0.14
32             |  (-2)  (-2)  (-2)  (-3)  (+4)  (+2)  (-2)
33      'p6'   | +0.22 -0.42 +0.00 -1.00 +0.17 +1.00 -0.11
34             |  (+2)  (-2)  (+1)  (-2)  (+2)  (+4)  (-2)
35      'p7'   | +0.22 -0.50 +0.00 +0.00 +0.78 +0.42 +1.00
36             |  (+2)  (-2)  (+1)  (-1)  (+3)  (+2)  (+4)
```

All outranking situations, qualified at stability level ± 1, are now put to an *indeterminate* status (see Listing 19.7, Lines 23–36 above). In the example here, three positive outrankings get dropped: between p3 and p7, between p7 and p3, and between p6 and p3, where the last situation is actually already doubtful because of a veto situation. Three negative outrankings get dropped as well: between p1 and p7, between p4 and p2, and between p7 and p4.

Notice by the way that outranking or outranked situations, although qualified at level ± 2 or ± 3, may nevertheless become doubtful because of considerable performance differences. We observe such a doubtful situation when comparing, for instance, alternatives p2 and p4 (see Listing 19.8, Lines 4–8).

Listing 19.8 Inspecting polarised outranking situations

```
1  >>> rg.showPolarisations()
2  *----  Negative polarisations ----*
3  number of negative polarisations : 3
4  1: r(p2 >= p4) = 0.33
5  criterion: en3
6  Considerable performance difference : -60.02
7  Veto discrimination threshold       : -60.00
8  Polarisation: r(p2 >= p4) = 0.33 ==> 0.00
9  2: r(p6 >= p3) = 0.06
10 criterion: ec8
11 Considerable performance difference : -62.31
12 Veto discrimination threshold       : -60.00
13 Polarisation: r(p6 >= p3) = 0.06 ==> 0.00
14 3: r(p6 >= p4) = -0.36
15 criterion: en3
16 Considerable performance difference : -68.27
17 Veto discrimination threshold       : -60.00
18 Polarisation: r(p6 >= p4) = -0.36 ==> -1.00
```

Despite being robust, the '*at least as well evaluated as*' situation between alternatives p2 and p4 becomes doubtful because of a considerable counter-performance (-60.02) of p2 observed on criterion en3, a negative difference which exceeds slightly the assumed veto discrimination threshold $v = 60.00$. The same polarisation to 0.00 (indeterminate) happens when comparing alternatives p6 and

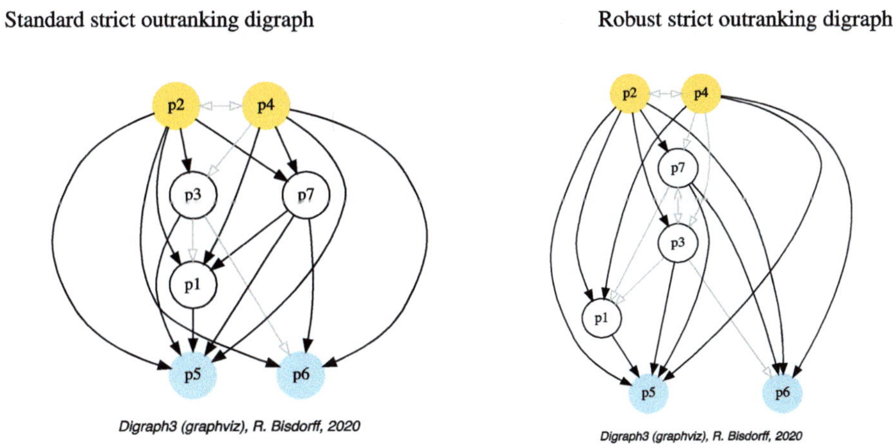

Fig. 19.1 *Standard* versus *robust* strict outranking digraphs oriented by their initial and terminal prekernels

p3 (Line 13). Notice the polarisation to −1.0 (certainly false) of the "*at least as well evaluated as*" situation between alternatives p6 and p4 (Line 18).

One may finally compare in Fig. 19.1 the *standard* and the *robust* versions of the corresponding strict outranking digraphs, both oriented by their respective identical initial and terminal prekernels.

The robust version (right in Fig. 19.1) drops two strict outranking situations: between p4 and p7 and between p7 and p1. The remaining 14 strict outranking (respectively, outranked) situations are now all verified at a stability level of +2 and more (respectively, −2 and less). They remain valid, hence, with all potential significance weights that are compatible with the given significance weights preordering (see Listing 19.2 on page 262).

To appreciate the apparent partial ordering of both the standard and the robust strict outranking digraphs shown in Fig. 19.1, let us have in Fig. 19.2 on the next page a final heatmap view on the underlying performance tableau ordered by the NetFlows ranking rule actually applied to the robust version of the outranking digraph (see the `outrankingModel='this'` flag in Line 4 below).

Listing 19.9 Computing a robust performance heatmap view

```
1 >>> rg.showHTMLPerformanceHeatmap(\
2 ...             Correlations=True,\
3 ...             colorLevels=5,\
4 ...             outrankingModel='this',\
5 ...             rankingRule='NetFlows')
```

As the initial prekernel {p2, p4} is in the robust outranking digraph validated at least at stability level ±2, recommending alternatives p4 and p2 as potential best choices appears robustly justified. Alternative p4 represents indeed an overall *best*

Heatmap of Performance Tableau 'robust_random3ObjectivesPerfTab'

criteria	ec4	so2	ec1	ec8	en9	en3	so7	en6	en5
weights	+8.00	+12.00	+8.00	+8.00	+6.00	+6.00	+12.00	+6.00	+6.00
tau(*)	+0.55	+0.52	+0.45	+0.38	+0.31	+0.29	+0.24	+0.05	+0.05
p4	72.26	49.05	85.19	44.62	50.92	80.81	55.57	66.21	49.69
p2	86.00	89.12	89.77	89.43	26.65	20.79	84.73	18.66	23.83
p3	68.70	58.85	80.19	91.94	14.85	46.78	15.49	1.65	27.25
p7	36.31	68.06	15.33	38.31	13.10	30.70	58.45	69.71	35.52
p1	22.65	22.43	38.11	77.02	44.37	58.16	28.41	11.41	31.40
p6	75.76	48.59	21.06	29.63	32.96	12.54	26.40	36.04	43.09
p5	8.96	12.36	46.75	35.91	9.83	31.05	44.92	31.22	29.52

Color legend:

quantile	20.00%	40.00%	60.00%	80.00%	100.00%

(*) **tau:** *Ordinal (Kendall) correlation between marginal criterion and global ranking relation*
Outranking model: **this**, *Ranking rule*: **NetFlows**
Ordinal (Kendall) correlation between global ranking and global outranking relation: **+0.960**
Mean marginal correlation (a) : **+0.338**
Standard marginal correlation deviation (b) : **+0.168**
Ranking fairness (a) - (b) : **+0.170**

Fig. 19.2 Robust heatmap of the random 3-objectives performance tableau ordered by the NetFlows ranking rule

compromise choice between all decision objectives, whereas alternative p2 gives an unanimous best choice with respect to two out of three decision objectives. It is up to the decision maker to make his final choice.

19.5 Characterising Unopposed Multiobjective Outranking Situations

When facing a performance tableau involving multiple decision objectives, the robustness level ± 3 may lead to distinguishing what we call *unopposed* outranking situations, like the one shown in the previous section between alternatives p4 and p1 ($r(p4 \succsim p1) = +0.78$, see Listing 19.4 on page 264, Line 11), namely preferential situations that are more or less validated or invalidated by all the decision objectives.

Definition 19.2 (Unopposed Outranking Situation)

- We say that decision alternative x *unopposedly outranks* decision alternative y when x positively outranks y on one or more decision objectives without x being positively outranked by y on any decision objective.

- Dually, we say that decision alternative x *is unopposedly outranked by* decision alternative y when x is positively outranked by y on one or more decision objectives without x positively outranking y on any decision objective.

Let us reconsider, for instance, the performance tableau pt with three decision objectives already seen in Listing 19.1 on page 261:

```
1 >>> pt.showObjectives()
2   *------ show objectives -------"
3    Eco: Economical aspect
4      ec1 criterion of objective Eco 8
5      ec4 criterion of objective Eco 8
6      ec8 criterion of objective Eco 8
7     Total weight: 24.00 (3 criteria)
8    Soc: Societal aspect
9      so2 criterion of objective Soc 12
10     so7 criterion of objective Soc 12
11    Total weight: 24.00 (2 criteria)
12   Env: Environmental aspect
13     en3 criterion of objective Env 6
14     en5 criterion of objective Env 6
15     en6 criterion of objective Env 6
16     en9 criterion of objective Env 6
17    Total weight: 24.00 (4 criteria)
```

We notice in this example three decision objectives of equal importance (see Lines 3, 13, and 17). What will be the outranking situations that are positively (respectively, negatively) validated for each one of the decision objectives taken individually?

Such unopposed multiobjective outranking situations may be obtained by operating an epistemic *average fusion* (see the symmetricAverage() method) of the marginal outranking digraphs restricted to the coalition of criteria supporting each one of the decision objectives (see Listing 19.10).

Listing 19.10 Computing unopposed outranking situations

```
>>> from outrankingDigraphs import\
...                          BipolarOutrankingDigraph
>>> geco = BipolarOutrankingDigraph(pt,\
...                          objectivesSubset=['Eco'])
>>> gsoc = BipolarOutrankingDigraph(pt,\
...                          objectivesSubset=['Soc'])
>>> genv = BipolarOutrankingDigraph(pt,\
...                          objectivesSubset=['Env'])
>>> from digraphs import FusionLDigraph
>>> objectiveWeights = \
...                  [pt.objectives[obj]['weight']\
...                          for obj in t.objectives]
>>> uopg = FusionLDigraph([geco,gsoc,genv],\
...                       operator='o-average',\
...                       weights=objectiveWeights)
>>> uopg.showRelationTable(ReflexiveTerms=False)
```

```
17  * ---- Relation Table -----
18    r   |  'p1'   'p2'   'p3'   'p4'   'p5'   'p6'   'p7'
19  -----|------------------------------------------------
20  'p1' |   -    +0.00  +0.00  -0.69  +0.39  +0.11  +0.00
21  'p2' | +0.00    -    +0.83  +0.00  +0.00  +0.00  +0.00
22  'p3' | +0.00  -0.33    -    +0.00  +0.50  +0.00  +0.00
23  'p4' | +0.78  +0.00  +0.61    -    +1.00  +1.00  +0.67
24  'p5' | -0.11  +0.00  +0.00  -0.89    -    +0.11  +0.00
25  'p6' | +0.00  +0.00  +0.00  -0.44  +0.17    -    +0.00
26  'p7' | +0.00  +0.00  +0.00  +0.00  +0.78  +0.42    -
27  Valuation domain: [-1.0; 1.0]
```

Positive (respectively, negative) $r(x \succsim y)$ characteristic values, like $r(p1 \succsim p5) = +0.39$ (see Listing 19.10, Line 20 above), show hence only outranking situations being validated (respectively, invalidated) by one or more decision objectives without being invalidated (respectively, validated) by any other decision objective.

For easily computing this kind of *unopposed multiobjective* outranking digraphs, the `outrankingDigraphs` module conveniently provides a corresponding `UnOpposedBipolarOutrankingDigraph` constructor .

Listing 19.11 Computing unopposed outranking digraphs

```
1  >>> from outrankingDigraphs import\
2  ...                  UnOpposedBipolarOutrankingDigraph
3  >>> uopg = UnOpposedBipolarOutrankingDigraph(pt)
4  >>> uopg
5    *------- Object instance description ------*
6     Instance class        : UnOpposedBipolarOutrankingDigraph
7     Instance name         : unopposed_outrankings
8     Actions             : 7
9     Criteria            : 9
10    Size                  : 13
11    Oppositeness (%)      : 43.48
12    Determinateness (%) : 61.71
13    Valuation domain      : [-1.00;1.00]
14    Attributes            : ['name', 'actions',
15                   'valuationdomain', 'objectives',
16                   'criteria', 'methodData',
17                   'evaluation', 'order', 'runTimes',
18                   'relation', ...
19                   'gamma', 'notGamma']
20 >>> uopg.computeOppositeness(InPercents=True)
21   {'standardSize': 23, 'unopposedSize': 13,
22    'oppositeness': 43.47826086956522}
```

The resulting *unopposed* outranking digraph keeps in fact 13 (see Listing 19.11, Lines 10 11) out of the 23 positively validated *standard* outranking situations, leading to a degree of *oppositeness* – preferential disagreement between decision objectives – of $(1.0 - 13/23) = 0.4348$.

We can now, for instance, verify the unopposed status of the outranking situation observed between alternatives `p1` and `p5`.

Listing 19.12 Example of unopposed multiobjective outranking situation

```
>>> uopg.showPairwiseComparison('p1','p5')
 *------------  pairwise comparison ----*
  Comparing actions : ('p1', 'p5')
  crit. wght.  g(x)   g(y)     diff    | ind   pref     r()
  ec1   8.00  38.11  46.75  -8.64    | 5.00  10.00   +0.00
  ec4   8.00  22.65  8.96  +13.69   | 5.00  10.00   +8.00
  ec8   8.00  77.02  35.91  +41.11   | 5.00  10.00   +8.00
  en3   6.00  58.16  31.05  +27.11   | 5.00  10.00   +6.00
  en5   6.00  31.40  29.52  +1.88    | 5.00  10.00   +6.00
  en6   6.00  11.41  31.22  -19.81   | 5.00  10.00   -6.00
  en9   6.00  44.37  9.83  +34.54   | 5.00  10.00   +6.00
  so2   12.00  22.43  12.36  +10.07 | 5.00  10.00  +12.00
  so7   12.00  28.41  44.92  -16.51 | 5.00  10.00  -12.00
  Valuation in range: -72.00 to +72.00;            -------
                             global concordance: +28.00
```

In Listing 19.12, we see that alternative p1 does indeed positively outrank alternative p5 from the economic perspective ($r(p1 \succsim_{Eco} p5) = +16/24$) as well as from the environmental perspective ($r(p1 \succsim_{Env} p5) = +12/24$), whereas, from the societal perspective, both alternatives appear incomparable ($r(p1 \succsim_{Soc} p5) = 0/24$).

When proportionally equal criteria significance weights per objective are given, these outranking situations appear hence *stable* with respect to all possible importance weights we could allocate to the decision objectives.

This gives way for computing multiobjective *Pareto efficient* choice recommendations.

19.6 Computing Pareto Efficient Multiobjective Choices

Indeed, best choice recommendations, computed from an *unopposed multiobjective* outranking digraph, deliver the case given *Pareto efficient* choices.

Listing 19.13 Pareto efficient multiobjective choice

```
>>> uopg.showBestChoiceRecommendation()
 Best choice recommendation(s) (BCR)
  (in decreasing order of determinateness)
  Credibility domain: [-1.00,1.00]
  === >> potential first choice(s)
  choice              : ['p2', 'p4', 'p7']
     independence        : 0.00
     dominance           : 0.33
     absorbency          : 0.00
     covering (%)        : 33.33
     determinateness (%) : 50.00
  === >> potential last choice(s)
  choice              : ['p3', 'p5', 'p6', 'p7']
```

```
14        independence          : 0.00
15        dominance             : -0.61
16        absorbency            : 0.11
17        covered (%)           : 33.33
18        determinateness (%)   : 50.00
```

Our previous *robust* best choice recommendation (p2 and p4, see Fig. 19.1 on page 271) remains, in this example here, *stable*. We recover indeed the best choice recommendation [p2, p4, p7] (see Listing 19.13 on the preceding page, Line 6). Yet, notice that decision alternative p7 appears to be at the same time a potential *first* as well as a potential *last* choice recommendation (see Line 13), a consequence of p7 being completely *incomparable* to the other decision alternatives when restricting the comparability to only unopposed strict outranking situations.

This kind of Pareto efficient result is shown in Fig. 19.3.

```
>>> (~(-uopg)).exportGraphViz(fileName = 'unopDigraph',\
...                           firstChoice = ['p2','p4'],\
...                           lastChoice = ['p3','p5','p6'])
 *---- exporting a dot file for GraphViz tools ----*
  Exporting to unopDigraph.dot
  dot -Grankdir=BT -Tpng unopDigraph.dot -o unopDigraph.png
```

In order to make now an eventual best unique choice, a decision maker will necessarily have to weigh, in a second stage of the decision aiding process, the relative importance of the individual decision objectives.

For concluding, let us mention that it is precisely again our bipolar-valued logical characteristic framework that provides us here with a first-order distributional dominance test for effectively qualifying the stability level ±2 robustness of an

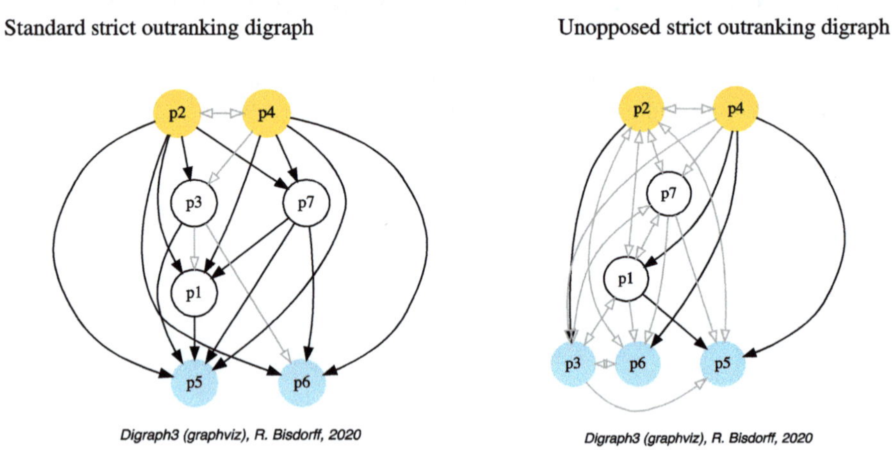

Fig. 19.3 Standard versus *unopposed* strict outranking digraphs oriented by first and last choice recommendations

outranking digraph when facing performance tableaux with criteria of only ordinal-valued significance. A real-world application of our stability analysis with such a kind of performance tableau may be consulted in Bisdorff (2015).

In a *social choice* context, where decision objectives would match different political parties, Pareto efficient best choice recommendations represent in fact *multipartisan* social choices that may judiciously temper plurality tyranny effects. This idea is developed in Chap. 20.

References

Bisdorff R (2004a) Concordant outranking with multiple criteria of ordinal significance. 4OR 2(4):293–308. http://hdl.handle.net/10993/23721

Bisdorff R (2004b) Preference aggregation with multiple criteria of ordinal significance. Annales du LAMSADE 3:25–44. http://hdl.handle.net/10993/42420

Bisdorff R (2014) On confident outrankings with multiple criteria of uncertain significance. In: Mousseau V, Pirlot M (eds) DAP'2014 from multiple criteria decision aid to preference learning, Ecole Centrale de Paris, pp 1–6. http://hdl.handle.net/10993/23910

Bisdorff R (2015) The EURO 2004 best poster award: choosing the best poster in a scientific conference. In: Bisdorff R, Dias L, Meyer P, Mousseau V, Pirlot M (eds) Evaluation and decision models with multiple criteria: case studies. Springer, Berlin, pp 117–166

Bisdorff R, Meyer P, Veneziano T (2009) Inverse analysis from a CONDORCET robustness denotation of valued outranking relations. In: Rossi F, Tsoukiàs A (eds) Algorithmic decision theory. Lecture notes in artificial intelligence (LNAI), vol 5783. Springer, Berlin, pp 180–191. http://hdl.handle.net/10993/23454

Bisdorff R, Meyer P, Veneziano T (2014) Elicitation of criteria weights maximising the stability of pairwise outranking statements. J Multi-Criteria Decision Analy (Wiley) 21:113–124. http://hdl.handle.net/10993/23701

Chapter 20
Tempering Plurality Tyranny Effects in Social Choice

The choice of a voting procedure shapes the democracy in which we live

— Baujard, Gavrel, Igersheim, Laslier, and Lebon (2013)

Contents

Abstract In a *social choice* context, where decision objectives would match different political parties, Pareto efficient choice recommendations represent in fact *multipartisan* social choices that may judiciously deliver the primary selection in a two-stage election system. Our bipolar-valued outranking model is based on approvals-disapprovals of '*at least as well evaluated as*' statements. A similar approach is put into practice with approval–disapproval voting systems. When converting such approval–disapproval voting ballots into corresponding performance records, we obtain a $(-1, 0, 1)$-valued evaluative voting system. We eventually show that in such an approval–disapproval voting system, the winner tends to be among the more or less multipartisan candidates.

20.1 Two-Stage Elections with Multipartisan Primary Selection

From the seminal work by *Bernard Roy*, tempering plurality tyranny effects is achieved in the outranking approach by taking into account considerable negative performance differences that render doubtful otherwise positive '*at least as well*

R. Bisdorff, *Algorithmic Decision Making with Python Resources*,
International Series in Operations Research & Management Science 324,
https://doi.org/10.1007/978-3-030-90928-4_20

evaluated as' situations (Benayoun et al. 1966). In the social choice context of general elections, such a polarisation is not feasible as the genuine uninominal voting ballots do not contain rich enough preferential information. Yet, when matching decision objectives with political parties, the Pareto efficient outranking situations seen in Sect. 19.6 correspond to multipartisan '*at least as well approved as*' situations when pairwisely comparing the eligible candidates. A best choice recommendation based on the corresponding multipartisan strict '*better approved as*' situations may, the case given, deliver a convincing primary selection in a two-stage election.

To compute multipartisan social choices, we need to, first, convert a given linear voting profile with pre-election polls into a corresponding performance tableau. We shall illustrate this point with a voting profile we discussed already in Chap. 7.

Listing 20.1 Example of a 3 parties voting profile

```
>>> from votingProfiles import RandomLinearVotingProfile
>>> lvp = RandomLinearVotingProfile(numberOfCandidates=15,
...                              numberOfVoters=1000,
...                              WithPolls=True,
...                              partyRepartition=0.5,
...                              other=0.1,
...                              seed=0.9189670954954139)
>>> lvp
  *------- VotingProfile instance description ------*
   Instance class    : RandomLinearVotingProfile
   Instance name     : randLinearProfile
   Candidates        : 15
   Voters            : 1000
   Attributes        : ['name', 'seed', 'candidates',
             'voters', 'WithPolls', 'RandomWeights',
             'sumWeights', 'poll1', 'poll2',
             'other', partyRepartition,
             'linearBallot', 'ballot']
>>> lvp.showRandomPolls()
  Random repartition of voters
   Party_1 supporters : 460 (46.0%)
   Party_2 supporters : 436 (43.6%)
   Other voters       : 104 (10.4%)
  *---------------- random polls ---------------
   Party-1(46.0%)  | Party-(43.6%)|  expected
   -----------------------------------------
    c06 : 19.91%  | c11 : 22.94%  | c06 : 15.00%
    c07 : 14.27%  | c08 : 15.65%  | c11 : 13.08%
    c03 : 10.02%  | c04 : 15.07%  | c08 : 09.01%
    c13 : 08.39%  | c06 : 13.40%  | c07 : 08.79%
    c15 : 08.39%  | c03 : 06.49%  | c03 : 07.44%
    c11 : 06.70%  | c09 : 05.63%  | c04 : 07.11%
    c01 : 06.17%  | c07 : 05.10%  | c01 : 05.06%
    c12 : 04.81%  | c01 : 05.09%  | c13 : 05.04%
    c08 : 04.75%  | c12 : 03.43%  | c15 : 04.23%
    c10 : 04.66%  | c13 : 02.71%  | c12 : 03.71%
    c14 : 04.42%  | c14 : 02.70%  | c14 : 03.21%
```

```
38      c05 : 04.01%    | c15 : 00.86%    | c09 : 03.10%
39      c09 : 01.40%    | c10 : 00.44%    | c10 : 02.34%
40      c04 : 01.18%    | c05 : 00.29%    | c05 : 01.97%
41      c02 : 00.90%    | c02 : 00.21%    | c02 : 00.51%
```

In this example (see Listing 20.1 on the facing page, Lines 19–41), we observe 460 Party-1 supporters (46%), 436 Party-2 supporters (43.6%), and 104 other voters (10.4%). Favorite candidates of Party-1 supporters, with more than 10%, are `c06` (19.91%), `c07` (14.27%), and `c03` (10.02%). Whereas for Party-2 supporters, favorite candidates are `c11` (22.94%), followed by `c08` (15.65%), `c04` (15.07%), and `c06` (13.4%).

We can convert this linear voting profile into a `PerformanceTableau` object where each political party matches a decision objective.

Listing 20.2 Converting a voting profile into a performance tableau

```
1  >>> lvp.save2PerfTab('votingPerfTab')
2  >>> from perfTabs import PerformanceTableau
3  >>> vpt = PerformanceTableau('votingPerfTab')
4  >>> vpt
5    *------- PerformanceTableau instance description ---*
6      Instance class     : PerformanceTableau
7      Instance name      : votingPerfTab
8      Actions            : 15
9      Objectives         : 3
10     Criteria           : 1000
11     Attributes         : ['name', 'actions', 'objectives',
12               'criteria', 'weightPreorder', 'evaluation']
13 >>> vpt.objectives
14   OrderedDict([
15     ('party0', {'name': 'other', 'weight': Decimal('104'),
16      'criteria': ['v0003', 'v0008', 'v0011', ... ']}),
17     ('party1', {'name': 'party 1', 'weight': Decimal('460'),
18      'criteria': ['v0002', 'v0006', 'v0007', ...]}),
19     ('party2', {'name': 'party 2', 'weight': Decimal('436'),
20       'criteria': ['v0001', 'v0004', 'v0005', ... ]})
21     ])
```

In Listing 20.2, the linear voting profile lvp is first stored in `PerformanceTableau` format (see Line 1). In Line 3, the PerformanceTableau class reloads this stored performance tableau data. The three parties of the linear voting profile represent three decision objectives and the 1000 voters are distributed as 1000 performance criteria according to the party they support.

In order to operate now a *primary multipartisan selection* of potential election winners, we compute the corresponding unopposed multiobjective outranking digraph (see Sect. 19.5).

Listing 20.3 Computing unopposed multiobjective outranking situations

```
1  >>> from outrankingDigraphs import \
2  ...        UnOpposedBipolarOutrankingDigraph
3  >>> uog = UnOpposedBipolarOutrankingDigraph(vpt)
4  >>> uog
5    *------- Object instance description ------*
6     Instance class       : UnOpposedBipolarOutrankingDigraph
7     Instance name        : unopposed_outrankings
8     Actions              : 15
9     Criteria             : 1000
10    Size                 : 34
11    Oppositeness (%)     : 67.31
12    Determinateness (%)  : 57.61
13    Valuation domain     : [-1.00;1.00]
14    Attributes           : ['name', 'actions', 'valuationdomain',
15                             'objectives', 'criteria',
                                   'methodData',
16                             'evaluation', 'order', 'runTimes',
17                             'relation',
                                   'marginalRelationsRelations',
18                             'gamma', 'notGamma']
```

From the potential 105 pairwise outranking situations, we keep 34 positively validated outranking situations, leading to a degree of *oppositeness* between political parties of 67.31% (see Listing 20.3 Line 11).

The corresponding bipolar-valued relation table can be shown by orienting the list of candidates with the help of the initial and terminal prekernels.

Listing 20.4 Computing unopposed multiobjective outranking situations

```
>>> uog.showPreKernels()
  *--- Computing preKernels ---*
 Dominant preKernels :
  ['c11', 'c06', 'c13', 'c15']
     independence :  0.0
     dominance    :  0.18
     absorbency   : -0.66
     covering     :  0.43
 Absorbent preKernels :
  ['c02', 'c04', 'c14', 'c03']
     independence :  0.0
     dominance    :  0.0
     absorbency   :  0.37
     covered      :  0.46
>>> orientedCandidatesList = ['c06','c11','c13','c15',\
...          'c01','c05','c07','c08','c09','c10','c12',\
...          'c02','c03','c04','c14']
>>> uog.showHTMLRelationTable(\
...     actionsList=orientedCandidatesList,\
...     tableTitle='Unopposed three-partisan outrankings')
```

Multipartisan outranking situations

r(x S y)	a06	a11	a13	a15	a01	a05	a07	a08	a09	a10	a12	a02	a03	a04	a14
a06	-	0.00	0.00	0.00	0.44	0.00	0.25	0.00	0.00	0.00	0.56	0.86	0.29	0.00	0.57
a11	0.00	-	0.00	0.00	0.00	0.55	0.00	0.18	0.59	0.51	0.39	0.80	0.00	0.42	0.47
a13	0.00	0.00	-	0.00	0.00	0.52	-0.27	0.00	0.00	0.00	0.00	0.77	0.00	0.00	0.16
a15	0.00	0.00	0.00		0.00	0.39	0.00	0.00	0.00	0.00	0.00	0.66	0.00	0.00	0.00
a01	-0.44	0.00	0.00	0.00	-	0.00	0.00	0.00	0.00	0.00	0.00	0.77	0.00	0.00	0.20
a05	0.00	-0.55	-0.52	-0.39	0.00	-	0.00	-0.47	0.00	-0.12	0.00	0.37	0.00	0.00	0.00
a07	-0.25	0.00	0.27	0.00	0.00	0.00	-	0.00	0.00	0.00	0.30	0.83	0.00	0.00	0.38
a08	0.00	-0.18	0.00	0.00	0.00	0.47	0.00	-	0.00	0.00	0.00	0.77	0.00	0.29	0.00
a09	0.00	-0.59	0.00	0.00	0.00	0.00	0.00	0.00	-	0.00	0.00	0.55	0.00	0.00	0.00
a10	0.00	-0.51	0.00	0.00	0.00	0.12	0.00	0.00	0.00	-	0.00	0.50	0.00	0.00	0.00
a12	-0.56	-0.39	0.00	0.00	0.00	0.00	-0.30	0.00	0.00	0.00	-	0.72	0.00	0.00	0.10
a02	-0.86	-0.80	-0.77	-0.66	-0.77	-0.37	-0.83	-0.77	-0.55	-0.50	-0.72	-	0.00	0.00	0.00
a03	-0.29	0.00	0.00	0.00	0.00	0.00	0.00	0.00	0.00	0.00	0.00	0.00	-	0.00	0.00
a04	0.00	-0.42	0.00	0.00	0.00	0.00	0.00	-0.29	0.00	0.00	0.00	0.00	0.00	-	0.00
a14	-0.57	-0.47	-0.16	0.00	-0.20	0.00	-0.38	0.00	0.00	0.00	-0.10	0.00	0.00	0.00	-

Valuation domain: [-1.00; +1.00]

Fig. 20.1 Relation table of multipartisan outranking digraph

In Fig. 20.1, we may notice that the dominating outranking prekernel {`c06`, `c11`, `c13`, `c15`} gathers in fact a multipartisan selection of potential election winners. It is worthwhile noticing that the majority margins obtained from a linear voting profile do verify the zero-sum rule: $\big(r(x \succsim y) + r(y \succsim x) \;=\; 0.0\big)$. To each positive outranking situation corresponds an equivalent negative converse situation and the resulting outranking and strict outranking digraphs are the same.

When restricting now, in a secondary election stage, the set of eligible candidates to this dominating prekernel, we may compute the actual best social choice.

Listing 20.5 Recommending the secondary election winner

```
>>> from outrankingDigraphs import BipolarOutrankingDigraph
>>> g2 = BipolarOutrankingDigraph(vpt,\
..             actionsSubset=['c06','c11','c13','c15'])
>>> g2.showRelationTable(ReflexiveTerms=False)
  * ---- Relation Table -----
      r    | 'c06'  'c11'  'c13'  'c15'
     ------|----------------------------
     'c06' |   -    +0.10  +0.48  +0.52
     'c11' | -0.10    -    +0.27  +0.29
     'c13' | -0.48  -0.27    -    +0.19
     'c15' | -0.52  -0.29  -0.19    -
     Valuation domain: [-1.0; 1.0]
>>> g2.computeCondorcetWinners()
  ['c06']
>>> g2.computeCopelandRanking()
  ['c06', 'c11', 'c13', 'c15']
```

Candidate `c06` appears clearly to be the winner of this election. Notice by the way that the restricted pairwise outranking relation, shown in Listing 20.5 Lines 8–11, models a linear ordering of the preselected candidates.

We can eventually check the quality of this best choice by noticing that Candidate `c06` represents indeed the *simple majority*, the *instant runoff*, the BORDA, as well as the CONDORCET winner of the initially given linear voting profile `lvp`.

```
>>> lvp.computeSimpleMajorityWinner()
  ['c06']
>>> lvp.computeInstantRunoffWinner()
  ['c06']
>>> lvp.computeBordaWinners()
  ['c06']
>>> from votingProfiles import MajorityMarginsDigraph
>>> cd = MajorityMarginsDigraph(lvp)
>>> cd.computeCondorcetWinners()
  ['c06']
```

In our example voting profile here, the multipartisan primary selection stage appears quite effective in reducing the number of eligible candidates to four out of a set of 15 candidates without by the way rejecting the actual winning candidate.

However, in a very *divisive* two major parties system, like in the USA, where preferences of the supporters of one major party are opposite to the preferences of the supporters of the other major party, the multipartisan outranking digraph will become nearly indeterminate.

In Listing 20.6 below, we generate such a divisive kind of linear voting profile with the help of the `DivisivePolitics` flag (see Lines 4 and 14–20). When now converting the voting profile into a performance tableau (Lines 20-21), we can compute the corresponding unopposed outranking digraph.

Listing 20.6 A divisive two-party example of a random linear voting profile

```
1  >>> from votingProfiles import RandomLinearVotingProfile
2  >>> lvp = RandomLinearVotingProfile(\
3  ...        numberOfCandidates=7,numberOfVoters=500,\
4  ...        WithPolls=True, partyRepartition=0.4,other=0.2,\
5  ...        DivisivePolitics=True, seed=1)
6  >>> lvp.showRandomPolls()
7    Random repartition of voters
8     Party-1 supporters : 240 (48.00%)
9     Party-2 supporters : 160 (32.00%)
10    Other voters       : 100 (20.00%)
11   *---------------- random polls -------------
12    Party_1(48.0%) | Party_2(32.0%) | expected
13    ------------------------------------------
14    c2 : 30.84%    | c1 : 30.84%   | c2 : 15.56%
15    c3 : 23.67%    | c4 : 23.67%   | c3 : 12.91%
16    c7 : 17.29%    | c6 : 17.29%   | c7 : 11.43%
17    c5 : 11.22%    | c5 : 11.22%   | c1 : 11.00%
18    c6 : 09.79%    | c7 : 09.79%   | c6 : 10.23%
19    c4 : 04.83%    | c3 : 04.83%   | c4 : 09.89%
20    c1 : 02.37%    | c2 : 02.37%   | c5 : 08.98%
```

```
21 >>> lvp.save2PerfTab('divisiveExample')
22 >>> dvp = PerformanceTableau('divisiveExample')
23 >>> from outrankingDigraphs import \
24 ...         UnOpposedBipolarOutrankingDigraph
25 >>> uodg = UnOpposedBipolarOutrankingDigraph(dvp)
26 >>> uodg
27   *------- Object instance description ------*
28    Instance class : UnOpposedBipolarOutrankingDigraph
29    Instance name  : divisiveExample
30    Actions        : 7
31    Criteria       : 500
32    Size           : 0
33    Oppositeness (%)     : 100.00
34    Determinateness (%)  : 50.00
35    Valuation domain     : [-1.00;1.00]
```

With an oppositeness degree of 100.0% (see Listing 20.6, Lines 32–33), the preferential disagreement between the political parties is complete, and the unopposed outranking digraph uodg becomes completely indeterminate as shown in the relation table below.

```
>>> uodg.showRelationTable(ReflexiveTerms=False)
  * ---- Relation Table -----
   r   | 'c1'   'c2'   'c3'   'c4'   'c5'   'c6'   'c7'
  -----|-----------------------------------------
  'c1' |   -   +0.00 +0.00 +0.00 +0.00 +0.00 +0.00
  'c2' | +0.00   -   +0.00 +0.00 +0.00 +0.00 +0.00
  'c3' | +0.00 +0.00   -   +0.00 +0.00 +0.00 +0.00
  'c4' | +0.00 +0.00 +0.00   -   +0.00 +0.00 +0.00
  'c5' | +0.00 +0.00 +0.00 +0.00   -   +0.00 +0.00
  'c6' | +0.00 +0.00 +0.00 +0.00 +0.00   -   +0.00
  'c7' | +0.00 +0.00 +0.00 +0.00 +0.00 +0.00   -
  Valuation domain: [-1.0; 1.0]
```

As a consequence, a multipartisan primary selection, computed with a `showBestChoiceRecommendation()` method, will keep the complete initial set of eligible candidates and, hence, become *ineffective* (see Listing 20.7, Line 6).

Listing 20.7 Example of ineffective primary multipartisan selection

```
1 >>> uodg.showBestChoiceRecommendation()
2   Rubis best choice recommendation(s) (BCR)
3    (in decreasing order of determinateness)
4    Credibility domain: [-1.00,1.00]
5    === >> ambiguous choice(s)
6     choice                : ['c1','c2','c3','c4','c5','c6','c7']
7     independence          : 0.00
8     dominance             : 1.00
9     absorbency            : 1.00
10    covered (%)           : 100.00
11    determinateness (%)   : 50.00
12     - most credible action(s) = { }
```

With such kind of divisive voting profile, there may indeed not always exist an obvious winner. In Listing 20.8, we see, for instance, that the simple majority winner is `c2` (Line 2), whereas the instant-run-off winner is `c6` (Line 4).

Listing 20.8 Example of non-obvious secondary selection

```
1  >>> lvp.computeSimpleMajorityWinner()
2    ['c2']
3  >>> lvp.computeInstantRunoffWinner()
4    ['c6']
5  >>> from votingProfiles import MajorityMarginsDigraph
6  >>> cg = MajorityMarginsDigraph(lvp)
7  >>> cg.showRelationTable(ReflexiveTerms=False)
8    * ---- Relation Table -----
9      r()   |  'c1'  'c2'  'c3'  'c4'  'c5'  'c6'  'c7'
10   ------|------------------------------------------
11     'c1'  |   -    -68   -90   -46   -68   -88   -84
12     'c2'  |  +68    -    -32   +80   +46    -6   -24
13     'c3'  |  +90   +32    -    +58   +46    +4    +8
14     'c4'  |   +4   -80   -58    -    -16   -68   -72
15     'c5'  |  +68   -46   -46   +16    -    -26   -64
16     'c6'  |  +88    +6    -4   +68   +26    -     -2
17     'c7'  |  +84   +24    -8   +72   +64    +2    -
18 Valuation domain: [-500;+500]
19 >>> cg.computeCondorcetWinners()
20   ['c3']
21 >>> lvp.computeBordaWinners()
22   ['c3','c7']
23 >>> cg.computeCopelandRanking()
24   ['c3', 'c7', 'c6', 'c2', 'c5', 'c4', 'c1']
```

But in our example here, we are lucky. When constructing with the pairwise majority margins digraph (Line 6), a CONDORCET winner, namely `a3`, becomes apparent (Lines 13 and 20), which is also one of the two BORDA winners (Line 22). More interesting even is to notice that the apparent majority margins digraph models in fact a linear ranking `[a3, a7, a6, a2, a5, a4, a1]` of all the eligible candidates, as shown with a COPELAND ranking rule (Line 24).

In Fig. 20.2, this linear ranking is shown with a graphviz drawing where all transitive arcs are dropped and the drawing is oriented with the CONDORCET winner `a3` and loser `a1` (Line 3 below).

```
1 >>> cg.closeTransitive(Reverse=True)
2 >>> cg.exportGraphViz('divGraph',\
3 ...                firstChoice=['c3'],lastChoice=['c1'])
4   *---- exporting a dot file for GraphViz tools ---------*
5    Exporting to divGraph.dot
6    dot -Grankdir=BT -Tpng divGraph.dot -o divGraph.png
```

Fig. 20.2 The linear ranking modelled by the majority margins digraph

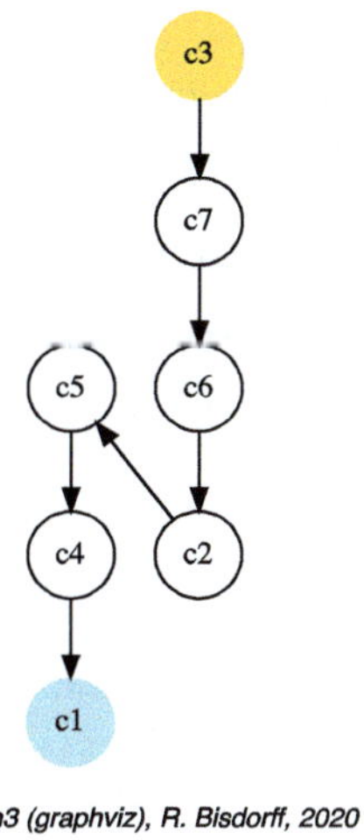

20.2 Bipolar Approval–Disapproval Voting Systems

Traditionally, most of the voting systems in use in the world do only collect approval votes and abstentions, dismissing potentially strong disapproval opinions, not to be assimilated to voting abstentions. Collecting both explicit approvals (+1) and explicit disapprovals (−1) essentially enriches the expression of voters' preferences.

In the `votingProfiles` module, we provide a `BipolarApprovalVotingProfile` class for handling voting results where, for each eligible Candidate `x`, the voters are invited to *approve* (+1), *disapprove* (−1), or *ignore* (0) the statement that Candidate `c` should win the election (Baujard et al. 2013).

File `bpApVotingProfile.py` contains an example of such an approval–disapproval voting profile concerning 100 voters and 15 eligible candidates.[1] We can inspect its content with the `BipolarApprovalVotingProfile` class:

Listing 20.9 Bipolar approval voting profiles

```
>>> from votingProfiles import \
...                     BipolarApprovalVotingProfile
>>> bavp = BipolarApprovalVotingProfile('bpApVotingProfile')
>>> bavp
  *------- VotingProfile instance description ------*
    Instance class    : BipolarApprovalVotingProfile
    Instance name     : bpApVotingProfile
    Candidates        : 15
    Voters            : 100
    Attributes        : ['name', 'candidates', 'voters',
                 'approvalBallot', 'netApprovalScores',
                 'ballot']
```

[1] The file `bpApVotingProfile.py` may be found in the `examples` directory of the DIGRAPH3 resources.

Beside the `candidates` and `voters` attributes, we discover in Listing 20.10 the `approvalBallot` attribute which gathers approval–disapproval votes. Its content is the following.

Listing 20.10 Inspecting an approval–disapproval ballot

```
1  >>> bavp.approvalBallot
2    {'v001':
3        {'c01': Decimal('0'),
4         ...
5         'c04': Decimal('1'),
6         ...
7         'c15': Decimal('0')
8        },
9     'v002':
10        {'c01': Decimal('-1'),
11         'c02': Decimal('0'),
12         ...
13         'c15': Decimal('1')
14        },
15        ...
16    'v100':
17       {'c01': Decimal('0'),
18        'c02': Decimal('1'),
19        ...
20        'c15': Decimal('1')
21       }
22     }
```

Let us denote by A_{v} the set of candidates approved by voter v. In the `approvalBallot` attribute, we hence record in fact the bipolar-valued truth characteristic values $r(\mathrm{c} \in A_{\mathrm{v}})$ of the statements that Candidate c *is approved* by voter v. In Listing 20.10 Line 5, we observe, for instance, that voter `v001` positively approves Candidate `c04`. And, in Line 10, we see that voter `v002` negatively approves, i.e. positively disapproves, Candidate `c01`.

The `showApprovalResults()` method and the `showDisapprovalResults()` method show how many approvals (respectively, disapprovals) each candidate receives.

```
>>> bavp.showApprovalResults()
    Approval results
     Candidate: 'c12' obtains 34 votes
     Candidate: 'c05' obtains 30 votes
     Candidate: 'c03' obtains 28 votes
     Candidate: 'c14' obtains 27 votes
     Candidate: 'c11' obtains 27 votes
     Candidate: 'c04' obtains 27 votes
     Candidate: 'c01' obtains 27 votes
     Candidate: 'c13' obtains 24 votes
     Candidate: 'c07' obtains 24 votes
     Candidate: 'c15' obtains 23 votes
     Candidate: 'c02' obtains 23 votes
     Candidate: 'c09' obtains 22 votes
```

```
15       Candidate: 'c08' obtains 22 votes
16       Candidate: 'c10' obtains 21 votes
17       Candidate: 'c06' obtains 21 votes
18      Total approval votes: 380
19      Approval proportion: 380/1500 = 0.25
20  >>> bavp.showDisapprovalResults()
21      Disapproval results
22       Candidate: 'c12' obtains 16 votes
23       Candidate: 'c03' obtains 22 votes
24       Candidate: 'c09' obtains 23 votes
25       Candidate: 'c04' obtains 24 votes
26       Candidate: 'c06' obtains 24 votes
27       Candidate: 'c13' obtains 24 votes
28       Candidate: 'c11' obtains 25 votes
29       Candidate: 'c02' obtains 26 votes
30       Candidate: 'c07' obtains 26 votes
31       Candidate: 'c08' obtains 26 votes
32       Candidate: 'c05' obtains 27 votes
33       Candidate: 'c10' obtains 27 votes
34       Candidate: 'c14' obtains 27 votes
35       Candidate: 'c15' obtains 27 votes
36       Candidate: 'c01' obtains 32 votes
37      Total disapproval votes: 376
38      Disapproval proportion: 376/1500 = 0.25
```

In Lines 3 and 22 above, we notice that, of all eligible candidates, it is Candidate `c12` who receives the highest number of approval votes (34) and the lowest number of disapproval votes (16). The total number of approval (respectively, disapproval) votes approaches more or less a proportion of 25% of the $100 \times 15 = 1500$ potential approval votes. About 50% of the latter remain hence ignored.

When operating now, for each Candidate x, the difference between the number of approval and the number of disapproval votes they receive, we obtain per candidate a corresponding *net approval* score, in fact, the bipolar truth characteristic value of the statement '*Candidate* x *should win the election*'.

$$r(\text{Candidate x should win the election}) \;=\; \sum_{\mathrm{v}} \big(r(\mathrm{x} \in A_{\mathrm{v}})\big)\,. \tag{20.1}$$

These bipolar characteristic values are stored in the `netApprovalScores` attribute and may be printed out with the `showNetApprovalScores()` method:

```
>>> bavp.showNetApprovalScores()
 Net Approval Scores
    Candidate: 'c12' obtains 18 net approvals
    Candidate: 'c03' obtains 6 net approvals
    Candidate: 'c05' obtains 3 net approvals
    Candidate: 'c04' obtains 3 net approvals
    Candidate: 'c11' obtains 2 net approvals
    Candidate: 'c14' obtains 0 net approvals
    Candidate: 'c13' obtains 0 net approvals
    Candidate: 'c09' obtains -1 net approvals
```

```
11       Candidate: 'c07' obtains -2 net approvals
12       Candidate: 'c06' obtains -3 net approvals
13       Candidate: 'c02' obtains -3 net approvals
14       Candidate: 'c15' obtains -4 net approvals
15       Candidate: 'c08' obtains -4 net approvals
16       Candidate: 'c01' obtains -5 net approvals
17       Candidate: 'c10' obtains -6 net approvals
```

We observe in Line 3 above that Candidate `c12`, with a net approval score of $34 - 16 = 18$, represents indeed the *best approved* candidate for winning the election. With a net approval score of $28 - 22 = 6$, Candidate `c03` appears second best approved. The net approval scores define hence a potentially weak ranking on the set of eligible election candidates, and the winner(s) of the election is(are) determined by the first-ranked candidate(s).

20.3 Pairwise Comparison of Approval–Disapproval Votes

The approval votes of each voter define now on the set of eligible candidates three ordered categories: his approved (+1), his ignored (0), and his disapproved (−1) ones. Within each of these three categories, we consider the voter's actual preferences as *not communicated*, i.e. as missing data. This gives for each voter a partially determined strict order which we find in the `ballot` attribute.

```
>>> bavp.ballot['v001']['c12']
   {'c02': Decimal('1'), 'c11': Decimal('1'),
    'c14': Decimal('1'), 'c04': Decimal('0'),
    'c06': Decimal('1'), 'c05': Decimal('1'),
    'c12': Decimal('0'), 'c13': Decimal('0'),
    'c15': Decimal('1'), 'c01': Decimal('1'),
    'c08': Decimal('1'), 'c07': Decimal('1'),
    'c09': Decimal('0'), 'c03': Decimal('1'),
    'c10': Decimal('0')}
```

For voter `v001`, for instance, the best approved Candidate `c12` is strictly preferred to Candidates: `c01`, `c02`, `c03`, `c05`, `c06`, `c07`, `c08`, `c11`, `c14`, and `c15`. No candidate is preferred to `c12`, and the comparison with `c04`, `c09`, `c10`, and `c13` is not communicated, hence indeterminate. Mind by the way that the reflexive comparison of `c12` with itself is, as usual, ignored, i.e. indeterminate. Each voter `v` defines thus a partially determined transitive strict preference relation denoted $\succ_{\text{v}}$ on the eligible candidates.

For each pair of eligible candidates, we aggregate the previous individual voter's preferences into a truth characteristic of the statement: 'Candidate x is *better approved than* Candidate y', denoted $r(x \succ y)$:

$$r(x \succ y) \;=\; \sum_{\text{v}} \big(r(x \succ_{\text{v}} y) \big) \,. \tag{20.2}$$

We say that Candidate x is *better approved than* Candidate y when $r(x \succ y) > 0$, i.e. there is a majority of voters who approve *more* and disapprove *less* x than y. Vice versa, we say that Candidate x is *not* better approved than Candidate y when $r(x \succ y) < 0$, i.e. there is a majority of voters who disapprove more and approve less x than y. This computation is achieved with the `MajorityMarginsDigraph` constructor.

```
1  >>> from votingProfiles import MajorityMarginsDigraph
2  >>> m = MajorityMarginsDigraph(bavp)
3  >>> m
4    *------- Digraph instance description ------*
5     Instance class       : MajorityMarginsDigraph
6     Instance name        : rel_bpApVotingProfile
7     Digraph Order        : 15
8     Digraph Size         : 97
9     Valuation domain     : [-100.00;100.00]
10    Determinateness (%) : 52.55
11    Attributes           : ['name', 'actions',
12                'criteria','ballot',
13                'valuationdomain', 'relation',
14                'order', 'gamma', 'notGamma']
```

The resulting digraph `m` contains 97 positively validated relations (see Line 8 above), and for all pairs (x, y) of eligible candidates, $r(x \succ y)$ takes value in a valuation range from -100.00 (unanimously rejected) to $+100.00$ (unanimously approved).

These pairwise $r(x \succ y)$ values can be inspected in a browser view with the `showHTMLRelationTable()` method:

```
>>> m.showHTMLRelationTable(relationName='r(x > y)')
```

In Fig. 20.3, it gets apparent that Candidate `c12` is a CONDORCET winner, i.e. the candidate who beats all the other candidates and, with the given voting profile `gavp`, should without doubt win the election. This strongly confirms the first-ranked result obtained with the previous net approval scoring.

Let us eventually compute, with the help of the NETFLOWS ranking rule,[2] a linear ranking of the 15 eligible candidates and compare the result with the net approval scores ranking.

Listing 20.11 Comparing the net approval and the NETFLOWS rankings

```
>>> from linearOrders import NetFlowsOrder
>>> nf = NetFlowsOrder(m,Comments=True)
>>> print('NetFlows versus Net Approval Ranking')
>>> print('Candidate\tNetFlows score\tNet Approval score')
>>> for item in nf.netFlows:
...     print( '%9s\t  %+.3f\t %+.1f' %\
...               (item[1], item[0],\
...                bavp.netApprovalScores[item[1]]) )
```

[2] See Sect. 8.3.

r(x > y)	c01	c02	c03	c04	c05	c06	c07	c08	c09	c10	c11	c12	c13	c14	c15
c01	–	1	-10	-10	-8	-2	-2	-3	-3	2	-5	-19	-4	-2	2
c02	-1	–	-5	-4	-9	-2	5	1	-7	3	-2	-20	-1	-4	-2
c03	10	5	–	1	4	6	5	10	7	11	0	-5	2	6	9
c04	10	4	-1	–	2	5	7	8	-1	4	0	-7	8	2	8
c05	8	9	-4	-2	–	4	6	7	4	2	1	-17	6	0	3
c06	2	2	-6	-5	-4	–	-1	2	-3	5	-4	-13	-1	-1	2
c07	2	-5	-5	-7	-6	1	–	7	-4	2	-5	-17	-2	-2	4
c08	3	-1	-10	-8	-7	-2	-7	–	-2	2	-2	-16	0	0	-1
c09	3	7	-7	1	-4	3	4	2	–	3	0	-18	-4	0	2
c10	-2	-3	-11	-4	-2	-5	-2	-2	-3	–	-6	-15	-4	-4	2
c11	5	2	0	0	-1	4	5	2	0	6	–	-15	4	0	5
c12	19	20	5	7	17	13	17	16	18	15	15	–	12	13	18
c13	4	1	-2	-8	-6	1	2	0	4	4	-4	-12	–	1	4
c14	2	4	-6	-2	0	1	2	0	0	4	0	-13	-1	–	-1
c15	-2	2	-9	-8	-3	-2	-4	1	-2	-2	-5	-18	-4	1	–

Valuation domain: [-100; +100]

Fig. 20.3 The bipolar-valued pairwise majority margins

```
9      NetFlows versus Net Approval Ranking
10     Candidate    NetFlows score   Net Approval score
11        'c12'          +410.000          +18.0
12        'c03'          +142.000           +6.0
13        'c04'           +98.000           +3.0
14        'c05'           +54.000           +3.0
15        'c11'           +34.000           +2.0
16        'c09'           -16.000           -1.0
17        'c14'           -20.000           +0.0
18        'c13'           -22.000           +0.0
19        'c06'           -50.000           -3.0
20        'c07'           -74.000           -2.0
21        'c02'           -96.000           -3.0
22        'c08'          -102.000           -4.0
23        'c15'          -110.000           -4.0
24        'c10'          -122.000           -6.0
25        'c01'          -126.000           -5.0
```

In Listing 20.11 on the previous page, we may notice that the NETFLOWS rule delivers a ranking that is very similar to the one previously obtained with the corresponding *Net Approval* scores. Only minor inversions do appear, like in the midfield, where Candidate `c09` advances before Candidates `c13` and `c14` (see Line 16), and Candidates `c06` and `c07` swap their positions 9 and 10 (see Line 19). The two last ranked Candidates `c10` and `c01` also swap their positions.

This confirms again the pertinence of the net approval scoring approach for finding the winner in an approval–disapproval voting system. Yet, voting by approving (+1), disapproving (−1), or ignoring (0) eligible candidates may also be seen as a performance evaluation of the eligible candidates on a $\{-1, 0, 1\}$-valued ordinal scale.

20.4 Three-Valued Evaluative Voting Systems

Following such an epistemic perspective, we may effectively convert the given `BipolarApprovalVotingProfile` instance into a `PerformanceTableau` instance, so as to get access to a corresponding outranking decision aiding approach.

Mind that, contrary to the majority margins of the '*better approved than*' relation, all voters consider now the approved candidates to be equivalent (+1). The same is true for the disapproved (−1) and, respectively, the ignored (0) candidates. The voter's marginal preferences model this time a complete preorder with three equivalence classes.

From the saved file `AVPerfTab.py` (see Line 1 below), we may construct an outranking relation on the eligible candidates with our standard `BipolarOutrankingDigraph` class constructor. The semantics of this outranking relation are the following:

- We say that Candidate x *outranks* Candidate y when there is a majority of voters who consider x *at least as well evaluated as* y.
- We say that Candidate x is *outranked by* Candidate y when there is a majority of voters who consider x *not at least as well evaluated as* y.
- Otherwise, the outranking situation is indeterminate.

Listing 20.12 Computing the outranking digraph

```
1  >>> bavp.save2PerfTab(fileName='AVPerfTab',valueDigits=0)
2    *--- Saving as performance tableau in AVPerfTab.py ---*
3  >>> from outrankingDigraphs import\
4  ...                                 BipolarOutrankingDigraph
5  >>> odg = BipolarOutrankingDigraph('AVPerfTab')
6  >>> odg
7    *------- Object instance description ------*
8     Instance class          : BipolarOutrankingDigraph
9     Instance name           : rel_AVPerfTab
10    Actions                 : 15
11    Criteria                : 100
12    Size                    : 210
13    Determinateness (%)     : 69.29
14    Valuation domain        : [-1.00;1.00]
15    Attributes              : ['name', 'actions', 'order,
16                           'criteria', 'evaluation', 'NA',
17                           'valuationdomain', 'relation',
18                           'gamma', 'notGamma', ...]
```

The size ($210 = 15 \times 14$) of the resulting outranking digraph odg, shown in Listing 20.12 on the preceding page Line 12 above, reveals that the corresponding '*at least as good evaluated as*' relation models actually a trivial *complete* digraph. All candidates appear to be *equally* at least as well evaluated and the strict '*better evaluated than*' (codual) outranking digraph becomes in fact empty. The converted performance tableau does apparently not contain sufficiently discriminatory performance evaluations for supporting any strict preference situations.

Yet, we may nevertheless try to apply again the NetFlows ranking rule to this complete outranking digraph odg and print side by side the corresponding NetFlows scores and the previous Net Approval scores.

Listing 20.13 Comparing the NetFlows and the Net Approval rankings

```
>>> from linearOrders import NetFlowsOrder
>>> nf = NetFlowsOrder(odg)
>>> print('NetFlows versus Net Approval Ranking')
>>> print('Candidate\tNetFlows Score\tNet Approval Score')
>>> for item in nf.netFlows:
...       print('%9s\t  %+.3f\t %+.0f' %\
...            (item[1],item[0],\
...             bavp.netApprovalScores[item[1]]) )
    NetFlows versus Net Approval Ranking
    Candidate     NetFlows score Net Approval score
       c12            +4.100           +18
       c03            +1.420            +6
       c04            +0.980            +3
       c05            +0.540            +3
       c11            +0.340            +2
       c09            -0.160            -1
       c14            -0.200             0
       c13            -0.220             0
       c06            -0.500            -3
       c07            -0.740            -2
       c02            -0.960            -3
       c08            -1.020            -4
       c15            -1.100            -4
       c10            -1.220            -6
       c01            -1.260            -5
```

Despite its apparent poor strict preference discriminating power, we obtain in Listing 20.13 here NetFlows scores that are directly proportional (divided by 100) to the scores obtained with the *better approved than* majority margins digraph m (see Listing 20.11 on page 291).

Encouraged by this positive result, we may furthermore compute a Rubis best choice recommendation (see Chap. 4).

Listing 20.14 Computing a best social choice recommendation

```
>>> odg.showBestChoiceRecommendation()
 Rubis best choice recommendation(s) (BCR)
  (in decreasing order of determinateness)
```

```
4    Credibility domain: [-1.00,1.00]
5   === >> ambiguous first choice(s)
6    * choice             : ['c01','c02','c03','c04','c05',
7                           'c06','c07','c08','c09','c10',
8                           'c11','c12','c13','c14','c15']
9     independence      : 0.06
10    dominance         : 1.00
11    absorbency        : 1.00
12    covering (%)      : 100.00
13    determinateness (%) : 61.13
14    - most credible action(s) = {
15        'c12': 0.44, 'c03': 0.34, 'c04': 0.30,
16        'c14': 0.28, 'c13': 0.24, 'c06': 0.24,
17        'c11': 0.20, 'c10': 0.20, 'c07': 0.20,
18        'c01': 0.20, 'c08': 0.18, 'c05': 0.18,
19        'c15': 0.14, 'c09': 0.14, 'c02': 0.06, }
20   === >> ambiguous last choice(s)
21    * choice             : ['c01','c02','c03','c04','c05',
22                           'c06','c07','c08','c09','c10',
23                           'c11','c12','c13','c14','c15']
24     independence      : 0.06
25     dominance         : 1.00
26     absorbency        : 1.00
27     covered (%)       : 100.00
28     determinateness (%) : 63.73
29     - most credible action(s) = {
30         'c13': 0.36, 'c06': 0.36, 'c15': 0.34,
31         'c01': 0.34, 'c08': 0.32, 'c07': 0.30,
32         'c02': 0.30, 'c14': 0.28, 'c11': 0.28,
33         'c09': 0.28, 'c04': 0.26, 'c10': 0.24,
34         'c05': 0.20, 'c03': 0.20, 'c12': 0.06, }
```

The strict outranking digraph ($\sim(-odg)$) being actually *empty*, we obtain a unique *ambiguous* – first as well as last – choice recommendation which trivially retains all fifteen candidates (see Listing 20.14 on the preceding page, Lines 6–8 above). Yet, the bipolar-valued best choice membership characteristic vector reveals that, among all the fifteen potential winners, it is indeed Candidate `c12` the most credible one with a 72% majority of voters' support (see Line 15, $(0.44 + 1.0)/2 = 0.72$), followed by Candidate `c03` (67%) and Candidate `c04` (65%). Similarly, Candidates `c13` and `c06` represent the most credible losers with a 68% majority voters' support (Line 30).

We observe here empirically that *evaluative* voting systems, using three-valued ordinal performance scales, match closely corresponding approval–disapproval voting systems. The latter systems model, however, more faithfully the very preferential information that is expressed with *approved*, *disapproved*, and *ignored* statements. The corresponding evaluation on a three-level scale, being value (numbers) based, cannot express the fact that in an approval–disapproval voting system there is no preferential information given concerning the pairwise comparison of all approved and, respectively, disapproved or ignored candidates.

Let us finally illustrate how approval–disapproval voting systems may favour multipartisan supported candidates. We shall therefore compare *approval–disapproval* versus *uninominal plurality* election results when considering a highly divisive and partisan political context.

20.5 Favouring Multipartisan Candidates

In modern democracy, politics are largely structured by political parties and activists. Let us so consider an approval–disapproval voting profile dvp where the random voter behaviour is simulated from two pre-electoral polls concerning a political scene with essentially two major highly competing parties, like the one existing in the USA.

Listing 20.15 A random approval–disapproval voting profile in a divisive political context

```
>>> dvp = RandomBipolarApprovalVotingProfile(\
...                 numberOfCandidates=15,\
...                 numberOfVoters=100,
...                 approvalProbability=0.25,
...                 disapprovalProbability=0.25,
...                 WithPolls=True,
...                 partyRepartition=0.5,
...                 other=0.05,
...                 DivisivePolitics=True,
...                 seed=200)
>>> dvp.showRandomPolls()
  Random repartition of voters
  Party-1 supporters : 45 (45.00%)
  Party-2 supporters : 49 (49.00%)
  Other voters       : 6 (06.00%)
  *---------------- random polls ----------------
  Party-1(45.0%) | Party-2(49.0%) |    expected
  -----------------------------------------------
  'c05' : 24.10% | 'c07' : 24.10% | 'c07' : 11.87%
  'c14' : 23.48% | 'c10' : 23.48% | 'c10' : 11.60%
  'c03' : 15.13% | 'c01' : 15.13% | 'c05' : 10.91%
  'c12' : 07.55% | 'c04' : 07.55% | 'c14' : 10.67%
  'c08' : 07.11% | 'c09' : 07.11% | 'c01' : 07.67%
  'c15' : 04.37% | 'c13' : 04.37% | 'c03' : 07.09%
  'c11' : 03.99% | 'c02' : 03.99% | 'c04' : 04.55%
  'c06' : 03.80% | 'c06' : 03.80% | 'c09' : 04.49%
  'c02' : 02.79% | 'c11' : 02.79% | 'c12' : 04.32%
  'c13' : 02.63% | 'c15' : 02.63% | 'c08' : 04.30%
  'c09' : 02.24% | 'c08' : 02.24% | 'c06' : 03.57%
  'c04' : 01.89% | 'c12' : 01.89% | 'c13' : 03.32%
  'c01' : 00.57% | 'c03' : 00.57% | 'c15' : 03.25%
  'c10' : 00.20% | 'c14' : 00.20% | 'c02' : 03.21%
  'c07' : 00.14% | 'c05' : 00.14% | 'c11' : 03.16%
```

In Listing 20.15 on the facing page, the divisive political situation is reflected by the fact that Party-1 and Party-2 supporters show strict opposite preferences. The leading candidates of Party-1 (`c05` and `c14`) are last choices for Party-2 supporters, and Candidates `c07` and `c10`, leading candidates for Party-2 supporters, are similarly the least choices for Party-1 supporters.

No clear winner may be guessed from these pre-election polls. As Party-2 shows however slightly more supporters than Party-1, the expected winner in a uninominal plurality or instant-run-off voting system will be Candidate `c07`, i.e. the leading candidate of majority Party-2 (see below).

```
>>> dvp.computeSimpleMajorityWinner()
  ['c07']
>>> dvp.computeInstantRunoffWinner()
  ['c07']
```

Now, in a corresponding approval–disapproval voting system, Party-1 supporters will usually approve their leading candidates and disapprove the leading candidates of Party-2. Vice versa, Party-2 supporters will usually approve their leading candidates and disapprove the leading candidates of Party-1. Let us consult the resulting approval votes per candidate.

```
>>> dvp.showApprovalResults()
     Candidate: 'c07' obtains 30 votes
     Candidate: 'c10' obtains 28 votes
     Candidate: 'c05' obtains 28 votes
     Candidate: 'c01' obtains 28 votes
     Candidate: 'c03' obtains 26 votes
     Candidate: 'c02' obtains 26 votes
     Candidate: 'c12' obtains 25 votes
     Candidate: 'c14' obtains 24 votes
     Candidate: 'c13' obtains 24 votes
     Candidate: 'c09' obtains 21 votes
     Candidate: 'c04' obtains 21 votes
     Candidate: 'c08' obtains 19 votes
     Candidate: 'c06' obtains 17 votes
     Candidate: 'c15' obtains 15 votes
     Candidate: 'c11' obtains 12 votes
    Total approval votes: 344
    Approval proportion: 344/1500 = 0.23
```

When considering only the approval votes, we find confirmed above that the leading candidate of Party-2 obtains in this simulation a plurality of approval votes. In uninominal plurality or instant-run-off voting systems, this candidate wins hence the election, quite to the despair of Party-1 supporters. As a foreseeable consequence, this election result will be more or less aggressively contested which leads to a loss of popular trust in democratic elections and institutions.

If we look however on the corresponding disapprovals, we discover that, not surprisingly, the leading candidates of both parties collect by far the highest number of disapproval votes.

```
>>> dvp.showDisapprovalResults()
    Candidate: 'c02' obtains 14 votes
    Candidate: 'c04' obtains 14 votes
    Candidate: 'c13' obtains 14 votes
    Candidate: 'c06' obtains 15 votes
    Candidate: 'c09' obtains 15 votes
    Candidate: 'c08' obtains 16 votes
    Candidate: 'c11' obtains 16 votes
    Candidate: 'c15' obtains 18 votes
    Candidate: 'c12' obtains 20 votes
    Candidate: 'c01' obtains 29 votes
    Candidate: 'c03' obtains 30 votes
    Candidate: 'c10' obtains 37 votes
    Candidate: 'c07' obtains 44 votes
    Candidate: 'c14' obtains 45 votes
    Candidate: 'c05' obtains 49 votes
   Total disapproval votes: 376
   Disapproval proportion: 376/1500 = 0.25
```

Balancing now approval against disapproval votes will favour the moderate, bipartisan supported, candidates.

```
>>> dvp.showNetApprovalScores()
   Net Approval Scores
    Candidate: 'c02' obtains 12 net approvals
    Candidate: 'c13' obtains 10 net approvals
    Candidate: 'c04' obtains 7 net approvals
    Candidate: 'c09' obtains 6 net approvals
    Candidate: 'c12' obtains 5 net approvals
    Candidate: 'c08' obtains 3 net approvals
    Candidate: 'c06' obtains 2 net approvals
    Candidate: 'c01' obtains -1 net approvals
    Candidate: 'c15' obtains -3 net approvals
    Candidate: 'c11' obtains -4 net approvals
    Candidate: 'c03' obtains -4 net approvals
    Candidate: 'c10' obtains -9 net approvals
    Candidate: 'c07' obtains -14 net approvals
    Candidate: 'c14' obtains -21 net approvals
    Candidate: 'c05' obtains -21 net approvals
```

Candidate `c02`, appearing in the pre-electoral polls in the midfield (in position 7 for Party-2 and in position 9 for Party-1 supporters, see Listing 20.15 on page 296), shows indeed the highest net approval score. The second highest net approval score obtains Candidate `c13`, in position 6 for Party-2 and in position 10 for Party-1 supporters.

Figure 20.4 on the facing page, showing the NETFLOWS ranked relation table of the '*better approved than*' majority margins digraph, confirms below this net approval scoring result.

```
>>> m = MajorityMarginsDigraph(dvp)
>>> m.showHTMLRelationTable(\
...        actionsList=m.computeNetFlowsRanking(),
...        relationName='r(x > y)')
```

r(x > y)	c02	c13	c04	c09	c12	c08	c06	c01	c11	c15	c03	c10	c07	c14	c05
c02	–	6	5	6	9	5	10	12	12	14	11	15	22	22	23
c13	-6	–	2	5	2	5	8	10	14	10	9	13	18	23	20
c04	-5	-2	–	0	2	2	3	7	7	8	11	13	18	21	18
c09	-6	-5	0	–	0	2	5	5	11	9	5	13	16	21	16
c12	-9	-2	-2	0	–	4	6	2	9	6	10	5	10	23	25
c08	-5	-5	-2	-2	-4	–	4	0	8	9	7	5	11	21	20
c06	-10	-8	-3	-5	-6	-4	–	-2	5	5	5	6	13	17	18
c01	-12	-10	-7	-5	-2	0	2	–	1	-1	3	8	11	9	13
c11	-12	-14	-7	-11	-9	-8	-5	-1	–	1	-2	7	14	13	16
c15	-14	-10	-8	-9	-6	-9	-5	1	-1	–	0	3	10	14	14
c03	-11	-9	-11	-5	-10	-7	-5	-3	2	0	–	-3	7	16	16
c10	-15	-13	-13	-13	-5	-5	-6	-8	-7	-3	3	–	3	6	7
c07	-22	-18	-18	-16	-10	-11	-13	-11	-14	-10	-7	-3	–	3	5
c14	-22	-23	-21	-21	-23	-21	-17	-9	-13	-14	-16	-6	-3	–	0
c05	-23	-20	-18	-16	-25	-20	-18	-13	-16	-14	-16	-7	-5	0	–

Valuation domain: [-100; +100]

Fig. 20.4 The pairwise *better approved than* majority margins

Candidate `c02` appears indeed *better approved than* any other candidate (CONDORCET winner), and the leading candidates of Party-1, `c05` and `c14`, are *less approved than* any other candidates (weak CONDORCET losers).

```
>>> m.computeCondorcetWinners()
  ['c02']
>>> m.computeWeakCondorcetLosers()
  ['c05','c14']
```

We see this result furthermore confirmed when computing the corresponding first and, respectively, last choice recommendations.

```
>>> m.showBestChoiceRecommendation()
 Rubis best choice recommendation(s) (BCR)
  (in decreasing order of determinateness)
  Credibility domain: [-100.00,100.00]
  === >> potential first choice(s)
   * choice              : ['c02']
     independence        : 100.00
     dominance           : 5.00
     absorbency          : -23.00
     covering (%)        : 100.00
     determinateness (%) : 52.50
     - most credible action(s) = { 'c02': 5.00, }
  === >> potential last choice(s)
   * choice              : ['c05', 'c14']
     independence        : 0.00
```

```
    dominance              : -23.00
    absorbency             : 5.00
    covered (%)            : 100.00
    determinateness (%)    : 50.00
    - most credible action(s) = { }
```

Candidate `c02`, being actually a CONDORCET winner, gives an initial prekernel of digraph m, whereas Party-1 leading Candidates `c05` and `c14`, both being weak CONDORCET losers, give together a terminal prekernel. They hence represent our *first choice* and, respectively, *last choice* recommendations for winning this simulated election.

It is worthwhile noticing again the essential structural and computational role, the ignored value is playing in approval–disapproval voting systems. This epistemic and logical *neutral* term is needed indeed for handling in a consistent and efficient manner *not communicated votes* and/or *indeterminate* preferential statements.

Let us conclude by predicting that, for leading political candidates in an aggressively divisive political context, the perspective to easily fail election with an approval–disapproval voting system might or will induce a change in the usual way of running electoral campaigns. Political parties and politicians, who avoid aggressive competitive propaganda and instead propose multipartisan collaborative social choices, will be rewarded with better election results than any kind of extremism. It could mean the end of sterile political obstructions and war like electoral battles. *Let us do it !*

The last part, Part V, of this monograph illustrates in three chapters computational resources for working with simple undirected graphs.

References

Baujard A, Gavrel F, Igersheim H, Laslier JF, Lebon I (2013) Approval voting, evaluation voting: an experiment during the 2012 French presidential election. Revue Économique (Presses de Sciences Po) 64(2):345–356

Benayoun R, Roy B, Sussmann B (1966) Electre: une méthode pour guider le choix en présence de points de vue multiples. Tech. Rep. 49, Société d'Economie et de Mathématique Appliqué, Direction Scientifique

Part V
Working with Undirected Graphs

The last part introduces Python resources for working with undirected graphs. Its aim is to eventually show the operational benefits one may get when implementing vertex adjacency with bipolar-valued characteristics. Several special graph models and algorithms are illustrated: Q-coloring, MIS and clique enumeration, line graphs and computing maximal matchings, grid graphs and computing the Ising model, n-cycle graphs and computing their non-isomorphic MISs, spanning tree graphs and graph forests, and generating and recognising split, interval, and permutation graphs.

Chapter 21
Bipolar-Valued Undirected Graphs

Contents

Abstract The chapter introduces bipolar-valued undirected graphs and illustrates several special graph models and algorithms like Q-coloring, MIS and clique enumeration, line graphs and maximal matchings, grid graphs, and n-cycle graphs with their non-isomorphic maximal independent sets of vertices.

21.1 Implementing Simple Graphs

In the DIGRAPH3 `graphs` module, the root `Graph` class provides a generic simple graph model, without loops and multiple links. A given object of this `Graph` type contains at least the following attributes:

1. `name`: usually the name of the stored instance
2. `vertices`: a dictionary with `name` and `shortName` attributes
3. `order`: the number of vertices
4. `valuationDomain`: a dictionary with three entries: the minimum (-1 means certainly no link), the median (0 means missing information), and the maximum characteristic value ($+1$ means certainly a link)
5. `edges`: a dictionary with *frozensets*[1] of pairs of vertices as keys carrying a characteristic value in the range of the previous valuation domain

[1] Python documentation.

R. Bisdorff, *Algorithmic Decision Making with Python Resources*,
International Series in Operations Research & Management Science 324,
https://doi.org/10.1007/978-3-030-90928-4_21

6. `size`: the number of positive edges
7. `gamma`: a dictionary containing the direct neighbours of each vertex, automatically added by the object constructor

Example Python Terminal Session

A random crisp graph instance can be generated with the `RandomGraph` class from the `graphs` module (see below).

Listing 21.1 Generating a random graph instance

```
>>> from graphs import RandomGraph
>>> g = RandomGraph(order=7,edgeProbability=0.5)
>>> g.save(fileName='tutorialGraph')
```

The saved `Graph` instance, named `tutorialGraph.py`, is encoded as shown in Listing 21.2.

Listing 21.2 Stored instance of a random graph

```
 # Graph instance saved in Python format
  from decimal import Decimal
  vertices = {
   'v1': {'shortName': 'v1', 'name': 'random vertex'},
   'v2': {'shortName': 'v2', 'name': 'random vertex'},
   'v3': {'shortName': 'v3', 'name': 'random vertex'},
   'v4': {'shortName': 'v4', 'name': 'random vertex'},
   'v5': {'shortName': 'v5', 'name': 'random vertex'},
   'v6': {'shortName': 'v6', 'name': 'random vertex'},
   'v7': {'shortName': 'v7', 'name': 'random vertex'},
  }
  valuationDomain = {
    'min':Decimal('-1'),'med':Decimal('0'),
    'max':Decimal('1')}
  edges = {
   frozenset(['v1','v2']) : Decimal('-1'),
   frozenset(['v1','v3']) : Decimal('-1'),
   frozenset(['v1','v4']) : Decimal('-1'),
   frozenset(['v1','v5']) : Decimal('1'),
   frozenset(['v1','v6']) : Decimal('-1'),
   frozenset(['v1','v7']) : Decimal('-1'),
   frozenset(['v2','v3']) : Decimal('1'),
   frozenset(['v2','v4']) : Decimal('1'),
   frozenset(['v2','v5']) : Decimal('-1'),
   frozenset(['v2','v6']) : Decimal('1'),
   frozenset(['v2','v7']) : Decimal('-1'),
   frozenset(['v3','v4']) : Decimal('-1'),
   frozenset(['v3','v5']) : Decimal('-1'),
   frozenset(['v3','v6']) : Decimal('-1'),
   frozenset(['v3','v7']) : Decimal('-1'),
   frozenset(['v4','v5']) : Decimal('1'),
   frozenset(['v4','v6']) : Decimal('-1'),
   frozenset(['v4','v7']) : Decimal('1'),
   frozenset(['v5','v6']) : Decimal('1'),
```

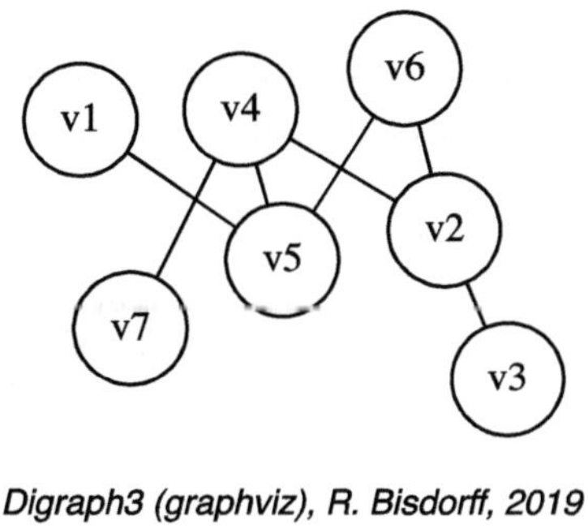

Fig. 21.1 Example simple graph instance

```
    frozenset(['v5','v7']) : Decimal('-1'),
    frozenset(['v6','v7']) : Decimal('-1'),
    }
```

The stored graph instance may be reloaded and plotted with the `exportGraphViz()` method (see Fig. 21.1):

```
>>> g = Graph('tutorialGraph')
>>> g.exportGraphViz()
 *---- exporting a dot file for GraphViz tools ---------*
  Exporting to tutorialGraph.dot
  fdp -Tpng tutorialGraph.dot -o tutorialGraph.png
```

Properties like the gamma function—the cover relation—, vertex degrees, and neighbourhood depths are shown with a `showShort()` method.

Listing 21.3 Inspecting a graph instance

```
>>> g.showShort()
 *---- short description of the graph ----*
  Name : 'tutorialGraph'
  Vertices :  ['v1','v2','v3','v4','v5','v6','v7']
  Valuation domain : {'min': -1, 'med': 0, 'max': 1}
  Gamma function   :
   v1 -> ['v5']
   v2 -> ['v6', 'v4', 'v3']
   v3 -> ['v2']
   v4 -> ['v5', 'v2', 'v7']
   v5 -> ['v1', 'v6', 'v4']
   v6 -> ['v2', 'v5']
   v7 -> ['v4']
  degrees       :  [0, 1, 2, 3, 4, 5, 6]
  distribution  :  [0, 3, 1, 3, 0, 0, 0]
  nbh depths    :  [0, 1, 2, 3, 4, 5, 6, 'inf.']
  distribution  :  [0, 0, 1, 4, 2, 0, 0, 0]
```

A `Graph` instance corresponds bijectively to a symmetric `Digraph` instance, and we can convert from one to the other with the `graph2Digraph()`, and vice versa, with the `digraph2Graph()` method (see Listing 21.4). Thus, all computing resources of the `Digraph` class suitable for symmetric digraphs become readily available, and vice versa.

Listing 21.4 Conversion between graphs and digraphs

```
>>> dg = g.graph2Digraph()
>>> dg.showRelationTable(ndigits=0,ReflexiveTerms=False)
  * ---- Relation Table -----
    S   |  'v1'  'v2'  'v3'  'v4'  'v5'  'v6'  'v7'
  -----|------------------------------------------
  'v1' |    -    -1    -1    -1     1    -1    -1
  'v2' |   -1     -     1     1    -1     1    -1
  'v3' |   -1     1     -    -1    -1    -1    -1
  'v4' |   -1     1    -1     -     1    -1     1
  'v5' |    1    -1    -1     1     -     1    -1
  'v6' |   -1     1    -1    -1     1     -    -1
  'v7' |   -1    -1    -1     1    -1    -1     -
>>> g1 = dg.digraph2Graph()
>>> g1
  *------- Digraph instance description ------*
   Instance class        : Digraph
   Instance name         : tutorialGraph
   Digraph Order         : 7
   Digraph Size          : 14
   Valuation domain      : [-1.00;1.00]
   Determinateness (%)   : 100.00
   Attributes            : ['name', 'order', 'actions',
                             'valuationdomain', 'relation',
                             'gamma', 'notGamma']
```

21.2 Q-Coloring of a Graph

With the `Q_Coloring` class, a 3-colouring of the tutorial graph `g` may be computed with a Gibbs sampler (Geman and Geman 1984) and plotted as shown in Listing 21.5.

Listing 21.5 Computing a 3-colouring of the random graph `g`

```
>>> from graphs import Q_Coloring
>>> qc = Q_Coloring(g)
  Running a Gibbs Sampler for 42 step !
  The q-coloring with 3 colors is feasible !!
>>> qc.showConfiguration()
    v5 lightblue
    v3 gold
    v7 gold
    v2 lightblue
    v4 lightcoral
```

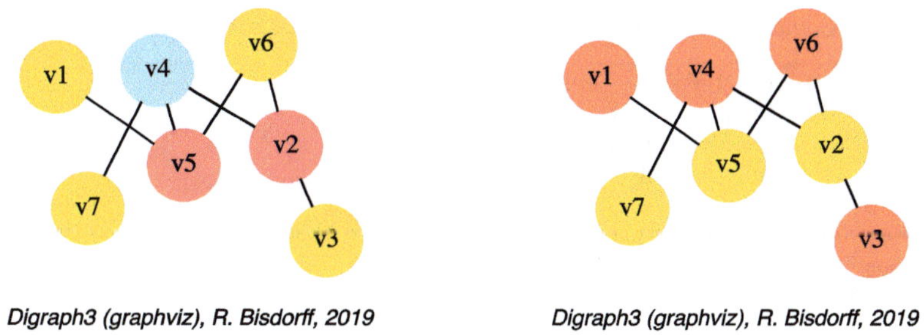

Fig. 21.2 3-Colouring and 2-colouring of the tutorial graph

```
    v1 gold
    v6 lightcoral
>>> qc.exportGraphViz('tutorial-3-coloring')
  *---- exporting a dot file for GraphViz tools
   Exporting to tutorial-3-coloring.dot
   fdp -Tpng tutorial-3-coloring.dot\
                     -o tutorial-3-coloring.png
```

Actually, with the given tutorial graph instance, a 2-colouring is already feasible (see Fig. 21.2).

```
>>> qc = Q_Coloring(g,colors=['gold','coral'])
  Running a Gibbs Sampler for 42 step !
  The q-coloring with 2 colors is feasible !!
>>> qc.showConfiguration()
    v5 gold
    v3 coral
    v7 gold
    v2 gold
    v4 coral
    v1 coral
    v6 coral
>>> qc.exportGraphViz('tutorial-2-coloring')
  Exporting to tutorial-2-coloring.dot
  fdp -Tpng tutorial-2-coloring.dot\
                     -o tutorial-2-coloring.png
```

21.3 MIS and Clique Enumeration

2-colouring define sets of independent vertices that are maximal in cardinality—the MISs. The `showMIS()` method computes and prints out such sets. The result is stored in a `misset` attribute (see Listing 21.6)

Listing 21.6 Computing and printing the maximal independent sets of graph g

```
>>> g = Graph('tutorialGraph')
>>> g.showMIS()
  *---  Maximal Independent Sets ---*
    ['v2', 'v5', 'v7']
    ['v3', 'v5', 'v7']
    ['v1', 'v2', 'v7']
    ['v1', 'v3', 'v6', 'v7']
    ['v1', 'v3', 'v4', 'v6']
    number of solutions:  5
    cardinality distribution
    card.:  [0, 1, 2, 3, 4, 5, 6, 7]
    freq.:  [0, 0, 0, 3, 2, 0, 0, 0]
    execution time: 0.00032 sec.
    Results in self.misset
>>> g.misset
    [frozenset({'v7', 'v2', 'v5'}),
     frozenset({'v3', 'v7', 'v5'}),
     frozenset({'v1', 'v2', 'v7'}),
     frozenset({'v1', 'v6', 'v7', 'v3'}),
     frozenset({'v1', 'v6', 'v4', 'v3'})]
```

A MIS in the dual `-g` of a graph instance `g` corresponds to a maximal *clique*, i.e. a maximal complete subgraph in `g`. Maximal cliques may be computed and printed with the `showCliques()` method. The result is stored in a `clique` attribute as shown in Listing 21.7.

Listing 21.7 Computing and printing the maximal independent sets of graph g

```
>>> g.showCliques()
  *---  Maximal Cliques ---*
    ['v2', 'v3']
    ['v4', 'v7']
    ['v2', 'v4']
    ['v4', 'v5']
    ['v1', 'v5']
    ['v2', 'v6']
    ['v5', 'v6']
   number of solutions:  7
   cardinality distribution
   card.:  [0, 1, 2, 3, 4, 5, 6, 7]
   freq.:  [0, 0, 7, 0, 0, 0, 0, 0]
   execution time: 0.00049 sec.
   Results in self.cliques
>>> g.cliques
  [frozenset({'v2', 'v3'}), frozenset({'v4', 'v7'}),
   frozenset({'v2', 'v4'}), frozenset({'v4', 'v5'}),
   frozenset({'v1', 'v5'}), frozenset({'v6', 'v2'}),
   frozenset({'v6', 'v5'})]
```

21.4 Line Graphs and Maximal Matchings

The `graphs` module also provides a `LineGraph` constructor. A *line graph* represents the adjacencies between edges of the given graph instance. In Listing 21.8, we compute, for instance, the line graph of the 5-cycle graph.

Listing 21.8 Computing the line graph of the 5-cycle graph

```
>>> from graphs import CycleGraph, LineGraph
>>> g = CycleGraph(order=5)
>>> g
  *------- Graph instance description ----*
   Instance class    : CycleGraph
   Instance name     : cycleGraph
   Graph Order       : 5
   Graph Size        : 5
   Valuation domain : [-1.00; 1.00]
   Attributes        : ['name', 'order',
             'vertices', 'valuationDomain',
             'edges', 'size', 'gamma']
>>> lg = LineGraph(g)
>>> lg
  *------- Graph instance description ------*
   Instance class    : LineGraph
   Instance name     : line-cycleGraph
   Graph Order       : 5
   Graph Size        : 5
   Valuation domain : [-1.00; 1.00]
   Attributes        : ['name', 'graph',
              'valuationDomain', 'vertices',
               'order', 'edges', 'size', 'gamma']
>>> lg.showShort()
  *---- short description of the graph ----*
   Name                : 'line-cycleGraph'
   Vertices            :   [frozenset({'v1', 'v2'}),
        frozenset({'v1', 'v5'}), frozenset({'v2', 'v3'}),
        frozenset({'v3', 'v4'}), frozenset({'v4', 'v5'})]
   Valuation domain :   {'min': Decimal('-1'),
              'med': Decimal('0'), 'max': Decimal('1')}
   Gamma function      :
    frozenset({'v1', 'v2'}) ->
        [frozenset({'v2', 'v3'}), frozenset({'v1', 'v5'})]
    frozenset({'v1', 'v5'}) ->
        [frozenset({'v1', 'v2'}), frozenset({'v4', 'v5'})]
    frozenset({'v2', 'v3'}) ->
        [frozenset({'v1', 'v2'}), frozenset({'v3', 'v4'})]
    frozenset({'v3', 'v4'}) ->
        [frozenset({'v2', 'v3'}), frozenset({'v4', 'v5'})]
    frozenset({'v4', 'v5'}) ->
        [frozenset({'v4', 'v3'}), frozenset({'v1', 'v5'})]
   degrees          :  [0, 1, 2, 3, 4]
   distribution :  [0, 0, 5, 0, 0]
   nbh depths      :  [0, 1, 2, 3, 4, 'inf.']
   distribution :  [0, 0, 5, 0, 0, 0]
```

Iterated line graph constructions are usually expanding, except for chordless cycles, where the same cycle is repeated, and for non-closed paths, where iterated line graphs progressively reduce one by one the number of vertices and edges and eventually become an empty graph.

Notice in Listing 21.9 that the MISs in a line graph provide *maximal matchings*—maximal sets of independent edges—of the original graph.

Listing 21.9 Computing the MISs of the line graph of the 8-cycle graph

```
1  >>> c8 = CycleGraph(order=8)
2  >>> lc8 = LineGraph(c8)
3  >>> lc8.showMIS()
4    *--- Maximal Independent Sets ---*
5     [frozenset({'v3', 'v4'}), frozenset({'v5', 'v6'}), frozenset({'v1', 'v8'})]
6     [frozenset({'v2', 'v3'}), frozenset({'v5', 'v6'}), frozenset({'v1', 'v8'})]
7     [frozenset({'v8', 'v7'}), frozenset({'v2', 'v3'}), frozenset({'v5', 'v6'})]
8     [frozenset({'v8', 'v7'}), frozenset({'v2', 'v3'}), frozenset({'v4', 'v5'})]
9     [frozenset({'v7', 'v6'}), frozenset({'v3', 'v4'}), frozenset({'v1', 'v8'})]
10    [frozenset({'v2', 'v1'}), frozenset({'v8', 'v7'}), frozenset({'v4', 'v5'})]
11    [frozenset({'v2', 'v1'}), frozenset({'v7', 'v6'}), frozenset({'v4', 'v5'})]
12    [frozenset({'v2', 'v1'}), frozenset({'v7', 'v6'}), frozenset({'v3', 'v4'})]
13    [frozenset({'v7', 'v6'}), frozenset({'v2', 'v3'}), frozenset({'v1',
          'v8'}),
14     frozenset({'v4', 'v5'})]
15    [frozenset({'v2', 'v1'}), frozenset({'v8', 'v7'}), frozenset({'v3', 'v4'}),
16     frozenset({'v5', 'v6'})]
17    number of solutions:  10
18    cardinality distribution
19    card.:  [0, 1, 2, 3, 4, 5, 6, 7, 8]
20    freq.:  [0, 0, 0, 8, 2, 0, 0, 0, 0]
21    execution time: 0.00029 sec.
```

The two last MISs of cardinality 4 (see Lines 13–16 13-16 in Listing 21.9) give perfect maximum matchings of the 8-cycle graph as shown in Fig. 21.3.

Fig. 21.3 *A perfect maximum matching of the 8-cycle graph*
Every vertex of the 8-cycle is adjacent to a matching edge. Odd cycle graphs do not admit any perfect matching

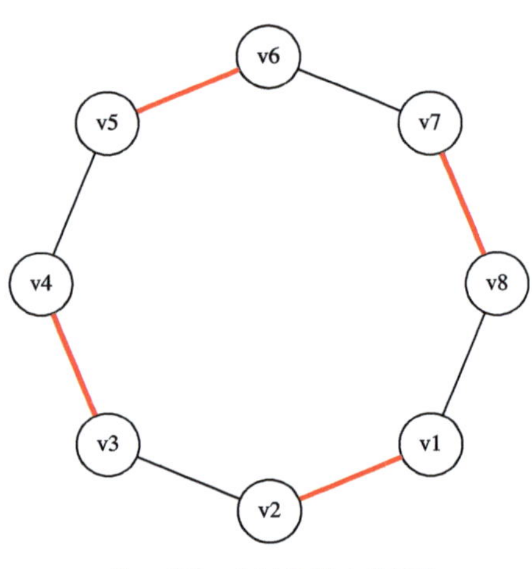

Listing 21.10 Computing maximum matchings in the 8-cycle graph

```
>>> maxMatching = c8.computeMaximumMatching()
>>> c8.exportGraphViz(fileName='maxMatchingcycleGraph',\
...                      matching=maxMatching)
 *---- exporting a dot file for GraphViz tools -----*
  Exporting to maxMatchingcyleGraph.dot
  Matching:  {frozenset({'v1', 'v2'}),
              frozenset({'v5', 'v6'}),
              frozenset({'v3', 'v4'}),
              frozenset({'v7', 'v8'}) }
  circo -Tpng maxMatchingcyleGraph.dot\
               -o maxMatchingcyleGraph.png
```

21.5 Grids and the ISING Model

Special classes of graphs, like the `GridGraph` class, provide $n \times m$ rectangular or triangular grids. With a *Gibbs* sampler[2], the `IsingModel` class can simulate in Fig. 21.4 on the following page an ISING model on, for instance, a 15×15 grid (Ising 1925).

Listing 21.11 Simulating an Ising model on a 15 × 15 rectangular grid

```
>>> from graphs import GridGraph, IsingModel
>>> g = GridGraph(n=15,m=15)
>>> g.showShort()
 *----- show short --------------*
  Grid graph    :  grid-6-6
  n             :  6
  m             :  6
  order         :  36
>>> im = IsingModel(g,beta=0.3,nSim=100000)
 Running a Gibbs Sampler for 100000 step !
>>> im.exportGraphViz(colors=['lightblue','lightcoral'])
 *---- exporting a dot file for GraphViz tools ---------*
  Exporting to grid-15-15-ising.dot
  fdp -Tpng grid-15-15-ising.dot -o grid-15-15-ising.png
```

21.6 Simulating METROPOLIS Random Walks

Finally, we provide the `MetropolisChain` class, a specialisation of the `Graph` class, implementing a generic Monte Carlo Markov Chain (MCMC) sampler for simulating random walks on a graph following given transition probabilities `probs`

[2] Geman and Geman (1984).

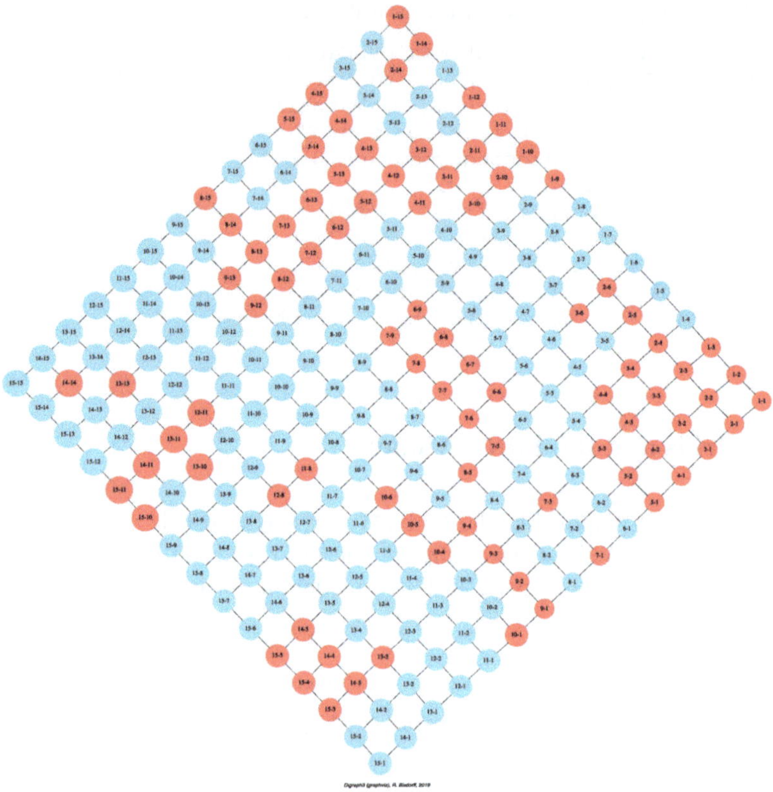

Fig. 21.4 Ising model of the 15 × 15 grid graph

= {`'v1''`: x, `'v2'`: y, ...} for visiting each vertex (see Listing 21.12) (Metropolis et al. 1953).

Listing 21.12 Simulating random walks on a graph

```
>>> from graphs import Graph
>>> g = Graph(numberOfVertices=5,edgeProbability=0.5)
>>> g.showShort()
  *---- short description of the graph ----*
   Name                : 'randomGraph'
   Vertices            :  ['v1', 'v2', 'v3', 'v4', 'v5']
   Valuation domain :  {'max': 1, 'med': 0, 'min': -1}
   Gamma function      :
    v1 -> ['v2', 'v3', 'v4']
    v2 -> ['v1', 'v4']
    v3 -> ['v5', 'v1']
    v4 -> ['v2', 'v5', 'v1']
    v5 -> ['v3', 'v4']
>>> # initialize a potential stationary probability vector
>>> n = g.order
```

```
>>> i = 0
>>> for v in g.vertices:
...     probs[v] = (n - i)/(n*(n+1)/2)
...     i += 1
```

The `checkSampling()` method of the `MetropolisChain` class (see below) generates a random walk of $nSim = 30{,}000$ steps on the given graph and records by the way the observed relative frequency with which each vertex is passed by. In Listing 21.13, we notice how well the random walk is matching the given stationary probability vector.

Listing 21.13 Checking the quality of the MCMC sampler

```
>>> from graphs import MetropolisChain
>>> met = MetropolisChain(g,probs)
>>> frequency = met.checkSampling(verticesList[0],nSim=30000)
>>> for v in verticesList:
...     print(v,probs[v],frequency[v])
    v1 0.3333 0.3343
    v2 0.2666 0.2680
    v3 0.2    0.2030
    v4 0.1333 0.1311
    v5 0.0666 0.0635
```

In this example, the stationary transition probability distribution (see Listing 21.13 above), shown by the `showTransitionMatrix()` method in Listing 21.14, is quite adequately simulated.

Listing 21.14 Printing the transition probability distribution

```
>>> met.showTransitionMatrix()
  * ---- Transition Matrix -----
    Pij  | 'v1'    'v2'    'v3'    'v4'    'v5'
    -----|-------------------------------------
    'v1' |  0.23    0.33    0.30    0.13    0.00
    'v2' |  0.42    0.42    0.00    0.17    0.00
    'v3' |  0.50    0.00    0.33    0.00    0.17
    'v4' |  0.33    0.33    0.00    0.08    0.25
    'v5' |  0.00    0.00    0.50    0.50    0.00
```

For the reader interested in algorithmic applications of Markov Chains, we recommend consulting (Häggström 2002).

21.7 Computing the Non-isomorphic MISs of the n-Cycle Graph

Due to the public success of our common work with Jean-Luc Marichal (Bisdorff and Marichal 2008), we present in this chapter an example Python session for computing the non-isomorphic maximal independent sets (MISs) of the 12-cycle graph, i.e. a `CirculantDigraph` class instance of order 12 and symmetric circulants 1 and −1.

```
>>> from digraphs import CirculantDigraph
>>> c12 = CirculantDigraph(order=12,circulants=[1,-1])
>>> c12 # 12-cycle digraph instance
 *------- Digraph instance description ------*
  Instance class      : CirculantDigraph
  Instance name       : c12
  Digraph Order       : 12
  Digraph Size        : 24
  Valuation domain  : [-1.0, 1.0]
  Determinateness     : 100.000
  Attributes          : ['name', 'order', 'circulants',
                          'actions', 'valuationdomain',
                          'relation', 'gamma', 'notGamma']
```

Such n-cycle graphs—see the 12-cycle graph in Fig. 21.5 on page 317—are also provided as undirected graph instances by the `CycleGraph` class.

```
>>> from graphs import CycleGraph
>>> cg12 = CycleGraph(order=12)
>>> cg12
 *------- Graph instance description ------*
  Instance class      : CycleGraph
  Instance name       : cycleGraph
  Graph Order         : 12
  Graph Size          : 12
  Valuation domain  : [-1.0, 1.0]
  Attributes          : ['name', 'order', 'vertices',
                          'valuationDomain', 'edges',
                          'size', 'gamma']
```

A non-isomorphic MIS of the 12-cycle graph corresponds in fact to a set of isomorphic MISs, i.e. an orbit of MISs under the automorphism group of the 12-cycle graph. In Listing 21.15, we are now first computing all maximal independent sets that are detectable in the 12-cycle digraph with the `showMIS()` method.

Listing 21.15 Computing the MISs of the 12-cycle graph

```
>>> c12.showMIS(withListing=False)
 *---  Maximal independent choices ---*
  number of solutions:   29
  cardinality distribution
  card.:   [0, 1, 2, 3, 4,  5,  6, 7, 8, 9, 10, 11, 12]
  freq.:   [0, 0, 0, 0, 3, 24,  2, 0, 0, 0,  0,  0,  0]
  Results in c12.misset
```

In the 12-cycle graph, we observe 29 labelled MISs: 3 of cardinality 4, 24 of cardinality 5, and 2 of cardinality 6. In case of n-cycle graphs with $n > 20$, as the cardinality of the MISs becomes big, it is preferable to use, as shown in

Listing 21.16, the `perrinMIS` shell command[3] compiled from C and installed along with all the DIGRAPH3 python modules for computing the set of MISs observed in the graph.

Listing 21.16 Computing the MISs with the `perrinMIS` shell command

```
...$ echo 12 | /usr/local/bin/perrinMIS
  # $---------------------------------- #
  # Generating MIS set of Cn with the   #
  # Perrin sequence algorithm.          #
  # Temporary files used.               #
  # even versus odd order optimised.    #
  # RB December 2006                    #
  # Current revision Dec 2018           #
  # ----------------------------------- #
  Input cycle order ? <-- 12
   mis 1 : 100100100100
   mis 2 : 010010010010
   mis 3 : 001001001001
     ...
     ...
     ...
   mis 27 : 001001010101
   mis 28 : 101010101010
   mis 29 : 010101010101
   Cardinalities:
   0 : 0
   1 : 0
   2 : 0
   3 : 0
   4 : 3
   5 : 24
   6 : 2
   7 : 0
   8 : 0
   9 : 0
   10 : 0
   11 : 0
   12 : 0
   Total: 29
   execution time: 0 sec. and 2 millisec.
```

[3] The `perrinMIS` shell command may be installed system-wide with the command `make installPerrin` from the main DIGRAPH3 directory. It is stored by default into `/usr/local/bin/`. This may be changed with the `INSTALLDIR` flag. The command `make installPerrinUser` installs it instead without sudo into the user's private `.bin` directory.

Reading in the result of the `perrinMIS` shell command, stored in a file called by default `curd.dat`, may be operated with the `readPerrinMisset()` method.

```
>>> c12.readPerrinMisset(file='curd.dat')
>>> c12.misset
    {frozenset({'5', '7', '10', '1', '3'}),
     frozenset({'9', '11', '5', '2', '7'}),
     frozenset({'7', '2', '4', '10', '12'}),
     ...
     ...
     ...
     frozenset({'8', '4', '10', '1', '6'}),
     frozenset({'11', '4', '1', '9', '6'}),
     frozenset({'8', '2', '4', '10', '12', '6'})
    }
```

For computing the corresponding non-isomorphic MISs, we actually need the automorphism group of the 12-cycle graph. The `Digraph` class therefore provides the `automorphismGenerators()` method which adds automorphism group generators to a `Digraph` class instance with the help of the external `dreadnaut` shell command from the *nauty* software package.[4]

Listing 21.17 Computing the automorphism group generators

```
>>> c12.automorphismGenerators()
      ...
    Permutations
    {'1':'1', '2':'12', '3':'11', '4':'10',
     '5':'9', '6':'8', '7':'7', '8':'6', '9':'5',
     '10':'4', '11':'3', '12':'2'}
    {'1':'2', '2':'1', '3':'12', '4':'11', '5':'10',
     '6':'9', '7':'8', '8':'7', '9':'6', '10':'5',
     '11':'4', '12':'3'}
>>> print('grpsize = ', c12.automorphismGroupSize)
   grpsize = 24
```

The 12-cycle graph's automorphism group, of size 24, is generated with both the permutations shown in Listing 21.17.

The `showOrbits()` method renders now the labelled representatives of each of the four orbits of isomorphic MISs observed in the 12-cycle graph (see Lines 7–10 below).

[4] The `automorphismGenerators` method uses the `dreadnaut` shell command from the *nauty* software package. See https://www3.cs.stonybrook.edu/~algorith/implement/nauty/implement.shtml. On Mac OS, there exist dmg installers, and on Ubuntu Linux or Debian, one may easily install it with `...$ sudo apt-get install nauty`.

Listing 21.18 Computing the MIS orbits of the 12-cycle graph

```
1  >>> c12.showOrbits(c12.misset,withListing=False)
2     ...
3    *---- Global result ----
4     Number of MIS:  29
5     Number of orbits :  4
6     Labelled representatives and cardinality:
7     1: ['2','4','6','8','10','12'], 2
8     2: ['2','5','8','11'], 3
9     3: ['2','4','6','9','11'], 12
10    4: ['1','4','7','9','11'], 12
11    Symmetry vector
12    stabilizer size: [1, 2, 3, ..., 8, 9, ..., 12, 13, ...]
13    frequency       : [0, 2, 0, ..., 1, 0, ...,  1,  0, ...]
```

The corresponding group stabilisers' sizes and frequencies, orbit 1 with 6 symmetry axes, orbit 2 with 4 symmetry axes, and orbits 3 and 4 both with one symmetry axis (see Lines 12–13 in Listing 21.18), are illustrated in the corresponding unlabelled graphs of Fig. 21.5.

The non-isomorphic MISs in the 12-cycle graph represent in fact all the ways one may write the number 12 as the circular sum of '2's and '3's without distinguishing opposite directions of writing. The first orbit corresponds to writing six times a '2', and the second orbit corresponds to writing four times a '3'. The third and fourth orbits correspond to writing two times a '3' and three times a '2'. There are two non-isomorphic ways to do this latter circular sum, separating the '3's either by one and two '2's or by zero and three '2's (Bisdorff and Marichal 2008).

Chapter 22 is devoted more specifically to tree graphs and graph forests.

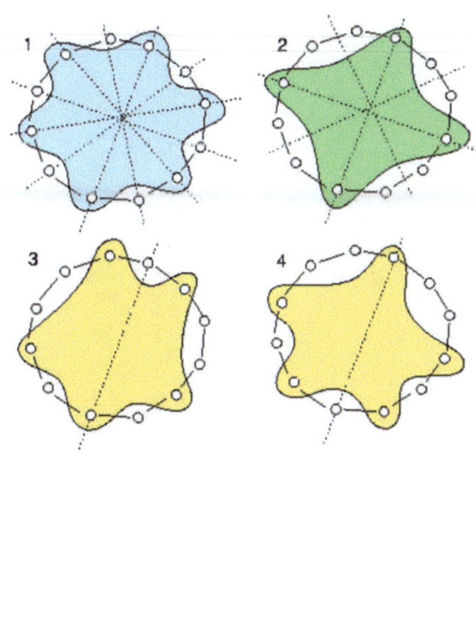

Fig. 21.5 Symmetry axes of the four non-isomorphic MISs of the 12-cycle graph

References

Bisdorff R, Marichal J (2008) Counting non-isomorphic maximal independent sets of the n-cycle graph. J Int Seq 11(Art. 08.5.7):1–16. https://cs.uwaterloo.ca/journals/JIS/VOL11/Marichal/marichal.html

Geman S, Geman D (1984) Stochastic relaxation, Gibbs distributions, and the Bayesian restoration of images. IEEE Trans Pattern Analy Mach Intell 6:721–741

Häggström O (2002) Finite Markov Chains and algorithmic applications. Cambridge University Press, Cambridge

Ising E (1925) Beitrag zur Theorie des Ferromagnetismus. Zeitschrift für Physik 31:253–258

Metropolis N, Rosenbluth A, Rosenbluth M, Teller A (1953) Equation of state calculations by fast computing machines. J Chem Phys 21(6):1087

Chapter 22
On Tree Graphs and Graph Forests

Contents

Abstract The chapter specifically addresses working with tree graphs and graph forests. We illustrate how to generate and recognise random tree graphs and how to compute the centres of a tree and draw a rooted and oriented tree. Finally, algorithms for computing spanning trees and forests are presented.

22.1 Generating Random Tree Graphs

Using the `RandomTree` class from the `graphs` module, we may, for instance, generate a random tree graph with 9 vertices.

Listing 22.1 Generating a random tree graph

```
>>> from graphs import RandomTree
>>> rt = RandomTree(order=9,seed=100)
>>> rt
  *------- Graph instance description ----*
   Instance class    : RandomTree
   Instance name     : randomTree
   Graph Order       : 9
   Graph Size        : 8
   Valuation domain : [-1.00; 1.00]
   Attributes    : ['name', 'order',
                   'vertices', 'valuationDomain',
                   'edges', 'prueferCode',
                   'size', 'gamma']
   *---- RandomTree specific data ----*
   Pruefer code  : ['v3','v8','v8','v3','v7','v6','v7']
```


R. Bisdorff, *Algorithmic Decision Making with Python Resources*,
International Series in Operations Research & Management Science 324,
https://doi.org/10.1007/978-3-030-90928-4_22

```
16 >>> rt.exportGraphViz('tutRandomTree')
17   *---- exporting a dot file for GraphViz tools --*
18    Exporting to tutRandomTree.dot
19    neato -Tpng tutRandomTree.dot -o tutRandomTree.png
```

In Fig. 22.1, one may notice that a tree graph of order $n > 2$ always contains $n-1$ edges (see Lines 7 and 8 in Listing 22.1 on the preceding page) and its structure is entirely characterised by a corresponding PRÜFER *code*, i.e. a list of vertices keys of length $n-2$. See, for instance, in Line 15 the code `['v3', 'v8', 'v8', 'v3', 'v7', 'v6', 'v7']` corresponding to our sample tree graph `rt`.

Each position of the code indicates the parent of the remaining leaf with the smallest vertex label. Vertex `v3` is thus the parent of `v1` and we drop leaf `v1`, `v8` is now the parent of leaf `v2` and we drop `v2`, vertex `v8` is again the parent of leaf `v4` and we drop `v4`, vertex `v3` is the parent of leaf `v5` and we drop `v5`, `v7` is now the parent of leaf `v3` and we may drop `v3`, `v6` becomes the parent of leaf `v8` and we drop `v8`, and `v7` becomes the parent of leaf `v6` and we may drop `v6`. The two eventually remaining vertices, `v7` and `v9`, give the last link in the reconstructed tree (Barthélemy and Guenoche 1991).

It is as well possible to, first, generate a random PRÜFER code of length $n-2$ from a set of n vertices and then construct the corresponding tree of order n by reversing the procedure illustrated above (see Listing 22.2 on the next page) . The resulting tree graph is shown in Fig. 22.2 on the facing page.

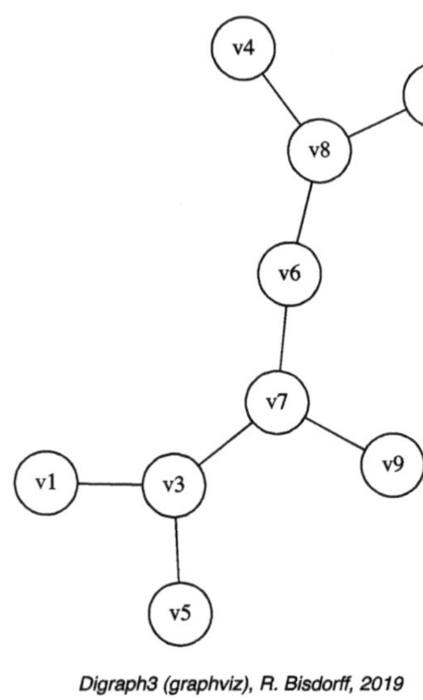

Fig. 22.1 *Random tree graph instance of order 9*. One may distinguish vertices like `v1`, `v2`, `v4`, `v5`, or `v9` of degree 1, called the *leaves* of the tree, and vertices like `v3`, `v6`, `v7`, or `v8` of degree 2 or more, called the *nodes* of the tree

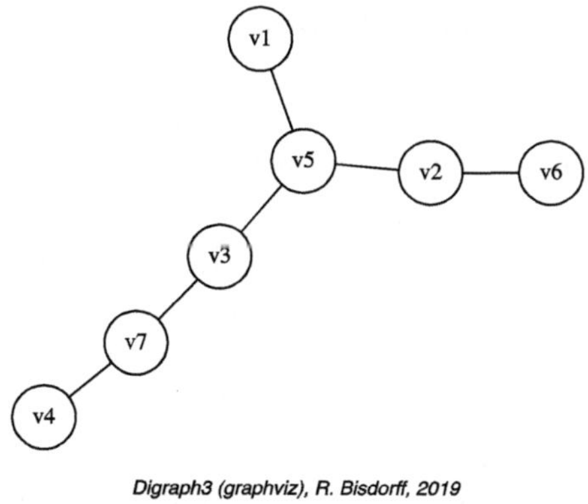

Fig. 22.2 Tree graph instance generated with the random PRÜFER code `['v5', 'v7', 'v2', 'v5', 'v3']`

Listing 22.2 Generating a tree graph with a random PRÜFER code

```
>>> verticesList = ['v1','v2','v3','v4','v5','v6','v7']
>>> n = len(verticesList)
>>> from random import seed, choice
>>> seed(101)
>>> code = []
>>> for k in range(n-2):
...     code.append( choice(verticesList) )
>>> print(code)
  ['v5', 'v7', 'v2', 'v5', 'v3']
>>> rt = RandomTree(prueferCode=code)
>>> rt
  *------- Graph instance description ------*
   Instance class    : RandomTree
   Instance name     : randomTree
   Graph Order       : 7
   Graph Size        : 6
   Valuation domain  : [-1.00; 1.00]
   Attributes : ['name', 'order', 'vertices',
                 'valuationDomain', 'edges',
                 'prueferCode', 'size', 'gamma']
  *---- RandomTree specific data ----*
   Pruefer code   : ['v5', 'v7', 'v2', 'v5', 'v3']
>>> rt.exportGraphViz('tutPruefTree')
  *---- exporting a dot file for GraphViz tools ---------*
   Exporting to tutPruefTree.dot
   neato -Tpng tutPruefTree.dot -o tutPruefTree.png
```

Following from the bijection between a labelled tree and its PRÜFER code, we actually know that there exist n^{n-2} different tree graphs with the same n vertices.

Given a genuine graph, how can we recognise that it is in fact a tree instance?

22.2 Recognising Tree Graphs

Given a graph `g` of order n and size s, the following 5 assertions `A1`, `A2`, `A3`, `A4`, and `A5` are all equivalent (see Barthélemy and Guenoche 1991):

`A1`: `g` is a tree.
`A2`: `g` is without (chordless) cycles and $n = s + 1$.
`A3`: `g` is connected and $n = s + 1$.
`A4`: Any two vertices of `g` are always connected by a unique path.
`A5`: `g` is connected and dropping any single edge will always disconnect `g`.

Assertion `A3`, for instance, gives a simple test for recognising a tree graph. In case of a *lazy evaluation* of the test in Listing 22.3, Line 3 below, it is opportune, from a computational complexity perspective, to first check the order and size of the graph, before checking its potential connectedness. We provide the `isTree()` method for computing both these tests (see Line 8).

Listing 22.3 Recognising a tree graph

```
1 >>> from graphs import RandomGraph
2 >>> g = RandomGraph(order=8,edgeProbability=0.3,seed=62)
3 >>> if g.order == (g.size +1) and g.isConnected():
4 ...     print('The graph is a tree ?', True)
5 ... else:
6 ...     print('The graph is a tree ?',False)
7  The graph is a tree ? True
8 >>> g.isTree()
9  True
```

The random graph of order 8 and edge probability 30%, generated with seed 62, is actually a tree graph instance, as confirmed by its graphviz drawing shown in Fig. 22.3 on the facing page.

```
>>> g.exportGraphViz('test62')
 *---- exporting a dot file for GraphViz tools ---*
  Exporting to test62.dot
  fdp -Tpng test62.dot -o test62.png
```

Yet, we still have to recover its corresponding PRÜFER code. Therefore, we can use the `tree2Pruefer()` method from the `TreeGraph` class. But, first, the instance class of graph `g` is changed to the `TreeGraph` type (see Line 2 below).

Listing 22.4 Computing the PRÜFER code of a tree graph instance

```
1 >>> from graphs import TreeGraph
2 >>> g.__class__ = TreeGraph
3 >>> g.tree2Pruefer()
4  ['v6', 'v1', 'v2', 'v1', 'v2', 'v5']
```

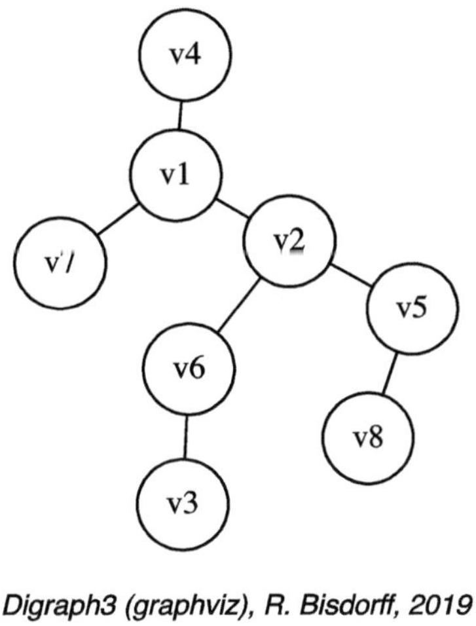

Fig. 22.3 *Recognising a tree graph*. One may notice that vertex `v2` is actually situated in the *centre* of the tree with a neighbourhood depth of 2

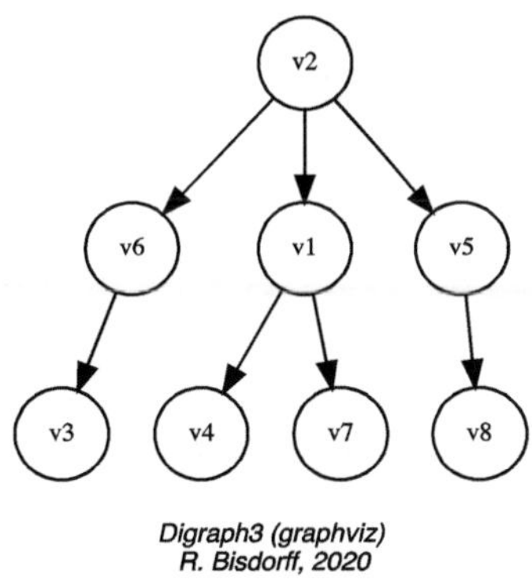

Fig. 22.4 Drawing an oriented tree rooted at its centre

In Fig. 22.3, we noticed that vertex `v2` is actually situated in the *centre* of the tree with a neighbourhood depth of 2. Centres of a graph are the vertices with minimal neighbourhood depth. For finding such centre(s), the `Graph` class provides the `computeGraphCentres()` method (see Line 1 in Listing 22.5). Knowing now the centre of graph `g`, we may draw a correspondingly rooted and oriented tree with the `exportOrientedTreeGraphViz()` method from the `TreeGraph` class (see Fig. 22.4).

Listing 22.5 Computing the centres of a tree and drawing a rooted and oriented tree

```
1 >>> g.computeGraphCentres()
2   {'v2': 2}
3 >>> g.exportOrientedTreeGraphViz(\
4 ...         fileName='rootedTree', root='v2')
5   *---- exporting a dot file for GraphViz tools
6    Exporting to rootedTree.dot
7    dot -Grankdir=TB -Tpng rootedTree.dot -o rootedTree.png
```

Let us now turn our attention towards a major application of tree graphs, namely *spanning trees* and *forests* related to graph traversals.

22.3 Spanning Trees and Forests

With the `RandomSpanningTree` class, we may generate, from a given connected graph `g` instance, *uniform* random instances of a spanning tree by using WILSON's algorithm (see Listing 22.6 and Fig. 22.5).

Listing 22.6 Generating uniform random spanning trees

```
>>> from graphs import RandomGraph,\
...                          RandomSpanningTree
>>> g = RandomGraph(order=9,\
...                     edgeProbability=0.4,seed=100)
>>> spt = RandomSpanningTree(g)
>>> spt
  *------- Graph instance description ------*
   Instance class     : RandomSpanningTree
   Instance name      : randomGraph_randomSpanningTree
   Graph Order        : 9
   Graph Size         : 8
   Valuation domain : [-1.00; 1.00]
   Attributes         : ['name','vertices','order',
                          'valuationDomain',
                          'edges','size','gamma',
                          'dfs','date', 'dfsx',
                          'prueferCode']
  *---- RandomTree specific data ----*
   Pruefer code   : ['v7','v9','v5','v1','v8','v4','v9']
>>> spt.exportGraphViz(fileName='randomSpanningTree',\
...                          WithSpanningTree=True)
  *---- exporting a dot file for GraphViz tools -----*
   Exporting to randomSpanningTree.dot
   [['v1','v5','v6','v5','v1','v8','v9','v3','v9','v4',
      'v7','v2','v7','v4','v9','v8','v1']]
   neato -Tpng randomSpanningTree.dot\
         -o randomSpanningTree.png
```

Fig. 22.5 Random spanning tree

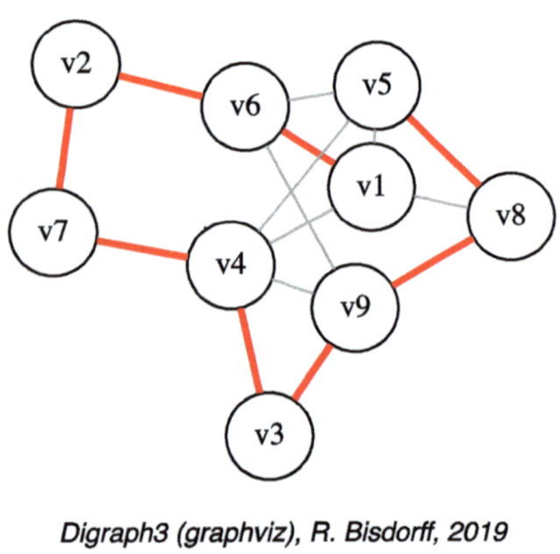

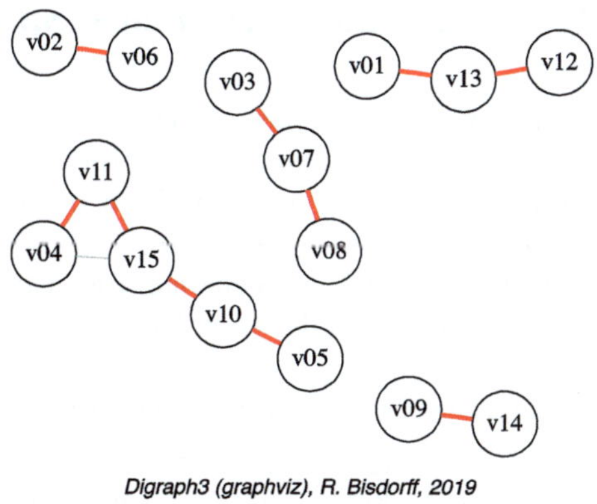

Fig. 22.6 Random spanning forest

WILSON's algorithm only works for connected graphs (Wilson 1996). More general, and in case of a not connected graph, the `RandomSpanningForest` class generates a not necessarily uniform random instance of a *spanning forest*—one or more random tree graphs—generated from a random *depth first search* of the graph components' traversals (see Listing 22.7 and Fig. 22.6).

Listing 22.7 Computing spanning forests over disconnected graphs

```
>>> g = RandomGraph(order=15,\
...                      edgeProbability=0.1,seed=140)
>>> g.computeComponents()
  [{'v12','v01','v13'}, {'v02','v06'},
   {'v08','v03','v07'}, {'v15','v11','v10','v04','v05'},
   {'v09','v14'}]
>>> spf = RandomSpanningForest(g,seed=100)
>>> spf.exportGraphViz(fileName='spanningForest',\
...                       WithSpanningTree=True)
  *---- exporting a dot file for GraphViz tools -----*
    Exporting to spanningForest.dot
    [['v03','v07','v08','v07','v03'],
     ['v13','v12','v13','v01','v13'],
     ['v02','v06','v02'],
     ['v15','v11','v04','v11','v15',
            'v10','v05','v10','v15'],
     ['v09','v14','v09']]
  neato -Tpng spanningForest.dot -o spanningForest.png
```

22.4 Maximum Determined Spanning Forests

In case of valued graphs—supporting weighted edges—we may finally construct a *most determined* spanning tree (or forest if not connected) using KRUSKAL's greedy minimum spanning tree algorithm on the dual valuation of the graph (Kruskal 1956).[1]

In Listing 22.8, we generate, for instance, a randomly valued graph with five vertices and seven edges bipolar-valued in $[-1.0; 1.0]$.

Listing 22.8 Generating randomly bipolar-valued graphs

```
>>> from graphs import RandomValuationGraph
>>> g = RandomValuationGraph(order=5,seed=2)
>>> g
 *------- Graph instance description ------*
  Instance class      : RandomValuationGraph
  Instance name       : randomGraph
  Graph Order         : 5
  Graph Size          : 7
  Valuation domain    : [-1.00; 1.00]
  Attributes          : ['name', 'order',
                   'vertices', 'valuationDomain',
                   'edges', 'size', 'gamma']
```

To inspect the edges' actual weights, we transform in Listing 22.9 the random graph `g` into a corresponding digraph `dg` and use the `showRelationTable()` method for printing its symmetric adjacency matrix.

Listing 22.9 Symmetric relation table

```
>>> dg = g.graph2Digraph()
>>> dg.showRelationTable()
 *---- Relation Table -----
   S   |  'v1'  'v2'  'v3'  'v4'  'v5'
  -----|------------------------------
  'v1' |  0.00  0.91  0.90 -0.89 -0.83
  'v2' |  0.91  0.00  0.67  0.47  0.34
  'v3' |  0.90  0.67  0.00 -0.38  0.21
  'v4' | -0.89  0.47 -0.38  0.00  0.21
  'v5' | -0.83  0.34  0.21  0.21  0.00
  Valuation domain: [-1.00;1.00]
```

To compute the most determined spanning tree or forest, we can use the `BestDeterminedSpanningForest` class constructor.

---

[1] KRUSKAL's algorithm is a minimum spanning tree algorithm which finds an edge of the least possible weight that connects any two trees in the forest.

Listing 22.10 Computing best determined spanning forests

```
>>> from graphs import\
                BestDeterminedSpanningForest
>>> mt = BestDeterminedSpanningForest(g)
>>> print(mt)
 *------- Graph instance description ------*
  Instance class    : BestDeterminedSpanningForest
  Instance name     : bdSpanningForest
  Graph Order       : 5
  Graph Size        : 4
  Valuation domain  : [-1.00; 1.00]
  Attributes        : ['name','vertices','order',
                       'valuationDomain',
                       'edges','size','gamma',
                       'dfs','date',
                       'averageTreeDetermination']
 *---- best determined spanning tree    ----*
  Depth first search path(s)  :
  [['v1','v2','v4','v2','v5','v2','v1','v3','v1']]
  Average determination(s)  :  [Decimal('0.655')]
```

The random graph g is connected and, hence, admits a single spanning tree of maximum mean determination = $(0.47 + 0.91 + 0.90 + 0.34)/4 = 0.655$ (see Lines 9, 6, and 10 in Listing 22.9 on the facing page) (Fig. 22.7).

```
>>> mt.exportGraphViz(\
...       fileName='bestDeterminedspanningTree',\
...       WithSpanningTree=True)
 *---- exporting a dot file for GraphViz tools ---*
  Exporting to spanningTree.dot
  [['v4','v2','v1','v3','v1','v2','v5','v2','v4']]
  neato -Tpng bestDeterminedSpanningTree.dot\
       -o bestDeterminedSpanningTree.png
```

One may easily verify that all other potential spanning trees, including instead the edges {v3, v5} and/or {v4, v5}, will show a lower average determination.

Chapter 23, the last on undirected graphs, is devoted to different models of BERGE or perfect graphs, namely split, interval, comparability, and permutation graphs.

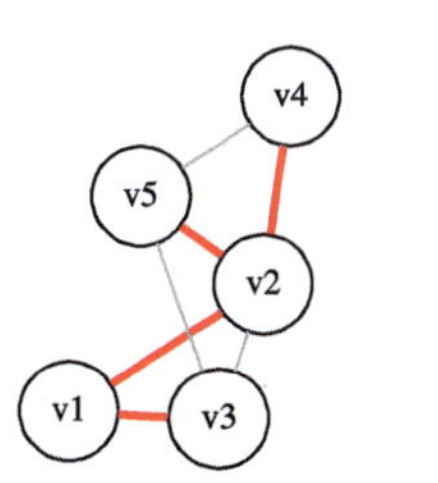

Fig. 22.7 Best determined spanning tree

References

Barthélemy J, Guenoche A (1991) Trees and Proximities Representations. Wiley, Hoboken

Kruskal J (1956) On the shortest spanning subtree of a graph and the traveling salesman problem. Proc Am Math Soc 7:48–50

Wilson DB (1996) Generating random spanning trees more quickly than the cover time. In: Proceedings of the twenty-eighth annual ACM symposium on the theory of computing (Philadelphia PA), ACM New York, pp 296–303

Chapter 23
About Split, Comparability, Interval, and Permutation Graphs

Contents

Abstract The last chapter of this book eventually presents some famous classes of perfect graphs, namely comparability, interval, permutation, and split graphs. We first present an example of a graph which is at the same time a triangulated, a comparability, a split, and a permutation graph. The importance to be an interval is illustrated with BERGE's mystery story. We discuss furthermore the generation of permutation graphs and close with how to recognise that a given graph is in fact a permutation graph.

23.1 A 'multiply' Perfect Graph

A graph `g` is called:

- BERGE or *perfect* when `g` and its dual `-g` both do not contain any chordless odd cycles of length greater than 3 (Berge 1963; Chudnovsky et al. 2006).
- *Triangulated* graph when `g` does not contain any *chordless cycle* of length 4 and more.

Definition 23.1 (Some Perfect Graph Classes) Following Martin Golumbic (see Golumbic 2004, p. 149), we call a given graph `g`:

- *Comparability* graph when `g` is *transitively orientable*.
- *Interval* graph when `g` is *triangulated* and its dual`-g` is a *comparability* graph.
- *Permutation* graph when `g` and its dual `-g` both are *comparability* graphs.
- *Split* graph when `g` and its dual `-g` both are *triangulated* graphs.

R. Bisdorff, *Algorithmic Decision Making with Python Resources*,
International Series in Operations Research & Management Science 324,
https://doi.org/10.1007/978-3-030-90928-4_23

To illustrate these *perfect* graph classes, we will generate from 8 intervals, randomly chosen in the default integer range [0, 10], a `RandomIntervalIntersectionsGraph` instance `g` (see Line 2 below).

```
1 >>> from graphs import\
2 ...                RandomIntervalIntersectionsGraph
3 >>> g = RandomIntervalIntersectionsGraph(\
4 ...                               order=8,seed=100)
5 >>> g
6   *------- Graph instance description ------*
7    Instance class    : RandomIntervalIntersectionsGraph
8    Instance name     : randIntervalIntersections
9    Seed              : 100
10   Graph Order       : 8
11   Graph Size        : 23
12   Valuation domain  : [-1.0; 1.0]
13   Attributes        : ['seed', 'name', 'order',
14         'intervals', 'vertices', 'valuationDomain',
15         'edges', 'size', 'gamma']
16 >>> print(g.intervals)
17  [(2,7), (2,7), (5,6), (6,8), (1,8), (1,1), (4,7), (0,10)]
```

With `seed = 100`, we obtain here an *interval graph*, in fact a *perfect graph*, which is *conjointly* a *triangulated*, a *comparability*, a *split*, and a *permutation* graph (see Listing 23.1, Lines 2, 6, 10, and 14 and Fig. 23.1 on the next page).

Listing 23.1 Testing perfect graph categories

```
1 >>> g.isPerfectGraph(Comments=True)
2   Graph randIntervalIntersections is perfect !
3 >>> g.isIntervalGraph(Comments=True)
4   Graph 'randIntervalIntersections' is triangulated.
5   Graph 'dual_randIntervalIntersections' is transitively orientable.
6   => Graph 'randIntervalIntersections' is an interval graph.
7 >>> g.isSplitGraph(Comments=True)
8   Graph 'randIntervalIntersections' is triangulated.
9   Graph 'dual_randIntervalIntersections' is triangulated.
10  => Graph 'randIntervalIntersections' is a split graph.
11 >>> g.isPermutationGraph(Comments=True)
12  Graph 'randIntervalIntersections' is transitively orientable.
13  Graph 'dual_randIntervalIntersections' is transitively orientable.
14  => Graph 'randIntervalIntersections' is a permutation graph.
15 >>> g.exportGraphViz('randomSplitGraph')
16   *---- exporting a dot file for GraphViz tools ---------*
17    Exporting to randomSplitGraph.dot
18    fdp -Tpng randomSplitGraph.dot -o randomSplitGraph.png
```

Notice however that the following four properties of a given graph `g`:

1. g is a *comparability* graph,
2. $-g$ is a comparability graph,
3. g is a *triangulated* graph,
4. $-g$ is a triangulated graph,

are all *independent* of one another (Golumbic 2004, p. 275).

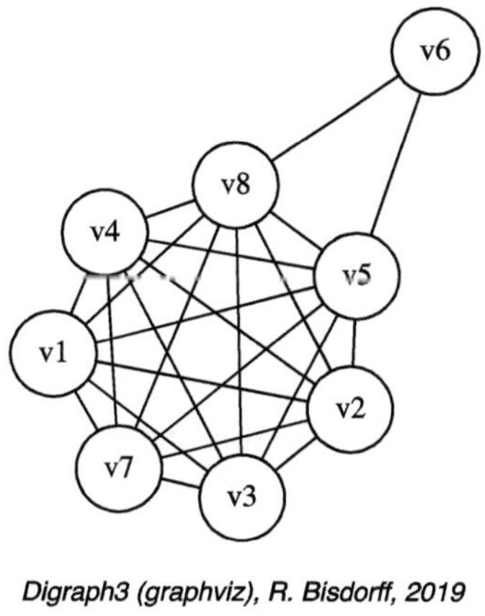

Fig. 23.1 *A conjointly triangulated, comparability, interval, permutation and split graph*
In the figure here, one may readily recognise the essential characteristic of split graphs, namely being always splittable into two disjoint sub-graphs: an *independent choice* {`v6`} and a *clique*—{`v1`, `v2`, `v3`, `v4`, `v5`, `v7`, `v8`}—which explains their name

23.2 Who Is the Liar?

Claude Berge's famous mystery story related by Golumbic (2004, p. 20) may well illustrate the importance of being an *interval* graph.

Suppose the file `berge.py`[1] contains the following `Graph` instance data:

```
vertices = {
  'A': {'name': 'Abe', 'shortName': 'A'},
  'B': {'name': 'Burt', 'shortName': 'B'},
  'C': {'name': 'Charlotte', 'shortName': 'C'},
  'D': {'name': 'Desmond', 'shortName': 'D'},
  'E': {'name': 'Eddie', 'shortName': 'E'},
  'I': {'name': 'Ida', 'shortName': 'I'},
  }
valuationDomain = {'min':-1,'med':0,'max':1}
edges = {
  frozenset(['A','B']) : 1,
  frozenset(['A','C']) : -1,
  frozenset(['A','D']) : 1,
  frozenset(['A','E']) : 1,
  frozenset(['A','I']) : -1,
  frozenset(['B','C']) : -1,
  frozenset(['B','D']) : -1,
  frozenset(['B','E']) : 1,
  frozenset(['B','I']) : 1,
  frozenset(['C','D']) : 1,
  frozenset(['C','E']) : 1,
  frozenset(['C','I']) : 1,
  frozenset(['D','E']) : -1,
  frozenset(['D','I']) : 1,
  frozenset(['E','I']) : 1,
  }
```

[1] The file `berge.py` is provided in the `examples` directory of the DIGRAPH3 resources.

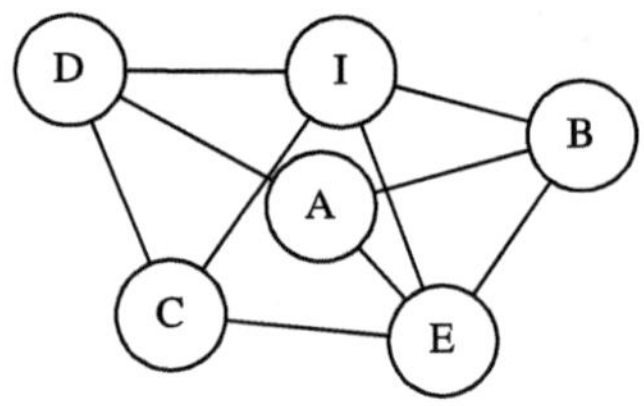

Digraph3 (graphviz), R. Bisdorff, 2019

Fig. 23.2 Graph representation of the testimonies of the professors

Six professors (labelled A, B, C, D, E, and I) had been to the library on the day that a precious document was stolen. Each entered once, stayed for some time, and then left. If two professors were in the library at the same time, then at least one of them saw the other. Detectives questioned the professors and gathered the testimonies that *Abe* saw *Burt* and *Eddie*; *Burt* saw *Abe* and *Ida*; *Charlotte* saw *Desmond* and *Ida*; *Desmond* saw *Abe* and *Ida*; *Eddie* saw *Burt* and *Ida*; and *Ida* saw *Charlotte* and *Eddie*. This data is gathered in the previous file, where each positive edge $\{x, y\}$ models the testimony that, either x saw y or y saw x.

```
>>> from graphs import Graph
>>> g = Graph('berge')
>>> g.showShort()
  *---- short description of the graph ----*
   Name                : 'berge'
   Vertices            :  ['A', 'B', 'C', 'D', 'E', 'I']
   Valuation domain :  {'min': -1, 'med': 0, 'max': 1}
   Gamma function      :
    A -> ['D', 'B', 'E']
    B -> ['E', 'I', 'A']
    C -> ['E', 'D', 'I']
    D -> ['C', 'I', 'A']
    E -> ['C', 'B', 'I', 'A']
    I -> ['C', 'E', 'B', 'D']
>>> g.exportGraphViz('berge1')
  *---- exporting a dot file for GraphViz tools ---------*
   Exporting to berge1.dot
   fdp -Tpng berge1.dot -o berge1.png
```

From graph theory, we know that time interval intersections graphs must in fact be interval graph instances, i.e. *triangulated* and *co-comparative* graphs. The testimonies graph shown in Fig. 23.2 should therefore not contain any chordless cycle of four and more vertices. Now, the presence or not of such chordless cycles in the testimonies graph can be checked with the `computeChordlessCycles()` method (Bisdorff 2010).

```
>>> g.computeChordlessCycles()
  Chordless cycle certificate: ['D','C','E','A','D']
  Chordless cycle certificate: ['D','I','E','A','D']
  Chordless cycle certificate: ['D','I','B','A','D']
  [(['D','C','E','A','D'],frozenset({'C','D','E','A'})),
   (['D','I','E','A','D'],frozenset({'D','E','I','A'})),
   (['D','I','B','A','D'],frozenset({'D','B','I','A'}))]
```

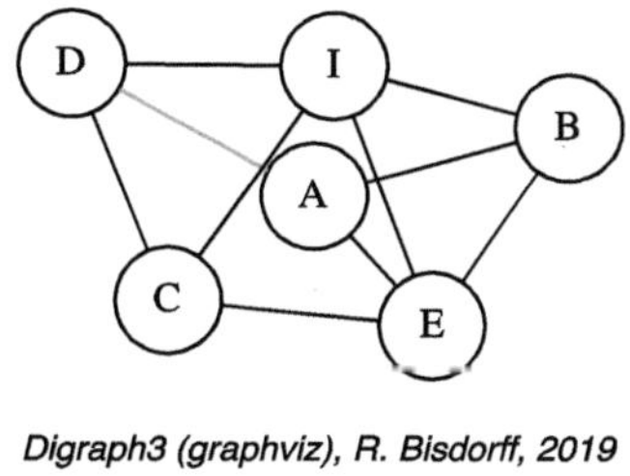

Fig. 23.3 The triangulated testimonies graph

We see in the listing above three intersection cycles of length 4, which is impossible to occur on the linear timeline. Obviously, one professor lied!

And it is *Desmond*. If we doubt his testimony that he saw *Abe* (see Line 1 below), we obtain indeed in Fig. 23.3 a triangulated graph instance whose dual is a comparability graph.

```
1  >>> g.setEdgeValue( ('D','A'), 0)
2  >>> g.showShort()
3    *---- short description of the graph ----*
4     Name               : 'berge'
5     Vertices           :  ['A','B','C','D','E','I']
6     Valuation domain   :  {'med': 0,'min': -1,'max': 1}
7     Gamma function     :
8      A -> ['B', 'E']
9      B -> ['A', 'I', 'E']
10     C -> ['I', 'E', 'D']
11     D -> ['I', 'C']
12     E -> ['A', 'I', 'B', 'C']
13     I -> ['B', 'E', 'D', 'C']
14 >>> g.isIntervalGraph(Comments=True)
15   Graph 'berge' is triangulated.
16   Graph 'dual_berge' is transitively orientable.
17   => Graph 'berge' is an interval graph.
18 >>> g.exportGraphViz('berge2')
19   *---- exporting a dot file for GraphViz tools ---------*
20    Exporting to berge2.dot
21    fdp -Tpng berge2.dot -o berge2.png
```

23.3 Generating Permutation Graphs

A graph is called a *permutation* or *inversion* graph if there exists a permutation of its list of vertices, like `[4,3,6,1,5,2]`, such that the graph is isomorphic to the inversions operated by the permutation in this list (see Golumbic 2004, Chap. 7, pp 157-170). This kind of graphs is also part of the class of BERGE graphs (Fig. 23.4).

```
>>> from graphs import PermutationGraph
>>> g = PermutationGraph(permutation = [4,3,6,1,5,2])
>>> g
```

Fig. 23.4 The [4,3,6,1,5,2] permutation graph

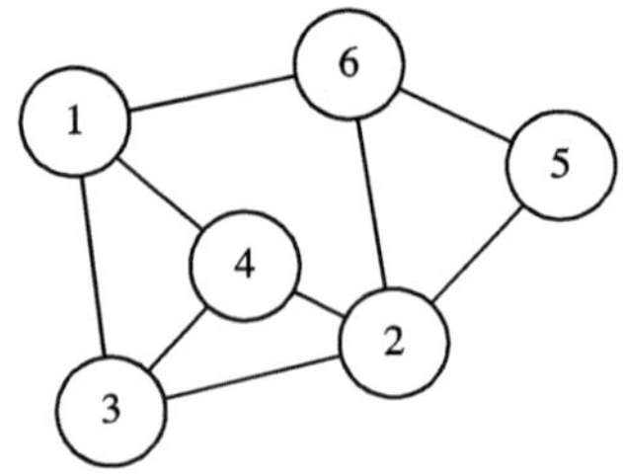

Digraph3 (graphviz), R. Bisdorff, 2019

```
  *------- Graph instance description ------*
   Instance class    : PermutationGraph
   Instance name     : permutationGraph
   Graph Order       : 6
   Permutation       : [4, 3, 6, 1, 5, 2]
   Graph Size        : 9
   Valuation domain  : [-1.00; 1.00]
   Attributes        : ['name', 'vertices', 'order',
                 'permutation', 'valuationDomain',
                 'edges', 'size', 'gamma']
>>> g.isPerfectGraph()
  True
>>> g.exportGraphViz(fileName='permutationGraph')
  *---- exporting a dot file for GraphViz tools ----*
   Exporting to permutationGraph.dot
   fdp -Tpng permutationGraph.dot\
                  -o permutationGraph.png
```

By using colour sorting queues, the minimal vertex colouring shown in Fig. 23.5 on the facing page is computable in $O\big(nlog(n)\big)$ (Golumbic 2004).

```
>>> g.computeMinimalVertexColoring(Comments=True)
    vertex 1: lightcoral
    vertex 2: lightcoral
    vertex 3: lightblue
    vertex 4: gold
    vertex 5: lightblue
    vertex 6: gold
>>> g.exportGraphViz(fileName='coloredPermutationGraph',\
...                  WithVertexColoring=True)
  *---- exporting a dot file for GraphViz tools -------*
   Exporting to coloredPermutationGraph.dot
   fdp -Tpng coloredPermutationGraph.dot\
       -o coloredPermutationGraph.png
```

The correspondingly coloured *matching diagram* of the nine inversions—the actual edges of the permutation graph –, which are induced by the given [4,3,6,1,5,2] permutation, may as well be drawn with the graphviz `neato` layout and explicitly positioned horizontal lists of vertices as shown in Fig. 23.6 on the next page.

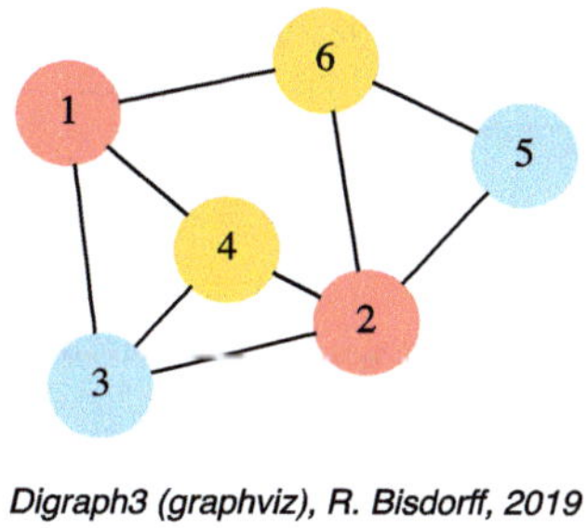

Fig. 23.5 Minimal vertex colouring of the permutation graph

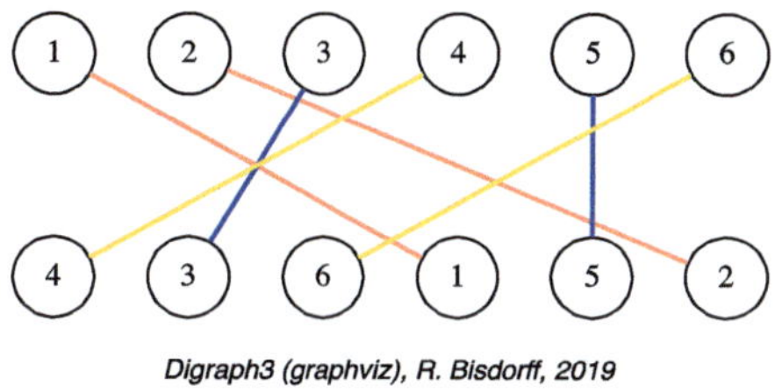

Fig. 23.6 Coloured matching diagram of the permutation [4,3,6,1,5,2]

```
>>> g.exportPermutationGraphViz('matchingDiagram',\
...                        WithEdgeColoring=True)
*---- exporting a dot file for GraphViz tools ----*
  Exporting to matchingDiagram.dot
  neato -n -Tpng matchingDiagram.dot\
          -o matchingDiagram.png
```

As mentioned before, a permutation graph and its dual are *transitively orientable*. The `transitiveOrientation()` method constructs from a given permutation graph a digraph where each edge of the permutation graph is converted into an arc oriented in an increasing alphabetic order of the adjacent vertices' keys (see Golumbic 2004). This orientation of the edges of a permutation graph is always transitive and delivers a *transitive ordering* of the vertices as shown in Fig. 23.7 on the following page.

```
>>> dg = g.transitiveOrientation()
>>> dg
  *------- Digraph instance description ------*
   Instance class   : TransitiveDigraph
   Instance name    : oriented_permutationGraph
   Digraph Order      : 6
   Digraph Size       : 9
   Valuation domain : [-1.00; 1.00]
   Determinateness  : 100.000
   Attributes       : ['name', 'order', 'actions',
                    'valuationdomain', 'relation',
                    'gamma', 'notGamma', 'size']
>>> print('Transitivity degree: %.3f' %\
...           dgd.computeTransitivityDegree() )
   Transitivity degree: 1.000
>>> dg.exportGraphViz(fileName='orientedPermGraph')
```

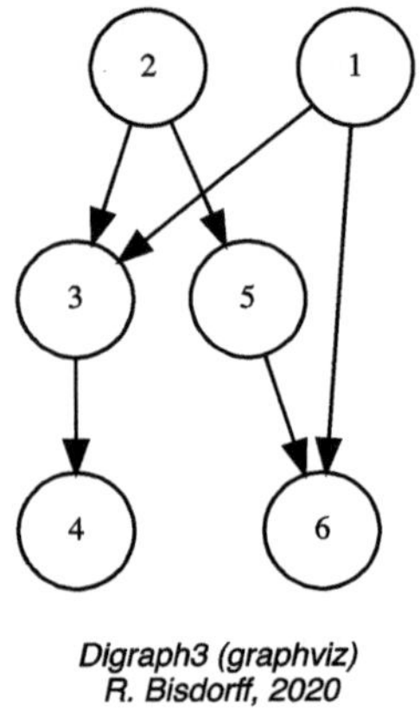

Fig. 23.7 The transitive orientation of the permutation graph

```
 *---- exporting a dot file for GraphViz tools -----*
  Exporting to orientedPermGraph.dot
  dot -Grankdir=TB -Tpng orientedPermGraph.dot\
                   -o orientedPermGraph.png
```

The dual of a permutation graph is again a permutation graph and as such also transitively orientable.

```
>>> dgd = (-g).transitiveOrientation()
>>> print('Dual transitivity degree: %.3f' %\
...               dgd.computeTransitivityDegree() )
   Dual transitivity degree: 1.00
```

23.4 Recognising Permutation Graphs

Now, a given graph `g` is a permutation graph if and only if both `g` **and** `-g` are transitively orientable. This property gives a polynomial test procedure (in $O(n^3)$ due to the transitivity check) for recognising permutation graphs.

Let us consider, for instance, the following random graph of order 8 generated with an edge probability of 40% and a random seed equal to 4335.

```
>>> from graphs import RandomGraph
>>> g = RandomGraph(order=8,\
...                 edgeProbability=0.4,seed=4335)
>>> g
 *------- Graph instance description ------*
  Instance class    : RandomGraph
  Instance name     : randomGraph
  Seed              : 4335
  Edge probability  : 0.4
  Graph Order       : 8
  Graph Size        : 10
  Valuation domain  : [-1.00; 1.00]
```

Fig. 23.8 Random graph of order 8 generated with edge probability 0.4

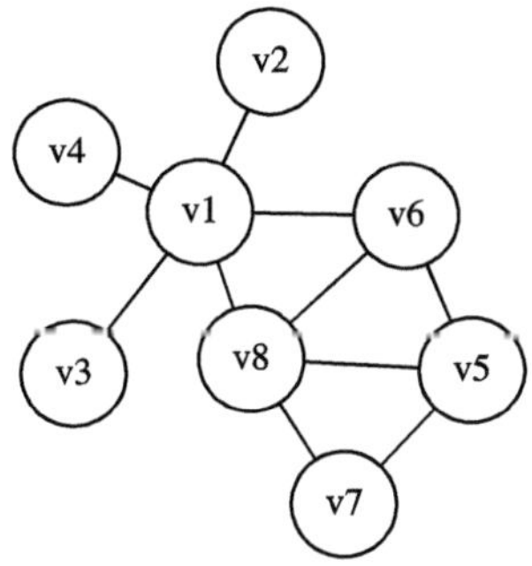

Digraph3 (graphviz), R. Bisdorff, 2019

```
  Attributes          : ['name', 'order', 'vertices',
                         'valuationDomain', 'seed',
                         'edges', 'size', 'gamma',
                         'edgeProbability']
>>> g.isPerfectGraph()
 True
>>> g.exportGraphViz(fileName='randomGraph4335')
 *---- exporting a dot file for GraphViz tools ----*
  Exporting to randomGraph4335.dot
  fdp -Tpdf randomGraph4335.dot -o randomGraph4335.pdf
```

If the random perfect graph instance g, shown in Fig. 23.8, is indeed a permutation graph, g and its dual -g are both *transitively orientable*, i.e. comparability graphs (Golumbic 2004). With the isComparabilityGraph() test, we may easily check this fact. This method proceeds indeed by trying to construct a transitive neighbourhood decomposition of a given graph instance and, if successful, stores the resulting edge orientations into an edgeOrientations attribute (Golumbic 2004, p.129-132).

```
>>> if g.isComparabilityGraph():
...     print(g.edgeOrientations)
 {('v1','v1'): 0, ('v1','v2'): 1, ('v2','v1'): -1, ('v1','v3'): 1,
  ('v3','v1'): -1, ('v1','v4'): 1, ('v4','v1'): -1, ('v1','v5'): 0,
  ('v5','v1'): 0, ('v1','v6'): 1, ('v6','v1'): -1, ('v1','v7'): 0,
  ('v7','v1'): 0, ('v1','v8'): 1, ('v8','v1'): -1, ('v2','v2'): 0,
  ('v2','v3'): 0, ('v3','v2'): 0, ('v2','v4'): 0, ('v4','v2'): 0,
  ('v2','v5'): 0, ('v5','v2'): 0, ('v2','v6'): 0, ('v6','v2'): 0,
  ('v2','v7'): 0, ('v7','v2'): 0, ('v2','v8'): 0, ('v8','v2'): 0,
  ('v3','v3'): 0, ('v3','v4'): 0, ('v4','v3'): 0, ('v3','v5'): 0,
  ('v5','v3'): 0, ('v3','v6'): 0, ('v6','v3'): 0, ('v3','v7'): 0,
  ('v7','v3'): 0, ('v3','v8'): 0, ('v8','v3'): 0, ('v4','v4'): 0,
  ('v4','v5'): 0, ('v5','v4'): 0, ('v4','v6'): 0, ('v6','v4'): 0,
  ('v4','v7'): 0, ('v7','v4'): 0, ('v4','v8'): 0, ('v8','v4'): 0,
  ('v5','v5'): 0, ('v5','v6'): 1, ('v6','v5'): -1, ('v5','v7'): 1,
  ('v7','v5'): -1, ('v5','v8'): 1, ('v8','v5'): -1, ('v6','v6'): 0,
  ('v6','v7'): 0, ('v7','v6'): 0, ('v6','v8'): 1, ('v8','v6'): -1,
  ('v7','v7'): 0, ('v7','v8'): 1, ('v8','v7'): -1, ('v8','v8'): 0}
>>> g.exportEdgeOrientationsGraphViz('transOrientGraph')
  *---- exporting a dot file for GraphViz tools ---------*
  Exporting to transOrientGraph.dot
  fdp -Tpng transOrientGraph.dot \
           -o transOrientGraph.png
```

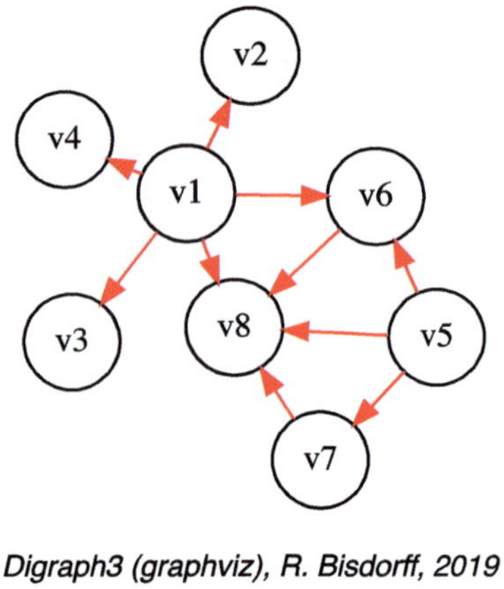

Fig. 23.9 Transitive neighbourhoods of the graph g

The resulting orientation of the edges of g, shown in Fig. 23.9, is indeed transitive. The same procedure applied to the dual graph -g gives as well a transitive orientation of its edges shown in Fig. 23.10 on the next page.

```
>>> gd = -g
>>> if gd.isComparabilityGraph():
...     print(gd.edgeOrientations)
  {('v1','v1'): 0, ('v1','v2'): 0, ('v2','v1'): 0, ('v1','v3'): 0,
   ('v3','v1'): 0, ('v1','v4'): 0, ('v4','v1'): 0, ('v1','v5'): 1,
   ('v5','v1'): -1, ('v1','v6'): 0, ('v6','v1'): 0, ('v1','v7'): 1,
   ('v7','v1'): -1, ('v1','v8'): 0, ('v8','v1'): 0, ('v2','v2'): 0,
   ('v2','v3'): -2, ('v3','v2'): 2, ('v2','v4'): -3, ('v4','v2'): 3,
   ('v2','v5'): 1, ('v5','v2'): -1, ('v2','v6'): 1, ('v6','v2'): -1,
   ('v2','v7'): 1, ('v7','v2'): -1, ('v2','v8'): 1, ('v8','v2'): -1,
   ('v3','v3'): 0, ('v3','v4'): -3, ('v4','v3'): 3, ('v3','v5'): 1,
   ('v5','v3'): -1, ('v3','v6'): 1, ('v6','v3'): -1, ('v3','v7'): 1,
   ('v7','v3'): -1, ('v3','v8'): 1, ('v8','v3'): -1, ('v4','v4'): 0,
   ('v4','v5'): 1, ('v5','v4'): -1, ('v4','v6'): 1, ('v6','v4'): -1,
   ('v4','v7'): 1, ('v7','v4'): -1, ('v4','v8'): 1, ('v8','v4'): -1,
   ('v5','v5'): 0, ('v5','v6'): 0, ('v6','v5'): 0, ('v5','v7'): 0,
   ('v7','v5'): 0, ('v5','v8'): 0, ('v8','v5'): 0, ('v6','v6'): 0,
   ('v6','v7'): 1, ('v7','v6'): -1, ('v6','v8'): 0, ('v8','v6'): 0,
   ('v7','v7'): 0, ('v7','v8'): 0, ('v8','v7'): 0, ('v8','v8'): 0}
>>> gd.exportEdgeOrientationsGraphViz('transOrientDualGraph')
   *---- exporting a dot file for GraphViz tools ---------*
    Exporting to transOrientDualGraph.dot
    fdp -Tpng transOrientDualGraph.dot \
             -o transOrientDualGraph.png
```

Let us recheck these facts by explicitly constructing transitively oriented digraph instances with the `computeTransitivelyOrientedDigraph()` method.

```
>>> og = g.computeTransitivelyOrientedDigraph(\
...                          PartiallyDetermined = True)
>>> print('Transitivity degree: %.3f' %\
...                          (og.transitivityDegree))
  Transitivity degree: 1.000
>>> ogd = (-g).computeTransitivelyOrientedDigraph(\
...                          PartiallyDetermined = True)
>>> print('Transitivity degree: %.3f' %\
...                          (ogd.transitivityDegree))
  Transitivity degree: 1.000
```

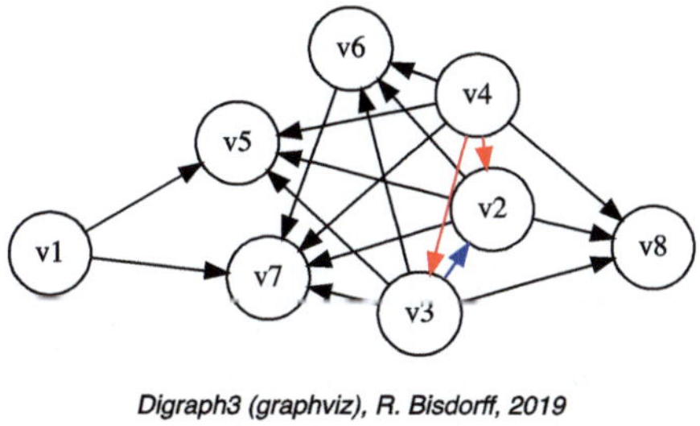

Fig. 23.10 *Transitive neighbourhoods of the dual graph* `-g`
It is worthwhile noticing that the orientation of `g` is achieved with a single neighbourhood decomposition, covering all the vertices, whereas the orientation of the dual graph `-g` here needs a decomposition into three subsequent neighbourhoods marked in black, red, and blue

The `PartiallyDetermined = True` flag in Lines 2 and 7 above is required here in order to orient only the actual edges of the graphs. Relations between vertices not linked by an edge are put to the indeterminate characteristic value 0. This allows us to compute, later on, convenient disjunctive digraph fusions.

As both graphs are indeed transitively orientable, we conclude that the given random graph `g` is actually a permutation graph instance. Yet, we still need to find now its corresponding permutation. We therefore implement a recipe given by (Golumbic 2004, p. 159).

We will first *fuse* both *og* and *ogd* orientations above with an *epistemic disjunction* operated with the symmetric `o-max` operator (see Sect. 2.5).

```
1 >>> from digraphs import FusionDigraph
2 >>> f1 = FusionDigraph(og,ogd,operator='o-max')
3 >>> s1 = f1.computeCopelandRanking()
4 >>> print(s1)
5   ['v5','v7','v1','v6','v8','v4','v3','v2']
```

With the `computeCopelandRanking()` method, we obtain the linear ordering [`v5`, `v7`, `v1`, `v6`, `v8`, `v4`, `v3`, `v2`] of the vertices (see Line 5 above).

We reverse now the orientation of the edges in `og` in order to generate, again by disjunctive fusion, the *inversions* that are produced by the permutation we are looking for (see `-og` in Line 1 below). Computing again a ranking with the COPELAND rule reveals the permuted list of vertices we are looking for (see Line 4 below).

```
1 >>> f2 = FusionDigraph((-og),ogd,operator='o-max')
2 >>> s2 = f2.computeCopelandRanking()
3 >>> print(s2)
4   ['v8','v7','v6','v5','v4','v3','v2','v1']
```

Vertex `v8` is put from position 5 to position 1, vertex `v7` is put from position 2 to position 2, vertex `v6` from position 4 to position 3, vertex `v5` from position 1 to position 4, etc. We generate these position swaps for all vertices and obtain thus the required permutation (see Line 5 below).

```
>>> permutation = [0 for j in range(g.order)]
>>> for j in range(g.order):
...     permutation[s2.index(s1[j])] = j+1
>>> print(permutation)
  [5, 2, 4, 1, 6, 7, 8, 3]
```

It is worthwhile noticing by the way that transitive orientations of a given graph and its dual are usually *not unique* and so may also be the resulting permutations. However, they all correspond to isomorphic graphs (Golumbic 2004). In our case here, we observe two different permutations and their reverses:

```
s1: ['v1', 'v4', 'v3', 'v2', 'v5', 'v6', 'v7', 'v8']
s2: ['v4', 'v3', 'v2', 'v8', 'v6', 'v1', 'v7', 'v5']
(s1 -> s2): [2, 3, 4, 8, 6, 1, 7, 5]
(s2 -> s1): [6, 1, 2, 3, 8, 5, 7, 4]
```

and

```
s3: ['v5', 'v7', 'v1', 'v6', 'v8', 'v4', 'v3', 'v2']
s4: ['v8', 'v7', 'v6', 'v5', 'v4', 'v3', 'v2', 'v1']
(s3 -> s4): [5, 2, 4, 1, 6, 7, 8, 3]
(s4 -> s3) = [4, 2, 8, 3, 1, 5, 6, 7]
```

The `computePermutation()` method does directly operate all these steps: computing transitive orientations, ranking their epistemic fusion, and delivering a corresponding permutation.

```
>>> g.computePermutation(Comments=True)
  ['v1', 'v2', 'v3', 'v4', 'v5', 'v6', 'v7', 'v8']
  ['v2', 'v3', 'v4', 'v8', 'v6', 'v1', 'v7', 'v5']
  [2, 3, 4, 8, 6, 1, 7, 5]
```

Let us finally check in Fig. 23.11 that the two permutations [2, 3, 4, 8, 6, 1, 7, 5] and [4, 2, 8, 3, 1, 5, 6, 7], observed above, will generate corresponding *isomorphic* permutation graphs.

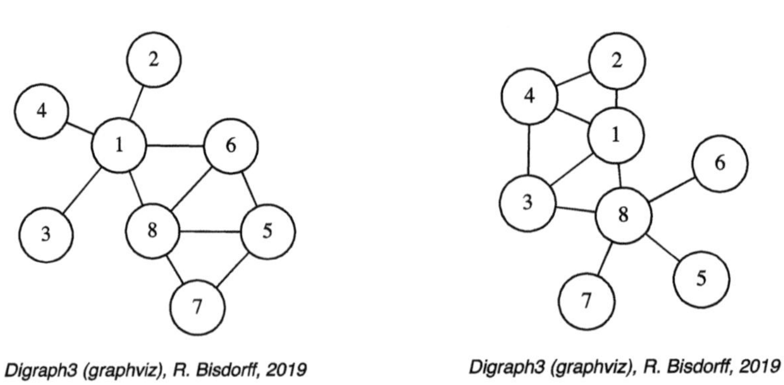

Fig. 23.11 Isomorphic permutation graphs

```
>>> gtesta = PermutationGraph(\
...                 permutation=[2, 3, 4, 8, 6, 1, 7, 5])
>>> gtestb = PermutationGraph(\
 ...                permutation=[4, 2, 8, 3, 1, 5, 6, 7])
>>> gtesta.exportGraphViz('gtesta')
>>> gtestb.exportGraphViz('gtestb')
```

And, we indeed recover two isomorphic copies of the original random graph (compare with Fig. 23.8 on page 337).

References

Berge C (1963) Perfect graphs. Six papers on graph theory. Indian Statistical Institute, Calcutta, pp 1–21

Bisdorff R (2010) Enumerating chordless circuits in directed graphs. In: ORBEL24-2010, 24th annual conference of the Belgian operational research society (ORBEL aka Sogesci-B.V.W.B.), January 28–29, Liège (BE), Université de Liège (BE), pp 1–12. http://hdl.handle.net/10993/23926

Chudnovsky M, Robertson N, Seymour P, Robin T (2006) The strong perfect graph theorem. Ann. Math. 164(1):51–229

Golumbic M (2004) Algorithmic graph theory and perfect graphs. Annals of discrete mathematics, vol 57, 2nd edn. Elsevier, Amsterdam

Index

R. Bisdorff, *Algorithmic Decision Making with Python Resources*,
International Series in Operations Research & Management Science 324,
https://doi.org/10.1007/978-3-030-90928-4